Cold Inflow-Free Solar Chimney

Md. Mizanur Rahman · Chi-Ming Chu
Editors

# Cold Inflow-Free Solar Chimney

## Design and Applications

*Editors*
Md. Mizanur Rahman
Department of Mechatronics Engineering
World University of Bangladesh
Dhaka, Bangladesh

Chi-Ming Chu
University Malaysia Sabah
Kota Kinabalu, Malaysia

ISBN 978-981-33-6833-0 ISBN 978-981-33-6831-6 (eBook)
https://doi.org/10.1007/978-981-33-6831-6



This Springer imprint is published by the registered company Springer Nature Singapore Pte Ltd.
The registered company address is: 152 Beach Road, #21-01/04 Gateway East, Singapore 189721, Singapore

# Contents

# Chapter 1
# Introduction of Cold Inflow Free Solar Chimney

**Md. Mizanur Rahman, Chi-Ming Chu, Sivakumar Kumaresen, and Shir Lee Yeoh**

Natural draft and forced draft chimneys are used in many industries to remove dust and dirt, hot gases and air from the process side to the atmosphere. Among them, the natural draft chimney is operated due to the effect of temperature difference between process side and ambient which is known as buoyancy force or stack effect. The process of flow is continuous as long as the buoyancy or stack effect is present. Solar chimney is a natural draft chimney that is used to generate electricity from solar energy; therefore, solar chimney is also known as solar updraft system. It is an economical and environmental friendly system to generate electricity and ventilation for houses or space. There are numerous works that have been found which discuss the enhancement of solar chimney power plant efficiency. The works also include the applications of solar chimney and feasibility study of hybrid systems. The researchers used experimental and simulation models for the study of solar chimney performance and its structural design. The purpose of this book is to provide information about the solar chimney for design. Solar chimney applications in many areas and incorporated in this book are drawn mainly from industries as dryers and households as natural ventilation systems.

Md. M. Rahman (✉)
Department of Mechatronics Engineering, World University of Bangladesh, 151/8, Green Road, Dhaka 1205, Bangladesh
e-mail: mizanur.rahman@mte.wub.edu.bd

C.-M. Chu · S. Kumaresen
Faculty of Engineering, Universiti Malaysia Sabah, Kota Kinabalu, Sabah, Malaysia

S. L. Yeoh
Faculty of Engineering, Universiti Malaysia Sabah, Kota Kinabalu, Sabah, Malaysia



Md. M. Rahman and C.-M. Chu (eds.), *Cold Inflow-Free Solar Chimney*,
https://doi.org/10.1007/978-981-33-6831-6_1

## Introduction

The energy consumption all over the world has gradually increased over the last century due to the modernization as well as industrialization (Cheng 2010). The demand of energy mainly depends on the energy conversion, levels of energy conversion and the standard of living as well as human expectation (Kreith et al. 2010). The population growth and development of cities as well living standard in the developing and developed countries resulted in global energy consumption averagely increased 2% every year. It is expected by the year 2025 the world population will reach about 8.4 billion and need a huge amount of energy (Golušin et al. 2013). To fulfill the demand of this energy, it will need to be extracted from the renewable and non-renewable sources. The reserve of non-renewable energy sources is now under threat because these sources are limited. In addition, the energy conversion technology from the non-renewable energy sources produces harmful greenhouse gases that are responsible for global warming (Prasad et al. 2017). The earth's temperature is increasing significantly every year, and the Arctic sea ice is melting due to climate change (Matishov et al. 2016). This is a warning sign from our mother Earth. One of the valid suitable options to reduce global warming is by using renewable energy, but it is still unable to contribute much toward the world energy supply. In a nutshell, clean and sustainable renewable energy is the key to the future. One of the promising renewable energy sources is solar energy (Zhou et al. 2010a). There are many ways electricity can be generated from solar energy. Among them, one of the suitable options is a solar chimney power generation system. Solar chimney power generation consists of solar collector and draft or chimney.

The history of a chimney was started from European when the house is heated with the fire to make it warm, and the roof hole is used to evacuate smoke and dust out from their house. In the seventeenth century, the industry started to build chimneys to remove unwanted gases out from the boiler or other fireplaces to maintain clean surrounding atmosphere inside the industry. Two types of chimneys named natural and forced draft are commonly used in the industries. In the forced draft chimney, either a fan or a driver is placed at the bottom or above the heat sources that are tube bundles in the cooling tower. The fan generates sufficient airflow that removes unwanted heat, smoke and dust particles from the system. In the natural draft chimney, the temperature difference between process sides and the ambient generates buoyancy force or stack effect. As a result, airflows through the chimney and removes waste heat from the system. This process will continue until the buoyancy or stack effect is present in the chimney. In the natural draft system, there are no mechanical appliances or devices used which makes it more feasible than forced draft in terms of operational safety and reliability (Arce et al. 2009; Fisher & Torrance 1999; Kumaresan et al. 2013; Chu-Hua 2007; Chu et al. 2012; Rahman et al. 2017; Damjakob & Tummers 2004). The solar chimney is also known as a solar updraft system that is an economical way to create natural flow used to generate electricity.

In 1903, Cabanyes came up with an idea of locating the wind blade to generate electricity inside the house. This was the origin of the hybrid system, which utilizes the usage of chimney, besides heating air. Lastly, the very first idea of a solar chimney power plant was proposed by Schlaich in the year 1968. The prototype of a solar chimney power plant was constructed in Manzanares, between the year 1981 and 1982 (Haaf 1984). Since then, the possibility of the solar chimney power plant is catching the researcher's attention. The concept of the chimney to generate electricity is using the natural convection with the support of thermal energy conversion to mechanical work by thermodynamics principles.

In addition, the solar chimney or solar draft natural convection process can also be used as alternative valid options to replace forced or mechanical cooling systems for houses, open spaces, etc.,(Chu 2002; Doyle & Benkly 1973). This is also called a passive or zero emission or green technology for power generation and ventilation. The solar chimney has two major components that are called solar collectors and chimney or draft. Not only power generation, the solar chimney can also be used for cooling space as well as for distilling seawater. In the solar chimney system, the air is heated due to the effect of solar radiations collected by solar collectors. The working principle of solar chimney is very simple to explain: The warm air rises up because of the buoyancy effect as it gets less dense, and it will exit through the chimney (Rahman et al. 2017; Koonsrisuk et al. 2010; Ahmed & Chaichan 2011; Verboom & Koten 2010). In the solar chimney, the air receives kinetic energy from solar radiation and becomes hot, resulting in initiated movement and air rises up. This process is known as buoyancy effect or stack effect of draft, the less dense air that leads to the draft from which it is exhaled. The efficiency of a solar chimney depends on the radius and design of the collector as well as the quality of collectors' material. In addition, the efficiency also depends on the physical shape of the chimney mainly height and diameter. The heat and mass transfer phenomena inside the natural draft chimney like solar chimney are very complex. It is very difficult to determine the velocity and the temperature distributions through solar chimneys under different environmental conditions (Zhou et al. 2007; Spencer et al. 2000; Gan & Riffat 1998).

After the solar chimney was introduced by the Cabanyes, there were a few patents that came up on this in Australia, Canada, Israel, the USA (Lucier 1981). The solar chimney power plant is started with the prototype presented by Schlaich and team, which has a height of 194.6 m, and the radius of the collector has 122 m. In addition, a single vertical axis four blades rotor turbine is used in the solar chimney prototype to generate electricity (Pasumarthi & Sherif 1998). This prototype model is able to generate electricity and to contribute 50 kW of peak power for almost eight years (Schlaich 1995). This is the first movement to prove that the contribution of the solar chimney power plant as sustainable renewable energy is possible. In Australia, the project of building 1000 m high with a 7000 m diameter collector solar chimney power plant was proposed and supported by the government. This plant was predicted to produce 200 MW of power, which is able to support over 200,000 households and a reduction of $CO_2$ gas emission by approximately 1,000,000 tonnes (Kasaeian et al. 2017; Dhahri & Omri 2013). In the Northwestern regions of China, there was a pilot solar chimney power plant set, which 200 m high with 500 m diameter solar collector, had the ability to produce 110 to 190 kW electric power on a monthly average (Dai et al. 2003). In 2008, a proposal was written, and the predicted result was analyzed in Mediterranean region. In this proposal, the size of the solar chimney power plant suggested to be 550 m high and 1250 m diameter of solar collector to produce about 2.8–6.2 MW of power (Nizetic et al. 2008). An analysis research had been completed in Arabian Gulf area, which showed that with a 500 m high and 1000 m diameter collector, the SCPP could produce about 8 MW of power in that region (Hamdan 2011). Another performance prediction research done in Adrar site estimated that the geometry capable of generating about 140 to 200 kW of power required the size to be 200 m high and 500 m diameter collector of SCPP (Larbi et al. 2010).

## Natural Convection

Convection is a mode of heat transfer for fluid. Convection could be categorized into two types, natural convection and forced convection. Natural convection happens when the buoyancy forces occur after the fluid absorbs heat. When the temperature difference is large enough, the internal energy of the air in the system has increased significantly. The air particles have received sufficient momentum due to change of internal energy, and as a result, the density of the air becomes lighter. The moment of air particles causes a significant amount of buoyancy force in the system that creates the movement of air or flow of air. According to Archimedes' principle, the buoyancy force is proportional to the density difference of the object.

The buoyancy or draft can be explained in another way since the cold air is denser than hot air. At higher temperature, the air particles have higher kinetic energy than the cold air. The hot air particles tend to vibrate due to change of kinetic energy and push these particles to the surrounding. Therefore, the air particles close to the heat source have higher kinetic energy, and due to its effect, the air particles vibrate and try to push the air particles to the surrounding. This natural phenomenon causes air circulation, and the solar chimney power plant is developed based on this concept as shown in Fig. 1.1. In the solar chimney power plant, the air temperature is increased due to heat energy received by the air at the solar collector. The hot air starts moving at the upward direction. A generator or wind turbine is placed at the entrance of the chimney to produce electricity (Cao et al. 2018).

To understand the fluid flow behavior in the solar chimney, the process can be simplified as the updraft of air from fire or hot source or heat exchanger. The heat sources of the solar chimney can also be considered as pure source. According to Byram and Nelson, partial verification of scaling laws for mass can be used to determine the updraft velocity from heat sources. The effects of surrounding environmental conditions such as cross wind velocity, geographical location are neglected in the Byram and Nelson mathematical correlation. A fully automated inspection and maintenance robot can also be used to measure the temperature and velocity distribution in a model solar chimney design for electricity generation, but it is not a cost effective system to understand the behavior of fluid flow and temperature distribution inside the solar chimney. The automated system in the solar chimney is able

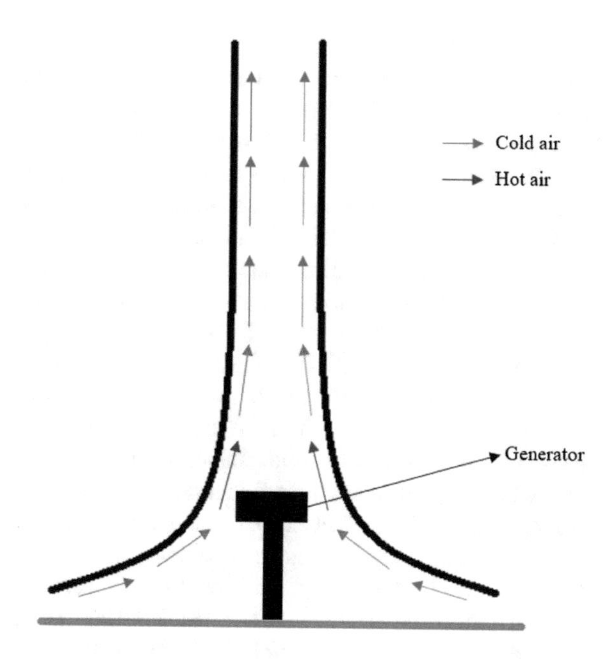


**Fig. 1.1** Airflow inside solar chimney by using natural convection concept

to enhance the performance of maintenance work as a relation operation and maintenance cost of the plant reduced significantly. In addition, the automated or robotic system is able to enhance operation safety and working environment of the service technician (Felsch et al. 2015). According to Nieuwenhuisen et al. 2017 discussed about a 80 cm flying robot for the purpose of inspection chimney. The robot integrated with a lightweight 3D laser scanner, cameras with an apex angle 122°. The proposed inspection robot can be used for inspection tall chimney that enhance operation and maintenance and reduce inspection cost as well down time (Nieuwenhuisen et al. 2017). The performance of the solar chimney as well as the effects of different operation parameters is studied numerically by the different researchers. The numerical study is also extended to evaluate the thermal performance of the solar chimney is also studied when integrated with building ventilation system. Sudprasert et al. 2016 constructed a numerical model to investigate the effect of humid air (RH 40 to 80%) on solar chimney thermal performance during summary when the ambient temperature and humidity are high. In this study, numerical analysis was done by using re-normalization group (RNG) turbulence model. In this model, the airflow at the solar collector is considered as turbulent. The Boussinesq approximation has been added with the dry air model when temperature and velocity distribution are estimated in the solar chimney model. The relative humidity significantly reduces the performance of the solar chimney up to 26.7% (Sudprasert et al. 2016). The effect of physical parameter mainly chimney inlet area on solar chimney thermal performance was studied by Bassiouny and Koura in the year 2008. FORTRAN software with relaxation iterative method is used to determine the temperature at different influencing points of the solar chimney. The results indicate significant difference between operation side and ambient temperature that act as driving force of the chimney (Bassiouny & Koura 2008). A three-dimensional quasi-steady CFD model was used to determine solar chimney performance in terms of air velocity and temperature. In addition, incompressible Navier–Stokes fluid flow equations can be used to estimate fluid dynamics and heat transfer mechanism in the solar chimney. The researchers are used simulation results to maximize the performance of the solar chimney model and validate the results with the outcome from the mathematical models or experimental outcomes. To understand the behavior or the fluid dynamics in the solar chimney, buoyancy force and radiation heat transfer models are used during CFD simulation (Abdeen et al. 2019; Rabehi et al. 2017; Fasel et al. 2013; Somsila et al. 2010; Gan & Riffat 1998). Numerical fan-model is also used to calculate averaged aerodynamics forces exerted in the solar chimney. This simulation model can be used to describe the thermal fluid behavior analysis of air-cooled heat exchangers. For better understanding about the fluid dynamics and heat transfer inside the chimney, incompressible fluid flow equations are combined with the CFD software. The effect of black body radiation was also united with this model for better understanding about the heat transfer phenomena in the solar chimney model (Dirkse et al. 2006). Numerical investigation instigated with atmospheric conditions to define the aerodynamics or detailed heat transfer characteristics in the solar chimney, but very limited number of experimental investigation was found about the aerodynamics of natural draft solar chimney system (Kitamura & Ishizuka 2004; Chu 2002; Lorenzini 2006).

## Solid Wall Chimney

Solid wall chimney is the conventional chimney we are familiar with, whose wall acts as a barrier to prevent the heat loss from hot air loss to the surrounding too quickly. In order to obtain sufficient buoyancy force to generate a significant amount of electricity, the height of the current design of solar chimney is often very tall. Numbers of research have been done to elucidate the governing factors of the natural convection process in the solid wall chimney. In 1942, Elenbaas analyzed natural convection phenomenon occurring between two parallel vertical plates. The process was considered an isothermal process, and the design optimized according to the maximum heat transfer rate. Then, in 1962, Bodoia & Osterle did the numerical analysis to determine the relation of the flow between plates with the temperature by using a finite difference method. In addition, in the year 1988, Sparrow et al. presented another numerical analysis by considering both natural convection and solid wall conduction (Sparrow et al. 1988; Bodoia & Osterle 1962; Elenbaas 1942).

## Cold Inflow

Cold inflow or flow inversion at chimney exit has been identified in cooling devices ranging from cooling tower, solar chimney to electronics vertical channels. Cold inflow is the most common problem that happens in the solid wall chimney exit. This is due to the unstable wind flow and downdraft occurring (Zhai & Fu 2006). Bouchair et al. conducted an experiment that proved that the reversal flow appeared at the chimney outlet. Kihm et al. (2013) also identified the reversal flow of the air through vertical isothermal channel walls. It was also found that cold air is liable to 'sink' into the glass tube shell under typical exit bulk velocities, of 3–5 m/s or below (Jörg & Scorer 1967; Chu & Rahman 2009). Modi & Torrance found a cold inflow phenomenon during experimental investigation of channel flow. Fisher & Torrance (1999) had quantified the results as cold inflow affects heat transfer performance by 4%. Zhou et al. (2010a) reported that the efficiency of a solar chimney power plant increases with the chimney height but the energy conversion performance is still poor.

Many researchers also noticed about the back flow in the solar chimney, but until today, very limited research has been conducted on it to determine the effects because of its complexity (Khanal & Lei 2012; Arce et al. 2009; Chen et al. 2003; Bouchair 1994). The research works on the effects of cold air inflow on performance not yet documented properly. Although cold inflow was recognized as a problem for air-cooled heat exchanger, but it also exits in solar chimney and cooling tower. These effects generate a complication to achieve its maximum efficiency (Arce et al. 2009; Meyer & Kröger 2004; Chen et al. 2003; Duvenhage et al. 1996; Bender et al. 1996; Duvenhage & Kröger 1996; Bouchair 1994). This phenomenon has been observed when the buoyant flow at the chimney exit is relatively slow, which introduce separation of fluid before the buoyant air leaves the tower.

The cold air inflow is an unresolved problem not only for air-cooled heat exchangers but also for solar chimneys or any other open top cylinder where natural convection takes place with weaker updraft flow. The effects of cold inflow still need further investigation (Dai et al. 2019). To minimize the impacts of cold air inflow, it is important to determine the magnitude of effects of cold air inflow on velocity and temperature inside the cooling tower. A screening device was applied by to prevent this cold air from sinking into a chimney duct to ensure the correct measurement of temperature for heat balance. The implications of this modification on the fluid flow and heat transfer have until now not been investigated, so the investigation was also extended to determine the effect of screening at top of the heat exchanger. By removing cold inflow, it was found that the chimney performance improves over that without cold inflow by using the screening device. (Chu et al. 2009, 2012). It was also found during the investigation by CFD that without cold inflow, adding the screening device would reduce the draft as normally expected. Once the adverse cold inflow is addressed by its removal, the other benefit is the introduction of stack effect of lazy plume, or plume-chimney (See chapter on lazy plume and its stack effect).

Thus, adverse cold inflow and the stack effect of lazy plume are mutually exclusive phenomena.

## Hybrid Solar Chimney Power Plant (SCPP)

Zuo et al. (2011) built a small-scale solar chimney power plant together with a seawater desalination system. The efficiency of this hybrid SCPP was more than a pure SCPP by 21.13%. However, this plant also involved with the fossil fuel system, it was still not considered as a fully green plant. According to Zhou et al. (2010a, b), the power output from the hybrid system of water desalination with SCPP was slightly less productive than the traditional SCPP system. Maia et al. (2009) used photovoltaic cells in their SCPP to conduct the performance analysis on agricultural crop drying. Cao et al. (2014) came out with a hybrid system of geothermal into SCPP. This study was conducted at Xi'an, and the performance of this plant was 26.3% more efficient than the classic SCPP. From all of the research, it could conclude that hybrid SCPP system is a possible idea which increased the purpose of the SCPP system other than electric generation (Zhou et al. 2010a, b, 2011).

## Conclusion

The solar chimney power plant is considered as future technology to generate electricity from renewable energy. It will also help to reduce the dependency on fossil fuel as well environmental pollution mainly air pollution. The performance of the solar chimney power plant depends on its location as well as the geometry of the solar chimney. The establishment of the thermal difference in the solar chimney is the key success to produce more power. The area where the solar radiation is limited, the hybrid system can be considered as a suitable valid option for power generation.

## References

Arce, J., Jiménez, M. J., Guzmán, J. D., Heras, M. R., Alvarez, G., & Xamán, J. (2009). Experimental study for natural ventilation on a solar chimney. *Renewable Energy, 34*(12), 2928–2934.

Abdeen, A., Serageldin, A. A., Ibrahim, M. G., El-Zafarany, A., Ookawara, S., & Murata, R. (2019). Solar chimney optimization for enhancing thermal comfort in Egypt: An experimental and numerical study. *Solar Energy, 180,* 524–536.

Ahmed, S. T., & Chaichan, M. T. (2011). A study of free convection in a solar chimney model. *Engineering and Technology Journal, 29*(14), 2986–2997.

Bassiouny, R., & Koura, N. S. (2008). An analytical and numerical study of solar chimney use for room natural ventilation. *Energy and Buildings, 40*(5), 865–873.

Bender, T. J., Bergstrom, D. J., & Rezkallah, K. S. (1996). A study on the effects of wind on the air intake flow rate of a cooling tower: Part 2. Wind wall study. *Journal of Wind Engineering and Industrial Aerodynamics, 64*(1), 61–72.

Bodoia, J. R., & Osterle, J. F. (1962). The development of free convection between heated vertical plates. *Journal of Heat Transfer, 84*(1), 40–43.

Bouchair, A. (1994). Solar chimney for promoting cooling ventilation in southern Algeria. *Building Services Engineering Research and Technology, 15*(2), 81–93.

Cao, F., Li, H., Ma, Q., & Zhao, L. (2014). Design and simulation of a geothermal-solar combined chimney power plant. *Energy Conversion and Management, 84,* 186–195.

Cao, F., Liu, Q., Yang, T., Zhu, T., Bai, J., & Zhao, L. (2018). Full-year simulation of solar chimney power plants in Northwest China. *Renewable Energy, 119,* 421–428.

Chen, Z. D., Bandopadhayay, P., Halldorsson, J., Byrjalsen, C., Heiselberg, P., & Li, Y. (2003). An experimental investigation of a solar chimney model with uniform wall heat flux. *Building and Environment, 38*(7), 893–906.

Cheng, J. (Ed.). (2010). *Biomass to renewable energy processes.* Raleigh, North Carolina: Taylor & Francis Group.

Zhang, C.-H. (2007). Thermodynamic analysis and calculation of large-scale solar chimney electricity generation plant. *Renewable Energy Resources, 2,* 2.

Chu, C. M., Rahman, M. M., & Kumaresan, S. (2012). Effect of cold inflow on chimney height of natural draft cooling towers. *Nuclear Engineering and Design, 249,* 125–131.

Chu, C. M. (2002). A Preliminary method for estimating the effective plume chimney height above a forced-draft air-cooled heat exchanger operating under natural convection. *Heat Transfer Engineering, 23*(3), 3–12.

Chu, C. C., & Rahman, M. M. (2009, January). A method to achieve robust aerodynamics and enhancement of updraft in natural draft dry cooling towers. In *Heat Transfer Summer Conference* (Vol. 43581, pp. 817–823).

Dai, Y., Huang, H., & Wang, R. (2003). Case study of solar chimney power plants in North-western regions of China. *Renewable Energy, 28*(8), 1295–1304. https://doi.org/10.1016/S0960-1481(02)00227-6.

Dai, Y., Kaiser, A. S., Lu, Y., Klimenko, A. Y., Dong, P., & Hooman, K. (2019). Addressing the adverse cold air inflow effects for a short natural draft dry cooling tower through swirl generation. *International Journal of Heat and Mass Transfer, 145,* 118738.

Damjakob, H., & Tummers, N. (2004, April), Back to the future of the hyperbolic concrete tower. In *Proceedings of the Fifth International Symposium on Natural Draught Cooling Towers* (pp. 3–21). Istanbul: AA Balkema Publishers.

Dirkse, M. H., van Loon, W. K., van der Walle, T., Speetjens, S. L., & Bot, G. P. (2006). A computational fluid dynamics model for designing heat exchangers based on natural convection. *Biosystems Engineering, 94*(3), 443–452.

Dhahri, A., & Omri, A. (2013). A review of solar chimney power generation technology. *International Journal of Engineering and Advanced Technology, 2*(3), 1–17.

Doyle, P. T., & Benkly, G. J. (1973). Use fanless air coolers. *Hydrocarbon Processing, 52*(7), 81–86.

Duvenhage, K. K. D. G., & Kröger, D. G. (1996). The influence of wind on the performance of forced draught air-cooled heat exchangers. *Journal of Wind Engineering and Industrial Aerodynamics, 62*(2–3), 259–277.

Duvenhage, K., Vermeulen, J. A., Meyer, C. J., & Kröger, D. G. (1996). Flow distortions at the fan inlet of forced-draught air-cooled heat exchangers. *Applied Thermal Engineering, 16*(8–9), 741–752.

Elenbaas, W. (1942). Heat dissipation of parallel plates by free convection. *Physica, 9*(1), 1–28. https://doi.org/10.1016/S0031-8914(42)90053-3.

Fasel, H. F., Meng, F., Shams, E., & Gross, A. (2013). CFD analysis for solar chimney power plants. *Solar Energy, 98,* 12–22.

Fisher, T. S., & Torrance, K. E. (1999). Experiments on chimney-enhanced free convection.

Felsch, T., Strauss, G., Perez, C., Rego, J. M., Maurtua, I., Susperregi, L., & Rodríguez, J. R. (2015). Robotized inspection of vertical structures of a solar power plant using NDT techniques. *Robotics, 4*(2), 103–119.

Golušin, M., Dodić, S., & Popov, S. (2013). *Sustainable energy management*. Academmic.

Gan, G., & Riffat, S. B. (1998). A numerical study of solar chimney for natural ventilation of buildings with heat recovery. *Applied Thermal Engineering, 18*(12), 1171–1187.

Haaf, W. (1984). Solar Chimneys. *International Journal of Solar Energy, 2*(2), 141–161. https://doi.org/10.1080/01425918408909921.

Hamdan, M. O. (2011). Analysis of a solar chimney power plant in the Arabian Gulf region. *Renewable Energy, 36*(10), 2593–2598. https://doi.org/10.1016/j.renene.2010.05.002.

Jörg, O., & Scorer, R. S. (1967). An experimental study of cold inflow into chimneys. *Atmospheric Environment, 1*(6), 645–654.

Kihm, K. D., Kim, J. H., & Fletcher, L. S. (2013). Onset of flow reversal and penetration length of natural convective flow between isothermal vertical walls. *Journal of Chemical Information and Modeling, 53*(9), 1689–1699.

Kasaeian, A. B., Molana, S., Rahmani, K., & Wen, D. (2017). A review on solar chimney systems. *Renewable and Sustainable Energy Reviews, 67*, 954–987.

Kitamura, Y., & Ishizuka, M. (2004). Chimney effect on natural air cooling of electronic equipment under inclination. *Journal of Electronic Packaging, 126*(4), 423–428.

Kreith, F., Kreider, J. F., & Krumdieck, S. (2010). *Principles of sustainable energy: Mechanical and Aerospace Engineering Series*. CRC Press.

Koonsrisuk, A., Lorente, S., & Bejan, A. (2010). Constructal solar chimney configuration. *International Journal of Heat and Mass Transfer, 53*(1–3), 327–333.

Khanal, R., & Lei, C. (2012). Flow reversal effects on buoyancy induced air flow in a solar chimney. *Solar Energy, 86*(9), 2783–2794.

Kumaresan, S., Rahman, M. M., Chu, C. M., & Phang, H. K. (2013). A chimney of low height to diameter ratio for solar crops dryer. In *Developments in Sustainable Chemical and Bioprocess Technology* (pp. 145–150). Springer.

Larbi, S., Bouhdjar, A., & Chergui, T. (2010). Performance analysis of a solar chimney power plant in the southwestern region of Algeria. *Renewable and Sustainable Energy Reviews, 14*(1), 470–477.

Lorenzini, G. (2006). Experimental analysis of the air flow field over a hot flat plate. *International Journal of Thermal Sciences, 45*(8), 774–781.

Lucier, R. E. (1981). U.S. Patent No. 4,275,309. Washington, DC: U.S. Patent and Trademark Office.

Maia, C., Ferreira, A., M. Valle, R., & F. B. Cortez, M. (2009). Analysis of the airflow in a prototype of a solar chimney dryer. *Heat Transfer Engineering, 30*.

Matishov, G. G., Dzhenyuk, S. L., Moiseev, D. V., & Zhichkin, A. P. (2016). Trends in hydrological and ice conditions in the large marine ecosystems of the Russian Arctic during periods of climate change. *Environmental Development, 17*, 33–45.

Meyer, C. J., & Kröger, D. G. (2004). Numerical investigation of the effect of fan performance on forced draught air-cooled heat exchanger plenum chamber aerodynamic behaviour. *Applied Thermal Engineering, 24*(2–3), 359–371.

Nizetic, S., Ninic, N., & Klarin, B. (2008). Analysis and feasibility of implementing solar chimney power plants in the Mediterranean region. *Energy, 33*(11), 1680–1690.

Nieuwenhuisen, M., Quenzel, J., Beul, M., Droeschel, D., Houben, S., & Behnke, S. (2017, June). ChimneySpector: Autonomous MAV-based indoor chimney inspection employing 3D laser localization and textured surface reconstruction. In *2017 International Conference on Unmanned Aircraft Systems (ICUAS)* (pp. 278–285). IEEE.

Pasumarthi, N., & Sherif, S. A. (1998). Experimental and theoretical performance of a demonstration solar chimney model—Part I: Mathematical model development. *International Journal of Energy Research, 22*(3), 277–288.

Prasad, P. V. V., Thomas, J. M. G., & Narayanan, S. (2017). Global warming effects. In *Encyclopedia of applied plant sciences* (pp. 289–299). https://doi.org/10.1016/B978-0-12-394807-6.00013-7.

Rabehi, R., Chaker, A., Aouachria, Z., & Tingzhen, M. (2017). CFD analysis on the performance of a solar chimney power plant system: Case study in Algeria. *International Journal of Green Energy, 14*(12), 971–982.

Rahman, M. M., Chu, C. M., Tahir, A. M., bin Ismail, M. A., bin Misran, M. S., & Ling, L. S. (2017). Experimentally identify the effective plume chimney over a natural draft chimney model. MS&E, *217*(1), 012002

Schlaich, J. (1995). The solar chimney: electricity from the sun. Edition Axel Menges.

Sparrow, E. M., Ruiz, R., & Azevedo, L. F. A. (1988). Experimental and numerical investigation of natural convection in convergent vertical channels. *International Journal of Heat and Mass Transfer, 31*(5), 907–915.

Somsila, P., Teeboonma, U., & Seehanam, W. (2010, June). Investigation of buoyancy air flow inside solar chimney using CFD technique. *In Proceedings of the International Conference on Energy and Sustainable Development: Issues and Strategies (ESD 2010)* (pp. 1–7). IEEE.

Sudprasert, S., Chinsorranant, C., & Rattanadecho, P. (2016). Numerical study of vertical solar chimneys with moist air in a hot and humid climate. *International Journal of Heat and Mass Transfer, 102,* 645–656.

Spencer, S., Chen, Z. D., Li, Y., & Haghighat, F. (2000). Experimental investigation of a solar chimney natural ventilation system. In *Air distribution in rooms* (pp. 813–818).

Verboom, G. K., & Van Koten, H. (2010). Vortex excitation: Three design rules tested on 13 industrial chimneys. *Journal of Wind Engineering and Industrial Aerodynamics, 98*(3), 145–154.

Zhai, Z., & Fu, S. (2006). Improving cooling efficiency of dry-cooling towers under cross-wind conditions by using wind-break methods. *Applied Thermal Engineering, 26*(10), 1008–1017.

Zhou, X., Yang, J., Xiao, B., & Hou, G. (2007). Experimental study of temperature field in a solar chimney power setup. *Applied Thermal Engineering, 27*(11–12), 2044–2050.

Zhou, X., Wang, F., & Ochieng, R. M. (2010). A review of solar chimney power technology. *Renewable and Sustainable Energy Reviews, 14*(8), 2315–2338.

Zhou, X., Xiao, B., Liu, W., Guo, X., Yang, J., & Fan, J. (2010). Comparison of classical solar chimney power system and combined solar chimney system for power generation and seawater desalination. *Desalination, 250*(1), 249–256.

Zuo, L., Zheng, Y., Li, Z., & Sha, Y. (2011). Solar chimneys integrated with seawater desalination. *Desalination, 276*.

# Chapter 2
# Theory of Natural Draft Chimney and Cold Inflow

**Md. Mizanur Rahman, Chi-Ming Chu, and Sivakumar Kumaresen**

This chapter will discuss the existing research on natural convection of fluid flow and its characteristics, temperature profile at the chimney exit as well as inside the chimney and effective plume chimney height above a hot plate or heat sources. The relation between the effective plume chimney height and natural convection processes will be discussed briefly in this chapter.

## Introduction to Heat Transfer

In the engineering thermodynamic, the heat transfer process takes place when the two thermal equilibrium states are having different values. It means that two thermal equilibrium states are having significant temperature differences. The heat transfer processes will continue as long as the temperature differences exist in the system. In thermodynamics, there is no indication of time that will discuss how long the processes will take. Therefore, when discussing the heat transfer process in any system in thermodynamics, it concerns conduction, convection and radiation. It is a heat transfer process that occurs in a solid or liquid object and depends on the heat transfer rate. The basic requirements of heat transfer by conduction in any system should have temperature difference between two surfaces. It can be mentioned that the temperature difference is the driving force for conduction heat transfer.

Md. M. Rahman (✉)
Department of Mechatronics Engineering, World University of Bangladesh, 151/8, Green Road, 1205 Dhaka, Bangladesh
e-mail: mizanur.rahman@mte.wub.edu.bd

C.-M. Chu · S. Kumaresen
Chemical Engineering Programme, Faculty of Engineering, Universiti Malaysia Sabah, Kota Kinabalu, Sabah, Malaysia


Md. M. Rahman and C.-M. Chu (eds.), *Cold Inflow-Free Solar Chimney*,
https://doi.org/10.1007/978-981-33-6831-6_2

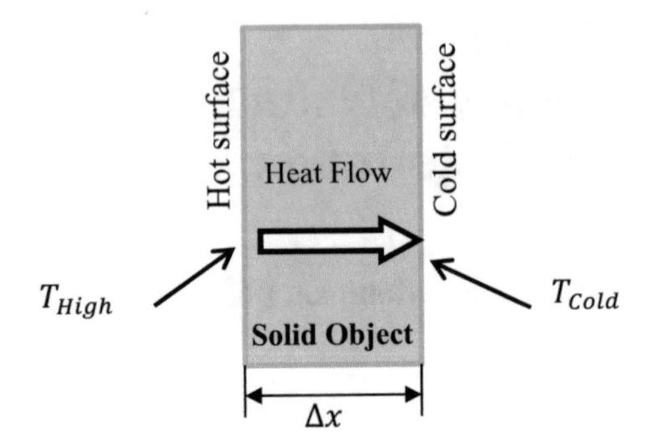


**Fig. 2.1** Conduction heat transfer

***Conduction***: The first one is known as heat transfer by conduction, which is defined as the process of heat transfer occur due to change of internal energy of the molecules of the solid without bulk motion of it. Consider a solid as shown in Fig. 2.1 has one hot surface and another one cold surface.

The hot surface of the solid has a higher temperature than the cold surface; therefore, conduction heat transfer occurs in the solid. According to the definition of the conduction, it is an energy transfer process from the particle having higher energy to the adjacent less energetic particle as a result of interaction due to vibration between the particles. Therefore, the conduction heat transfer is happened due to the energy interaction in the particles of the objects. The rate of heat conduction in the solid object can be determined experimentally. The rate of heat transfer due to conduction can be defined as the ratio to the temperature difference between hot and cold surfaces and to the heat transfer area. The rate of heat transfer due to conduction is inversely proportional to the thickness of the solid object wall (Lienhard & John, 2000).

***Convection***: In the convection heat transfer process, the heat is transferred from the solid surface to the fluid. In another word, the heat transfer takes place due to interaction between a solid or liquid surface and an adjacent dynamic fluid. A good example of convection heat transfer is boiling of water in a pot or airflow over a hot plate (solar chimney). The convection heat transfers also cover the thermal interaction between two fluids. To understand more about the heat transfer process due to convection, consider a hot flat plate at the open space as shown in Fig. 2.2. Let consider, the hot plate surface temperature and heat flux are $T_s$ and $q_s''$ respectively. The ambient temperature is $T_a$ and the stream velocity of air is $V_a$.

The heat energy is stored in the hot plate. The fluid immediately adjacent to the hot plate creates a thin layer which is a slowdown region named the boundary layer. From the hot plate, heat is transfer to this layer by conduction. The hot particles from

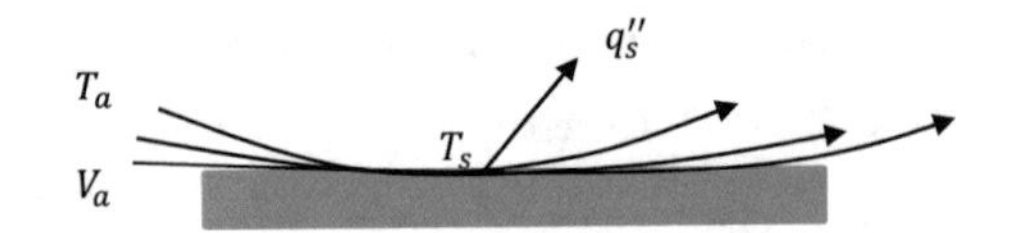


**Fig. 2.2** Hot flat plate

the boundary layer sweep away and at the downstream mix up with fluid particles at adjacent streamline. In this process, the air particles are carrying heat from the hot plate and cool down. Therefore, heat is dissipated and transferred to the air from the surface of the hot plate by convection process, if radiation is neglected. The cooling of the hot plate can be determined by using the following empirical relation.

$$\frac{\mathrm{d}T_{\text{Hot body}}}{\mathrm{d}t} \propto T_{\text{Hot body}} - T_{\infty} \tag{2.1}$$

where $T_{\infty}$ is the initial temperature of the fluid that is flowing over the hot plate. This equation indicates the energy is dissipated from the hot body. If the total energy of the hot plate is constant, then the hot plate temperature reduces.

The hot plate will loss the temperature faster, if

- The hot plate is placed in front of the fan. The air will move faster that will increase the heat transfer rate.
- Change of working fluid from air to a higher heat flux fluid. So the rate of heat transfer will be faster.
- Increase the surface area of the hot plate. The bigger surface area will provide a larger contact surface or air; as a result, the heat transfer rate increases.

From the above discussion, it can be concluded that convection is a physical process and three factors such as fluid motion, fluid nature and surface geometry play major roles in the convection heat transfer.

The convection heat transfer is divided into force convection process and natural convection process. In the force convection process, a mechanical system is introduced that enhances the flow rate as well as increases convection heat transfer. On the other hand, in the natural convection process, there is no mechanical device to create the flow over the hot surface. The natural convection process will be discussed further, later on in this chapter since the operation of solar chimney is based on it (Jiji 2006; Lienhard & John, 2000).

***Radiation***: In the radiation heat transfer process, the energy is transferred from one place to another place in the form of an electromagnetic wave which is known as thermal radiation. This phenomenon is used commonly to transfer heat from one source to another that can be feet around us every day. The best example of radiative heat transfer is the sun energy travels to the earth every day in the form of an electromagnetic wave. It is important for our daily life because it is a dominant factor for several natural and human-made phenomena. The development of ocean waves and wind flow is the most common phenomenon cause due to solar radiation. In the real life, all objects emit electromagnetic waves due to the energy level of atoms and molecules (Howell et al. 2010).

## Natural Convection Process

Natural convection is a process used in the solar chimney to generate power and can be an important part of the power generation sector. In the solar chimney power plant, the heat transfer process occurs by natural convection to the surrounding environment. During natural convection process, the absorber extract heat energy as much as possible from the thermodynamic cycle and reject it to the surrounding. Therefore, the efficiency improvement of a solar chimney depends on the intake air velocity and the temperature difference between processes sides and the ambient (Chu, 2002; Smrekar et al., 2006). The windy condition reduced more than 40% of the total power generation capacity (Zhai & Fu, 2006). Therefore, it is very important to know the basic working principle of natural convection processes before discussing the operation principle of the solar chimney power plant.

Generally, the natural convection is a process that is caused by buoyancy forces because of density differences due to temperature variations in the different layer of fluid or air. During heating, the density of a fluid or air changes due to temperature difference and develops a boundary layer that causes the fluid to flow in an upward direction. The hot air is replaced by cooler air and the process continues until the equilibrium condition is reached. This continuous process is known as free or natural convection and the phenomena are called stack effect.

In the convection processes, airflow takes place through both diffusion and Brownian motion of individual particles and resulted in bulk scale motion of currents in the air. In another way during natural convective processes, a bulk amount of air circulates as effects of buoyancy from the variation in density between two layers of air and creates movement of air. This large air acceleration is generated as result of gravitational force and air drives due to convection. Therefore, the convection occurs when air contacts with a hot surface and departs from it freely. This process continues until both layer of fluid or air having the same density or pressure.

If the heat source is continuous as like in a heat exchanger, then the natural convection process continues as long as the air receives thermal energy from the source. A vacuum pocket is generated by the upward movement of the hot or warm air that is replaced by cool air. The continuous heat flow to the cool air makes it warm and rise up, so that a steady flow of air over a heat source is observed as long as the cool air is available or unaffected by any other heat sources. The density difference between process sides with ambient results in a pressure difference that causes air movement in the upward direction. Therefore, the natural draft or convection cooling system starts when the following occurs.

- The process side temperature is higher than ambient temperature.
- The steady flow of hot air in the upward direction from process side,
- The upward movement of air produce negative pressure at the air inlet of the heat exchanger,
- Positive pressure is developed at the exit of the outlet of the heat exchanger,

- Hot air flows out through the chimney and cold air replaces the position of hot air from ambient. The flow reaches a steady state when the heat exchangers act as a continuous heat source.

In the fluid mechanics, the dimensionless number called Rayleigh number (Ra) is associated with the flow that is caused due to the buoyancy force or to explain the characteristic of the non-uniform mass density fluid flow. The effects of gravity force and temperature are considered in this dimensionless number; therefore, it is suitable to explain the characteristic of fluid flow region where density difference is a dominant factor. The application of the Rayleigh number is not only limited to fluid mechanics, it can also be used to explain the characteristic of solidifying alloys, the porous media and earth's mantle. In addition, study showed that high value of Rayleigh number is used to explain laminar or turbulent natural convection flow in a specific area (Dubief & Terrapon, 2020; Nemati et al., 2020; Dixit & Babu, 2006; Dol & Hanjalić, 2001; Ravi et al., 1994; Goldstein et al., 1973).

According to the definition of Rayleigh number, it is a product of Grashof number and Prandtl number. The Grashof number is a function of buoyancy force and viscosity of a working fluid, whereas the Prandtl number is a function of momentum and thermal diffusivity. If the value of Ra number is less than $10^8$, the flow in the natural convection region is considered as a laminar, and when it is more than $10^{10}$, the flow is considered as turbulent. The flow between Rayleigh number $10^8$ and $10^{10}$ is known as mix flow (Dirkse et al, 2006). In natural convection processes, the magnitude and the rate of flow depend mainly on the temperature difference between process side or heat exchanger and the ambient and modified by environmental conditions. The equation of free convection could be used for estimating the air velocity due to natural convection (Vlasov et al., 2002; Dirkse et al., 2006). The equation of free convection could be written as

$$v = K\left(\frac{(2gH\Delta\rho_m)}{\rho_m}\right)^{0.5} \tag{2.2}$$

where $K$ is empirical coefficient depending on the size and height of the cooling tower chimney and the total hydraulic drag of the cooling tower and the value of $K$ was assumed at approximately 0.5.

The natural convection concept was used to explain the free boundary flow phenomena of the column growth above the high land, forest fire and plume emitted from stacks (Khan et al., 2020; Wong & Chu, 2017; Scorer, 1954; Byram & Nelson, 1974). Generally, the cloud column growth was observed over high land during afternoon and low ground during night. The heat energy received by the high land is greatest in daytime compare to the low ground. Therefore, the loss of heat in the air between low and high lands is comparable although both grounds have similar capacity to heat absorption. The air acquires temperature in the high land of the mountain is greater than acquires temperature in the low land as a result the column of air above the high land become more warmer than the air over the low land, resulted a trend of air circulation from low land to high land is initiated (Scorer, 1954). Fire also generates upward airflow or plume because of natural convection process. In the plume, the heat is transferred from one layer to another by the circulation of air resulted from density difference. The intensity and spreading velocity of plume depends on natural convection process (Byram & Nelson, 1974; Wong & Chu 2017; Kondrashov et al., 2017).

## Natural Convection from Finite Heat Source

The plume or airflow over a pure heat source or heat exchanger due to natural convection is very hard to explain by using simple natural convection processes because the plume depends on the nature of heat sources as well as surrounding environmental conditions. Two general types, free and wall natural convection plume, were identified. The free plume is distinguished by the buoyant flow above a horizontal cylinder or flat plate, whereas natural convection wall plume resulted from a vertical plate heat source. Both types of plume from a line and point source have been well studied. Leu and Jang (1994) found in the plume above a point source embedded in a non-Darcian porous medium; when the centerline velocity decreases, the centerline temperature increased but when the thermal dispersion coefficient increased, both centerline temperature and velocity are decreased. Study also showed that dispersion or mixing of plume depends on surrounding environmental conditions. Baughman et al. (1994) mentioned that in the close condition also, plume-mixing time would vary with the conditions within the close area as well as the nature and intensity of plume. Degan et al. (1995) conducted study on plume from point heat source buried in a saturated fluid porous medium and results indicated that for a fixed Rayleigh number, an increase in the inertial effects reduces the convective flow in the upward direction.

In addition, Lushi and Stockie (2010) developed a method to estimate the plume from multi-point source and used Gaussian plume model to solve problem and to measure the particulate material deposited at ground level from lead–zinc smelting operation in Trail, British Columbia, Canada. Mehdizadeh and Rifai (2004) evaluated the traveling pattern of industrial plume at high altitudes, where industrial source complex was considered as a point source of plumes. Meteorological information was also considered during calculate effective plume velocity and its dispersion (Mehdizadeh & Rifai, 2004). Hunt and Kaye (2006) conducted a series of experiments to measure the entrainment rate of airflow into plumes from a large area source with low initial momentum flux. It was found that immediately below the source, the plume contracts until it reaches a neck of diameter that was approximately half of the source diameter. After the neck, the plume started to expand and form a classic pure plume with linear radial growth rate (Kaye & Hunt, 2009). Wang (1996) used a nonlinear plume kinematic equation to describe plume advection under the influence of wind speed shear effects and determine the speed of wind as well as plume diffusion from low-level point sources (Wang, 1996).

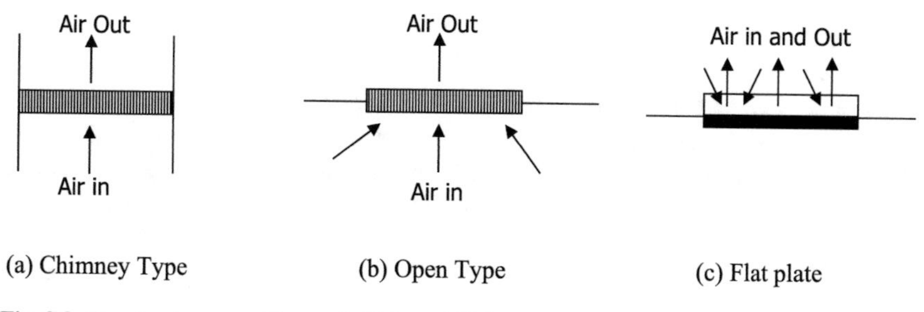


(a) Chimney Type (b) Open Type (c) Flat plate

**Fig. 2.3** Pure heat sources (Byram & Nelson, 1974)

Apart from this, Byram and Nelson (1974) defined fire heat source as a pure heat source without any kind of combustion products except carbon dioxide and water vapor and transferred heat to the surrounding air or atmosphere by natural thermal convection. To simply understand the velocity or movement of air before and after the pure heat source, it was considered as hot porous disk or hot screens mounted in the a vertical tube or chimney or hot screens mounted in an opening in a horizontal surface or hot flat plate with zero emissivity (Fig. 2.3). Partial verification of scaling laws for mass (or stationary) fires was used to calculate velocity at the entrance and exit of the hot porous disk that can be compared with chimney. From the above discussion, it can be considered that the plume rise and dispersion depend on surrounding environmental conditions as well as intensity of heat sources; therefore, thermodynamics laws and energy equation can be used to determine the mean plume velocity from the source.

## *Energy Balance Over Finite Heat Source*

Applying to the first law of thermodynamics, when a hot surface or heat source put in a hollow tube or a system, the air very near to the source receives heat energy from pure heat source. This heat energy is transferred from one layer to another layer of the air that causes natural thermal convection flow. Hot air continuously flows out and the cold air enters into the system. The amount of hot air exiting from the system depends on the hot surface area, its temperature and heat transfer coefficient or amount of heat generated by the pure heat source per unit time. If $m$ is the mass of air that received $\mathrm{d}Q$ amount of heat from the heat source, then according to the first law of thermodynamics when air is an ideal gas and received thermal energy at constant pressure, the internal energy is

$$\mathrm{d}E = mc_p\mathrm{d}T = \mathrm{d}Q - p\mathrm{d}V \tag{2.3}$$

where

$c_p$= The specific heat at constant pressure,

$P$ = Pressure,
$T$ = Absolute temperature and
$\mathrm{d}v$ = Change of volume due to received heat energy from the pure heat source.

The amount of heat rejected from the pure heat source described in Fig. 2.1a, b at constant pressure depends on specific heat, temperature difference between before and after pure heat source and mass of air that can be described mathematically as

$$Q = mc_p(T - T_0) \tag{2.4}$$

where
$c_p$ = Symbolized the specific heat at constant pressure and
$T$ and $T_0$= Air temperature before and after leaving the heating source.

For the ideal gas, the change of volume is

$$V - V_0 = 1/P(\gamma - 1/\gamma)Q = Q/\rho_0 c_p T_0 \tag{2.5}$$

where
$\rho_0$ = ambient air density and
$\gamma = c_p / c_v$ for air.

So the volumetric rate of flow is

$$\frac{\mathrm{d}V}{\mathrm{d}t} - \frac{\mathrm{d}V_0}{\mathrm{d}t} = 1/P(\gamma - 1/\gamma)\frac{\mathrm{d}Q}{\mathrm{d}t} = 1/\rho_0 c_p T_0 \times \mathrm{d}Q/\mathrm{d}t \tag{2.6}$$

For low velocity of air, the pressure $P$ can be considered as a constant, then the ideal gas law become

$$V_0 = V(T_0/T) \tag{2.7}$$

From the Eqs. 2.6 and 2.7, the equation for volumetric flow rate can be developed as

$$\frac{\mathrm{d}V}{\mathrm{d}t} = 1/P(T/T - T_0)(\gamma - 1/\gamma)\frac{\mathrm{d}Q}{\mathrm{d}t} \tag{2.8}$$

If $I_a$ is the rate of heat released from unit area of the pure heat source, then

$$I_a = {}^{\mathrm{d}Q/\mathrm{d}t}/_{A}$$

and

$$Q = mc_p(T_1 - T_2)$$

According to Byram and Nelson (1974), linear velocity of air at inlet and outlet of the pure heat source is

$$u = \frac{1}{p}\left(\frac{T}{T - T_0}\right)\left(\frac{\gamma - 1}{\gamma}\right)I_a = \left(\frac{T}{T - T_0}\right)\frac{I_a}{\rho_0 C_p T_0} \tag{2.9}$$

$$u_0 = \frac{1}{p}\left(\frac{T_0}{T - T_0}\right)\left(\frac{\gamma - 1}{\gamma}\right)I_a = \left(\frac{T_0}{T - T_0}\right)\frac{I_a}{\rho_0 C_p T_0} \tag{2.10}$$

The net velocity of air departs from the pure heat source

$$u - u_0 = \frac{1}{p}\left(\frac{\gamma - 1}{\gamma}\right)I_a = \frac{I_a}{\rho_0 C_p T_0} \tag{2.11}$$

where

$c_p$ = Specific heat at constant pressure,

$T$ and $T_0$ = Temperature of the air before entering and just after leaving the heat source and $\rho$ = density of air.

It was also found that the net velocity of air from pure heat source is independent of the temperature of pure heat source but the information about the structure of the plume and distribution of temperature in the zone of flow establishment above the pure heat source could not be explained properly by using Byram and Nelson (1974). Byram and Nelson (1974) mentioned that pure heat sources are a source of heat that is not affected by geographical and surrounding environmental conditions that why will not produce any combustion products during generation of heat (Byram & Nelson, 1974).

## Theory of Plume Chimney

Most of the research work was conducted related to plume on stack effluent dispersion, surrounding environmental effects on cooling tower and cooling devices designed for electronic goods by modeling them with point or line source. Limited research work on plumes near exit rising from finite sources was available (Chu, 1986; Hilst, 1957; Al-Waked 2010; Preez & Kroger, 1993; Kratjig et al., 1998). Although study showed that plume studies had been carried out by Zeldovich and Schmidt in the 1930s and 1940s (Chu, 1986). Therefore, the plume chimney height above the heat exchanger or over a hot plate in the absence of cold air inflow and crosswind can be calculated from the buoyancy equation. Due to the effect of buoyancy the gases move upward direction and create a vertical plume (Fig. 2.4). The buoyancy balance equation over a forced draft air-cooled heat exchanger is

$$\Delta P_{\text{tot}} = \rho_i (h_o - h_b) g_n - \left[\rho_o h_o g_n + \frac{\rho_i + \rho_o}{2} h_b g_n\right] \tag{2.12}$$

where $h_b$ = Bundle height,

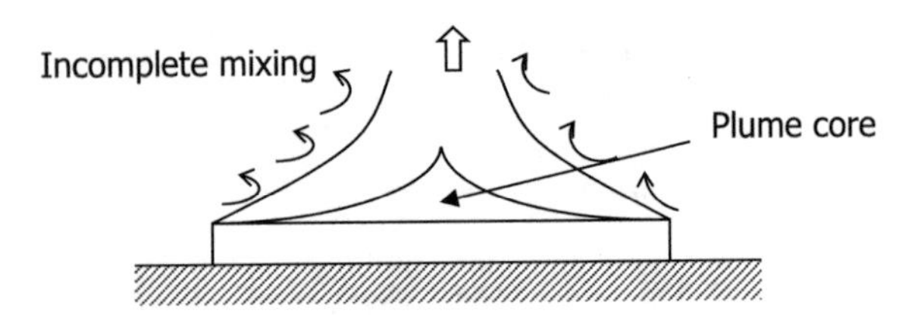


**Fig. 2.4** Effect of face dimensions on effective plume height

$h_o$ = Height over which the buoyancy difference is causing the draught or effective plume height.

A normal chimney or cooling tower hot air from the heat source is shielded by chimney or tower wall for the chimney to create a buoyancy difference to cause a flow (Fig. 2.5). In the fanless forced or induced draft air cool heat exchanger, the effluent plume exiting a stack is separated from surrounding cold air by a boundary layer. An incomplete mixing zone is established surrounding the plume. Constituting a partial wall was not 100% impervious to gas. This was named as plume–chimney (Chu, 2002).

To estimate the effective plume chimney height over a heat exchanger from the experimental results of effective plume over a hot flat plate, the energy per unit volume of the plume above a hot flat plate denoted by subscript 1 and a forced draft air-cooled heat exchanger of the same dimension and hot flat plate, denoted by subscript 2, the energy per unit volume of the plume or core was driven by Chu 2002.

$$\frac{\rho_a - \rho_1}{\rho_1} = \frac{\Delta T_1}{T_a} \text{and} \frac{\rho_a - \rho_2}{\rho_2} = \frac{\Delta T_2}{T_a} \tag{2.13}$$

Chu (2002) mentioned that the effective plume height appeared to be negligibly small for hot flat plate of face dimension 0.457 m × 0.457 m, but when simulating a forced draft air-cooled heat exchanger with large face dimensions can lead to a very conservative result.

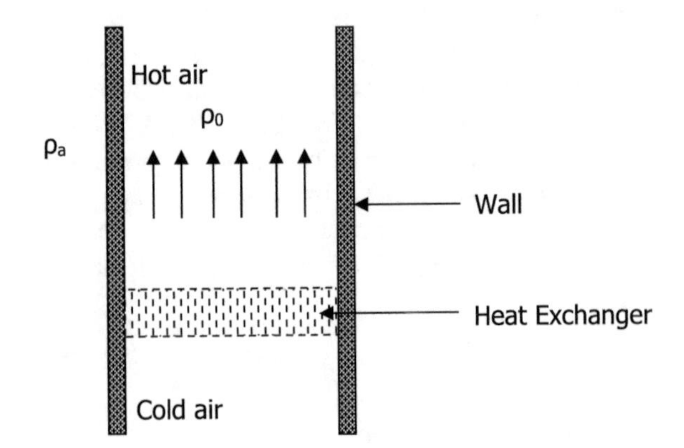


**Fig. 2.5** Airflow pattern in a normal chimney

## Estimation of Effective Plume Chimney Height

The concept of effective plume was used in the field of design cooling system electronic devices, safety equipment such as the safety of gas-cooled reactor. It was found that lower chimney height cooling tower with an equal performance duty, compare to taller chimney height reduce over design and construction cost. Enhancing the performance of the natural cooling tower as well as overcome the problem of over design effective plume chimney height could be used as bundle depth of the cooling tower (Nielsen & Tao, 1965; Zinoubi, et al, 2005; Zhou et al., 2009).

A recently developed method for effective plume chimney height was introduced by Chu in (2002). In this method, the effective plume chimney height is denoted by $h_o$ which can be defined as the height of the plume rising above the hot flat plate or heat exchanger due to the buoyancy effects or density difference resulted from thermal different between heat exchanger and surrounding when cooling tower or heat exchanger operates under natural convection. This mathematical correlation is developed to estimate effective plume chimney height for turbulence flow. Heat and mass balance equation is used to calculate total pressure drop as well as buoyancy force or pumping force in the fanless-induced draft air-cooled heat exchanger. To estimate effective plume chimney height and simulate the natural convection performance of air-cooled heat exchangers requires the solution of the buoyancy and the heat transfer equations. The volumetric buoyancy force $f_g$ (Nm$^{-3}$) for natural convection can be described by using Boussinesq approximation (Dirkse et al., 2006).

$$f_g = \rho \times g \times \beta \times (T - T_\alpha) \quad (2.14)$$

where $\rho$ = Density of fluid (Kg m$^{-3}$),

$g$ = Acceleration due to gravity = 9.81 m$^2$s$^{-1}$,

$\beta$ = Thermal expansion coefficient (K$^{-1}$),

$T$ = Local temperature ($K$),

$T_\alpha$ = Ambient temperature respectively ($K$).

This equation is suitable for moderate temperature because the pressure and the temperature differences are nonlinear over large temperature range. So, this equation is not suitable for natural convection processes where maintaining large temperature range.

Study showed that an electrically heated air-cooled heat exchanger is a face area of 457 × 457 mm$^2$ does not have a significant effective on plume height but this can be supported by data obtained from an industrial size heat exchanger of 2.0 × 3.1 m$^2$ when it was fitted with hardware chimney above the bundle face but without screen at chimney exit (Chu, 2002). It has been noticed that a large forced draft bundle should have a significant effective plume height. It was experimentally determined using a laboratory size air-cooled heat exchanger that $h_o$ was negligibly small (Fig. 2.6). Data from a full-size industrial heat exchanger of 2.0 × 3.1 m$^2$ face area and fitted with three chimney heights ($h_p$ = 0.30, 1.22 and 2.44 m) appeared to support the laboratory result reasonably well (Chu 2002).

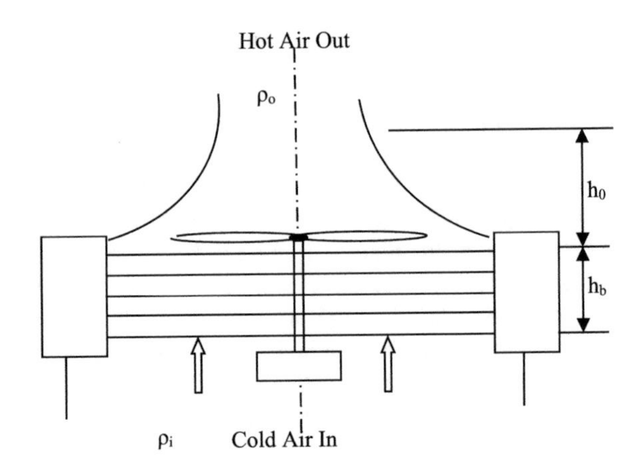


**Fig. 2.6** Induced draft air-cooled heat exchanger

The effective plume chimney height is to be distinguished from the maximum height of the plume that it can travel due to the initial momentum resulted from buoyancy forces. It is described as the height of plume that would act as a solid hardware chimney would except it is invisible (Chu, 2006). It also mentioned the parametric effects of bundle breadth and the temperature difference between processes side and the ambient on effective plume chimney height.

## Temperature Profile in the Zone of Flow Establishment

The zone of flow establishment above the hot plate or heat exchanger depends on the plume or jet characteristic. According to the definition of plume it is a free flow of hot gases in vertical direction over a constant heat source due to the effect of initial buoyancy force whereas jet is a flow of hot gases maintains due to initial momentum force of the gas molecules. The main difference between plume and jet is the characteristic of driving force. Plume is driven by buoyancy but jet is driven by momentum (Chu, 1986). Thus, plume flows only upward or downward direction depending on environmental conditions, whereas jet may be aimed in any direction. In addition, in the jet, the centerline velocity drops vary rapidly compare to plume with distance. A plume is created because of buoyancy force; therefore, momentum increased continuously and the decay of velocity is very slow but the zone of flow establishment is where the centerline velocity in a jet remains at the source value before it falls off due to momentum. In a plume, it is where the centerline temperature remains at the source value before it falls off due to mixing and loss of buoyancy as shown in Figs. 2.7 and 2.8. Zinoubi, et al., in (2005), conducted a study on plume–thermosiphon interaction but discussed temperature profile inside a cylinder. Although the aim of the research was to improve the industrial chimney efficiency but under this research, fundamental flow characteristic determine was determined. The air is loaded into the system from the bottom and moving out from the top due

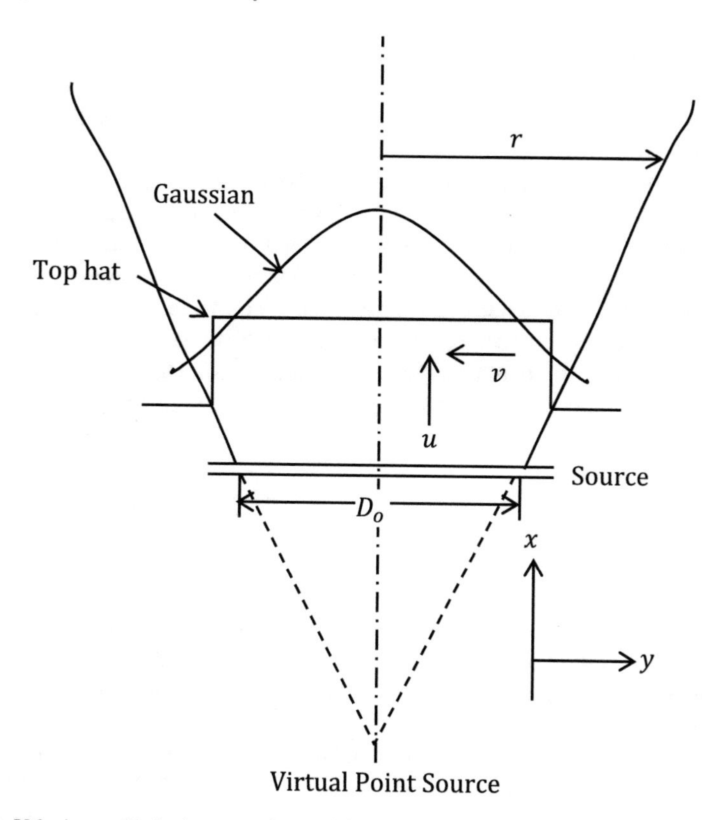


Fig. 2.7 Velocity profile in the zone of establish flow (ZFE)

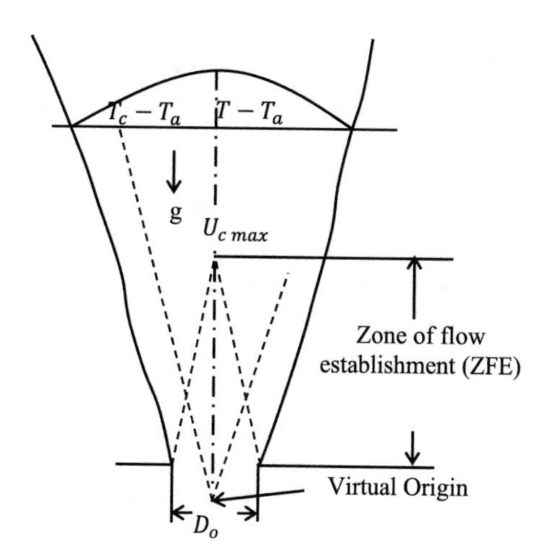


Fig. 2.8 Temperature profile in the ZFE (Henderson, 1983)

to natural convection process. The cooled air contact with hot source and received kinetic energy to flow through cylinder placed at different height from the source. Zinoubi et al. observed an isothermal field in the cylinder during experiments. The results showed that when cylinder at near to heat source, the temperature profile has three minimum downward apexes and has two maximum points centered at the plume axis. The average temperature falls down, on the plume axis due to cold air penetration coming from outside. On the other hand, when the cylinder comes closer to the heat source, the maximum temperature point in the temperature profile at near to the heat source comes closer to the vertical axis. For a higher level, the temperature profile was noticeably uniform that indicated the establishment of the turbulence and no penetration of cold air inflow (Zinoubi et al., 2005). The cold air penetration could be defined as the amount of air moving downward direction from the top to the system.

## Cold Inflow

Wind conditions and air temperature significantly influence the performance as well as the general operation of the cooling tower. This strong wind can reduce up to 40% of the power generation capacity. Most of the study was done to determine the effect of crosswind. The crosswinds 14 and 19 $ms^{-1}$ drop the airflow rate significantly 13.47–24.12% and the windbreak wall was effective for windward flow but it is less effective in the leeward flow (Bender et al., 1996; Zhai & Fu, 2006). Limited research was conducted to determine the effect of cold air inflow.

Modi and Torrance (1987) reported the effect of cold inflow when conducting studies of the separation of a smooth attached buoyant flow from the inner wall of a duct and experience on cold inflow in a turbulent flow and laminar flow region. The study showed that the influence of Re and Fr on the structure of the cold inflow region at the wall near to exit was experienced at moderate Reynolds number (Re = 200 to 500 and Fr = 1 to 5). However, in the laminar flow, cold inflow leads to premature separation of wall boundary layer and it merges into the buoyant jet or plume above the duct exit.

Although effective plume height is more noticeable in the large dimension air-cooled heat exchanger, cold air inflow disrupts the plume above the heat exchanger. Jörg and Scorer (1967) investigate the effect of cold inflow and found that exit dimension is responsible for penetration of cold air in the cooling tower resulting in decreased buoyancy. It was found that the penetration of cold air in the system depends upon the penetration depth ratio and critical velocity. Fisher and Torrance (1999) mentioned that the theory of cold air inflow not yet established properly but Modi and Torrance (1987) suggested Froude number is related to the presence of cold inflow by

$$\mathrm{Fr} = \frac{\rho_c U^2}{g r_H(\rho_c - \rho_h)} = \frac{U^2}{g\beta r_H(T_h - T_c)} \tag{2.15}$$

where U is the mean velocity of fluid in the chimney, the subscripts h and c represent the hot and cold fluids and $r_H$ is the chimney's hydraulic radius. It was also suggested by Fisher and Torrance (1999) that the cold inflow occurs when Fr values are less than a critical value that typically of the order $Fr_{cric} \sim 1$. Hence, cold inflow restricted the exit cross-section and it decreases the overall flow rate as well as cooling the chimney wall resulting in the reduction of the effective chimney height.

## *Roles of Wire Meshes in Fluid Flow and Cold Inflow*

Many applications of wire meshes were noted in different fields; among them, one of the most important applications in the research field was, wire meshes were used in the high-temperature air combustion furnace to characterize a radiative converter (Shiozaki et al., 2005). In addition, Kolodziej and Lojewska (2009) mentioned that for years, catalytic wire gauzes were used for ammonia oxidation. Stainless wire mesh was used by Ismail et al., (2007) during the synthesis and characterization of a highly porous and well adhere immobilization. Therefore, analysis of flow through wire meshes is an important factor before it is used in any system. In an induced draft air-cooled heat exchanger, the cold inflow disrupts the hot air rise. Chu (1986) used in the experiment wire screens to minimize the penetration of cold air inflow in the system.

The analysis of airflow through wire mesh can be comparable with fluid flow in the porous media. The pore size can be of the order of a few millimeter size or the order of centimeter or large. So the single-phase flow of air through the porous material depends on permeability tensor, which is a characteristic of porous material that indicates ability of fluid flow. When the gas pressure is very low and pores size is very small, then the mean free path of the gas molecules is affected by the pore sizes, and therefore, velocity slip occurs. Darcy–Weisbach equation depends on Fanning friction factor and can estimate the flow resistance when air is flowing through wire mesh (Kolodziej & Lojewska, 2009). Another mathematical correlation described by Kolodziej and Lojewska (2009) is that of Ergun (1952). The Ergun mathematical correlation described pressure drop through packed beds. The problem with Ergun model is that the void fraction should not exceed 50% which is usually more for wire meshes.

The new model was proposed by Kolodziej and Lojewska (2009) to estimate total pressure drop when air flows through wire mesh. The new model is the combination of Blake–Kozeny's laminar flow pressure drop mathematical correlation and Burke–Plumer's turbulent flow pressure drop mathematical correlation.

### *Boundary Layers Around Cylinder*

Wire mesh is made of round shape thin stainless steel wire that could be compared with a thin cylinder. The wires in the wiremesh are crossing one another and create small holes. When the fluid flow these holes, it modify the fluid flow characteristic depends on Reynolds number. Therefore, it is important to understand the modified fluid flow pattern in the presence of wire meshes. To understand the fluid flow pattern through wire meshes, at first it was considered that a thin layer region near to the cylinder surface was created due to viscous effects when fluid flows through it that is known as boundary layer. The Bernoulli equation was used to predict the pressure variation outside the boundary layer. In this case, the local pressure can be determined if the local-free stream velocity is known. Study showed that in the natural convection processes, the face velocity of the wire meshes was very low varying from 0.01 to 0.03 m/s which is very difficult to measure accurately (Nag & Datta, 2007; Kreith, 1999).

The flow of fluid at high Reynolds number, the boundary layer of the fluid become turbulent and seperation take places at the rear end of the object but at low Reynolds number the flow always encounter with drag and friction coefficient. Therefore, the fluid flow across a single circular cylinder is frequently encountered in practice but putting the magnitude of the drag and heat transfer coefficient is very complicated because complexity of the flow pattern is visualized around the cylinder. At high pressure, the boundary layer of a fluid over the surface of a cylinder separates, and beyond this point, the direction of fluid flow near to surface is opposite direction to the mainstream and that causes turbulent eddies. Both side eddies extended downstream to produce a turbulent wake at the rear of the cylinder as shown in Fig. 2.9. Fluid particles are striking the cylinder at the stagnation points and the pressure at that point raises equal to the velocity head $\frac{\rho u^2}{2g}$ over. This pressure is more than the free stream pressure. The boundary layer created is along the surface of the cylinder by the divided flow and it allows the velocity to reach a maximum on both sides and also experiences zero velocity at the stagnation point in the rear.

The pattern of flow around a circular cylinder undergoes a series of changes depends on Reynolds number and rate of heat transfer depends on flow pattern. At low Reynolds number which is less than equal to one, there is no flow separation was observed and the flow adheres to the surface. The flow streamlines maintain the same pattern that predicted from potential flow theory. The inertia force is very small that can be negligible and the drug is caused by viscous forces. If the Reynolds number increase and reached approximately 10 (Re=10) the inertia force in the fluid flow

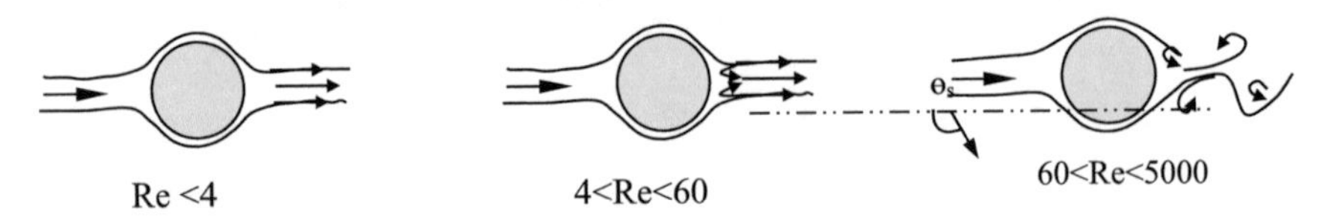


**Fig. 2.9** Effects of Reynolds number across a wire mesh screen

**Table 2.1** Flow characteristic and separation angle at a circular cylinder (incompressible flow)

| Reynolds number regime | Flow characteristic | Separation angle |
|---|---|---|
| Re = 0 | Steady, no wake | – |
| 3–4 < Re > 30–40 | Steady, symmetric separation | $130^\circ \langle\theta_s\rangle 180^\circ$<br>Re = 35 Re = 5 |
| 30–40 < Re > 80–90 | Laminar unstable wake | $115^\circ \langle\theta_s\rangle 130^\circ$<br>Re = 90 Re = 35 |
| 80–90 < Re > 150–300 | Karman vortex street | |
| 150–300 < Re > $10^5$ | Laminar with vortex street instabilities | $\theta_s \approx 80^\circ$ |

becomes appreciable and two weak eddies forms in the rear of the circular cylinder. The pressure drag becomes half of the total drug. When the Reynolds number reaches approximately 100, the flow characteristic changes and vortices separate alternatively from the both side of the cylinder and the pressure drag start dominates. The velocity of flow increases and When the Reynolds number lies from $10^3$ to $10^5$, the skin frictions drag force becomes negligible compared to pressure drag force caused by turbulent eddies in the wake. In this flow the constant drag coefficient observed because the boundary layer remains laminar from the leading edge to the point of separation. At the value Reynolds number, more than $10^5$ the flow in the boundary layer becomes turbulent and the separation point moves toward the rear (Nag & Datta, 2007; Baehr & Stephan, 1998). Therefore, fluid is separation is observed with increase of fluid velocity of Reynolds number. The separation angle decrease with increase of Reynolds number. The fluid separation angle as shown in Table 2.1 can be estimated from Reynolds number suggested by Schlichting and Gersten (2016).

After the separation, the boundary layers form both surfaces of the body initially laminar, then transition to turbulent flow. The boundary layer thickness caused the displacement of external flow lines and separation occurred in the region of increasing pressure on the rear of the cylinder. After separation, boundary layer fluid no longer remains in contact with the surface and forms the viscous wake behind the object that makes the flow field typical.

## Scale of Experimental in Different Studies

Usually, the air-cooled heat exchanger is very large in size and shape that is an obstacle to conduct experiment to measure velocity and differential pressure; therefore, most of the researchers used different mathematical empirical correlations to estimate the airflow rate and pressure in the system. Mainly, in the power industries, the size of the cooling tower depends on its capacity but compared to other systems in the power plant, the size of the cooling tower is quite large. Sometimes, the whole structure of the cooling tower can be as big as 500 m high and the platform area approximately 35,000 m$^2$ (Schreüder & Du Plessis, 1989) or tower height 100 m and base diameter about 70 m (Vlasov et al., 2002). Therefore, small-scale model was considered as an alternative approach to conduct experiment on air-cooled heat exchangers. To implement the model results to full-scale air-cooled heat exchanger, the emphasis is on the similarity of Reynolds number and Froude number. In the real tower, the Reynolds number is about $10^7$; to achieve this Reynolds number in the model's scales, the velocity of air needs to be about 1 km/s which is supersonic velocity of air (Zhai & Fu, 2006). To develop this velocity in the laboratory scale and conduct the experiment is very difficult and impractical. Therefore, Reynolds number and Froude number for small-scale model cannot be satisfied to exchange results from model heat exchanger to original size (Wei et al., 1995; Zhai & Fu, 2006). Without this, Garmize et al. (1994) mentioned that if model is used to represent the original object, then the model should have some closeness of the physical similarity numbers like Rayleigh (Ra) and Prandtl (Pr) numbers. A model was proposed by Garmize et al. (1994). In this model, from the physical point of view, there was a similarity in the Prandtl number where the Rayleigh number had significant difference in magnitude. The Rayleigh number for the proposed model was found within range from $10^8$ to $10^9$ whereas in the actual cooling tower, the Rayleigh number for 100 m tower height was calculated nearly equal to $10^{15}$. Besides this, to conduct experiment on the laboratory scale and to measure the external flow pattern, Bender et al. (1996) used 1: 25 scale double cell cooling tower model. The external dimension of the cooling tower was 0.366 m wide, 0.469 m in height and 0.732 m in length. To calculate the effective plume chimney height, a hot flat plate face of $0.457 \times 0.457$ m$^2$ is used by Chu (2002). Although this face area showed no significant effective plume chimney height, results used for the industrial size heat exchanger with face area 2.4 m $\times$ 6.0 m had significant effective plume chimney height (Chu, 2002). So it can be concluded that the industrial size chimney supported the laboratory results reasonably well although the face dimension of laboratory model heat exchanger was small.

Garmize et al. (1994) developed experimental facilities as a model to conduct experiment on chimney-type evaporative cooling tower where warm water interacted with cold air. The height of the model was 0.5 m and the diameter of the bottom was 0.5 m. In addition, Meyer and Kroger (1998) conducted a series of experiments by using different fan and heat exchanger sizes to represent actual situation of the industrial forced draft air-cooled heat exchanger that widely used in the chemical and processes industry where low-quality waste heat has to be rejected. The aim of the research was to investigate the forced draft heat exchanger characteristics as well as the flow losses in the plenum chamber (area between the fan and the heat exchanger) and its geometry. Therefore, to a heat exchanger, elliptical tube and rectangular or elliptical fin with frontal areas 1.6 $\times$ 1.9 m, 1.64 $\times$ 1.9 m and 1.9 $\times$ 1.9 m were used, respectively, and the tube length 1.6 m (Meyer & Kroger, 1998). To study the resulting flow of plume–thermosiphon interaction: application to chimney problem, Zinoubi et al. (2005) was conducted experiment by creating thermal plume over a 7 cm diameter flat disk. Electric heater is used to heat the plate at 300 °C. From the center to the end of the disk, the temperature difference observed 5 °C (Zinoubi et al. 2005). However, Schreüder and Du Plessis in 1989 studied an air-cooled heat exchanger in a power plant that received low-pressure steam directly from low-pressure turbine and increased fin temperature is about 50 °C. During study, the ambient temperature was 20 °C and the exit, air temperature from fin tube heat exchanger was almost the same as fin temperature (Schreüder & Du Plessis, 1989).

To analyze the average velocity above a hot metallic flat place, Lorenzini (2006) conducted experiment to determine the fluid flow field over a hot flat plate by using base plate having side of 350 mm and height of 30 mm and maintain the plate temperature of 65–85 °C. To investigate the influences of the air-cooled heat exchanger geometry mainly the orientation of the heat exchanger fin surface, Meyer and Kroger (2001) conducted experiment with nine different heat exchangers had frontal area of 600 × 600 $mm^2$. The outcome of the investigation showed that the inlet airflow losses are independent of the average air velocity.

Hayashi et al. (1976) published a research paper after analysis of cross-flow in the cooling tower. A rectangular box-type cooling tower having dimensions base, width and height 300 mm, 300 mm and 900 mm, respectively. The model is made from polyvinylchloride plates. Study also showed that the thermal and dehumidifier behavior was the same as standard cross-flow-type heat exchanger. Saman and Alizadeh (2001) investigated dehumidifier or cooler both experimentally and numerically. Air is blown with maximum velocity of 1.5 m/s through the standard dimension heat exchanger breadth width and height 600 mm, 600 mm and 600 mm, respectively. The aim of the research was to investigate the effect of ambient air condition on the heat exchanger performance. The summary information about the size and shape of air-cooled heat exchanger is mentioned in Table 2.2. From the table, it can be seen that the dimension of the hot flat plate that is used for investigation of the flow temperature profile is quite small relative to full-size air-cooled heat exchanger.

Hence, the actual size and shape of the cooling tower were too big; therefore, it was difficult and not cost-effective to conduct intensive research on actual dimension cooling tower. On the other hand, to conduct the parametric study on natural draft air-cooled heat exchanger, different researchers used a different shape of model to conduct experiment on laboratory scale. This also made the situation difficult to compare experimental results with real-world cooling towers. The cooling tower's performance depend on different environmental and physical conditions such as crosswind, geographic location of chimney, surrounding temperature and different orientations of fins and tubes surface deforms airflow patterns inside heat exchanger (Zhai & Fu, 2006; Meyer & Kroger, 2001; Wei et al., 1995) that makes the situation more complicated to give a suitable dimensions of a model chimney cooling tower. In addition it was assumed that the exit air velocity of heat exchanger is uniform but still now this profile is very hard to determine.

Chu (1986) found in the literature review that the real plumes are highly influenced by the geometrical and topographical condition but to conduct experiment on plume, the main drawback was no one can produce in the laboratory exactly the real condition.

The exit air velocity of the heat exchanger is estimated to be as low as 0.01 $ms^{-1}$. The giant size and shape of the heat exchanger make the system difficult to determine

**Table 2.2** Face dimension of experimental models heat exchanger used by researchers

| Researchers | Year | Aim of research | Heat exchanger or hot plate size |
|---|---|---|---|
| Lorenzini | 2006 | To determine the fluid flow field over a hot flat plate | Side 0.35 m and height 0.03 m |
| Zinoubi, Rejeb and Ali | 2005 | To determine the temperature profile over a hot flat plate | 0.07 m diameter circular hot plate |
| Meyer and Kroger | 2001 | To investigate the influences of the air-cooled heat exchanger geometry | 0.6 m × 0.6 m |
| Saman and Alizadah | 2001 | To investigate the effect of ambient air condition on the heat exchanger performance | 0.6 m × 0.6 m |
| Mayer and Kroger | 1998 | To determine the plenum chamber losses in forced draught air-cooled heat exchanger | 1.6 m × 1.9 m<br>1.64 m × 1.9 m<br>1.9 m × 1.9 m |
| Bender et al. | 1996 | To investigate the effect of different wind wall configuration | Model scale 1: 25 |
| Garmize et al. | 1994 | To carry out a quantitative study of the main aero-thermal processes | Diameter of model was 0.5 m |
| Chu C. M., Farrant, P. E. and Bolt, T. R. | 1988 | To determine the effective plume chimney height over a model air-cooled heat exchanger | 0.457 m × 0.457 m<br>2.0 m × 3.1 m |
| Hayashi, Hirai and Ito | 1976 | An analysis of cross-flow cooling tower | Rectangular Box Cooling Tower 300 × 300 × 900 $mm^2$ |

the pressure loss inside the fanless force draft air-cooled heat exchanger experimentally under normal environmental condition (Rahman & Chu, 2007; Chu 2002; Zhai & Fu, 2006; Meyer & Kroger, 2001; Wei et al., 1995).

## Instrumentation in Natural Convection

Differential pressure, temperature, flow rate and velocity data were recorded during the fluid flow processes through a system to ensure the safety and quality control, so different measurement techniques like direct, indirect, electronic, electromagnetic and optical were used to measure the flow parameter with high accuracy.

## Temperature Measurement

Thermocouple is a device used to measure the temperature from the system. It senses temperature changes based on voltage due to temperature gradient in two wires of different materials jointed at one end. The magnitude of developed voltage depends on the temperature at the junction point. Different types of thermocouple are available in the market. The use of thermocouple depends on the temperature range therefore different researchers used different thermocouples for experiments. To measure the temperature in the 7 cm flat hot plate during developing a hot plume, Al–Cr type thermocouples were used to measure the temperature by Zinoubi et al. (2005). Laguerre et al. (2005) conducted an experiment on convection heat transfer processes for low-velocity air used for stack objects cooling. During experiment, the temperature at the entrance was measured with T-type calibrated thermocouples. Garmize et al. (1994) used thermocouples along the vertical axis to measure the temperature in four different points to record the temperature of the vapor–air mixture along and across the direction of the wind stream in the chimney-type evaporated cooling tower laboratory model to enhance heat and mass transfer processes.

## Velocity Measurement

Generally, Pitot tube is used to measure the velocity of fluid flow. The Pitot tube is used Bernoulli's energy principle which is one of the most accurate methods to measure the velocity. In the Pitot tube, an error of few percentages occurs if the tube has misalignment less than $15^0$. In addition, flow meter or hot wire anemometer is used to measure the velocity of air. But in the natural convection processes, the velocity is too small ranging from 0.01 to 0.03 m/s that are very difficult to measure by using traditional flow meter or hot wire anemometer. To measure the flow rate due natural convection. A new technique is proposed by Lorenzini, 2006. In this technique a very light weight polyethylene coil is placed above the hot plate that start rotate due to the effect of natural convection flow. This technique was published as a research article and discuss the mechanism of air velocity measurement due to natural convection. In this process, the real force of air is developed by mechanical energy created from the conservation of thermal energy available in the hot flat plate. To measure the axial velocity of air over hot flat plate, a lightweight polyethylene coil is used as tracer that is able to rotate around a rod fixed over the hot flat plate. In this experiment, the weight of the coil is an important factor because if it is too heavy, then buoyancy is insufficient to create angular velocity for it; therefore, very lightweight material is used to construct it. Without this, to conduct this experiment, high thermal conductivity and diffusivity material is chosen for hot plate will create uniform temperature and quick transient states to plate temperature variations (Lorenzini, 2006).

The rotating coil that is used to measure the updraft air velocity over the hot flat plate is disturbed with surrounding air velocity (crosswind) and temperature difference between exits and surrounding. Meyer and Kroger mentioned that the velocity profile in the plume just above the hot flat plate is uniform which is impossible in the real world (Meyer & Kroger, 1998).

The surrounding air pressure always disturbs the natural plume; therefore, only a shallow plume core forms near to the flat surface of the horizontal plate where the air has uniform velocity. The height of this plume core is as small as near to zero that could be neglected for small hot flat plate. The main aim of the Lorenzini, 2006, experiment was to determine the correlation between the thermal stratification of the air due to horizontal hot flat plate not to determine the air velocity.

The experimental results showed that the trend of the relationships between weight of the coil and angular velocity is appeared irregular but theoretically, the heavy coil should be affected by frictional force at the end of the coil support and coil contact point. The irregular velocity trend observed between the coil weight lies 1 g to less than 4 g but trend is right for the coil weight 4–6 g (Lorenzini, 2006). This is because, for very small coil weight values, the dynamic resistance of air on moving coil produces different angular rotation from what is usual. Meyer and Kroger (1998) used spaced propeller anemometers across the heat exchanger to determine the velocity profile in the force draft air-cooled heat exchangers exit. Hotwired anemometer was used by Zinoubi et al. (2005) to determine velocity of air above hot flat plate when covered with a thermosiphon. In this experiment, very small hot surface was used which length surface was only 7.0 cm. In addition, Laguerre et al. (2005) used Annubar probe to measure the airflow rate during experiment on model air-cooled heat exchanger.

## *Differential Pressure Measurement*

The measurement of air pressure is required to determine the velocity of air because the relationship between velocity and pressure is well mentioned in the energy equation. The static pressure of air can explain velocity when it is undisturbed by the measurement. Piezometer is used to measure the static pressure when the flow is parallel and the pressure variation is hydrostatic normal to the steam line. The opening for the piezometer should be very small, and the length should be more than double of its diameter. When the piezometer is fixed with system, it should normal with surface. The main problem with piezometer was small misalignment or roughness of the opening may cause error in the measurement. So for the rough surface, static tube is used that consists of a tube which is directed upstream with the end closed. The tube has radial hole and it needs to calibrate before use because it may read too high or too low. If the static tube is not able to read true static pressure, the pressure head difference is equal to the square of the velocity of flow. The alignment of static tube is very simple so that error may occur only few percentages.

Pressure is often measured on U tube manometers or micromanometer. With a micromanometer, it is possible to detect the change of pressure head about 0.001 in of fluid. So corresponding pressure change depends on the specific gravity of fluid. On the other hand, the same sensitiveness can be achieved by using electric transducers. It can sense the pressure changes by electrically measuring the displacement of sensor (Streeter, 1971; Fox et al. 2004). Laguerre et al. (2005) conducted an experiment on convection heat transfer processes for low-velocity air used for stack objects cooling. During experiment, a differential pressure transmitter model FC0352 was used to measure the pressure drop in the stack that has precision ±0.25%.

## References


Al-Waked, R. (2010). Crosswinds effect on the performance of natural draft wet cooling towers. *International Journal of Thermal Sciences, 49*(1), 218–224.

Baehr, H. D., & Stephan, K. (1998). *Heat and mass transfer*.

Baughman, A. V., Gadgil, A. J., & Nazaroff, W. W. (1994). Mixing of a point source pollutant by natural convection flow within a room. *Indoor Air, 4*(2), 114–122.

Bender, T. J., Bergstrom, D. J., & Rezkallah, K. S. (1996). A study on the effects of wind on the air intake flow rate of a cooling tower: Part 2 wind wall study. *Journal of Wind Engineering and Industrial Aerodynamics, 64,* 61–72.

Byram, G. M., & Nelson, R. M. (1974). Buoyancy characteristics of a fire heat source. *Fire Technology, 10,* 68–79.

Chu, C. C. M. (1986). *Studies of the Plumes above Air Cooled Heat Exchangers operating under Natural Convection.* Ph.D. Thesis. University of Birmingham.

Chu, C. M. (2002). A Preliminary method for estimating the effective plume chimney height above a forced-draft air-cooled heat exchanger operating under natural convection. *Heat Transfer Engineering, 23*(3), 3–12.

Chu, C. M. (2006). Use of Chilton - Colburn analogy to estimate effective plume chimney height of a forced draft air- colled heat exchanger. *Heat Transfer Engineering, 27*(9), 81–85.

Degan, G., Vasseur, P., & Bilgen, E. (1995). Convective heat transfer in a vertical anisotropic porous layer. *International Journal of Heat and Mass Transfer, 38*(11), 1975–1987.

Dirkse, M. H., van Loon, W. K., van der Walle, T., Speetjens, S. L., & Bot, G. P. (2006). A computational fluid dynamics model for designing heat exchangers based on natural convection. *Biosystems Engineering, 94*(3), 443–452.

Dixit, H. N., & Babu, V. (2006). Simulation of high Rayleigh number natural convection in a square cavity using the lattice Boltzmann method. *International Journal of Heat and Mass Transfer, 49*(3–4), 727–739.

Dol, H. S., & Hanjalić, K. (2001). Computational study of turbulent natural convection in a side-heated near-cubic enclosure at a high Rayleigh number. *International Journal of Heat and Mass Transfer, 44*(12), 2323–2344.

Dubief, Y., & Terrapon, V. E. (2020). Heat transfer enhancement and reduction in low-Rayleigh number natural convection flow with polymer additives. *Physics of Fluids, 32*(3), 033103.

Ergun, S. (1952). Fluid flow through packed columns. *Chemical Engineering Progress, 48*, 89–94.

Fisher, T. S., & Torrance, K. E. (1999). Experiments on Chimney Enhanced Free Convection. *Journal of Heat Transfer, 121*, 603–609.

Fox, R., Mcdonald, A., & Pritchard, P. (2004). *Introduction to Differential Analysis of Fluid Motion: Newtonian Fluid, Navier-Stokes Equations.* _______ Introduction fluids mechanic (Vol. 6ª, p. 213). John Wiley & Sons.

Garmize, L Kh, Dashkov, G. V., Solodukhin, A. D., & Fisenko, S. P. (1994). Laboratory modeling of the enhancement of heat and mass transfer processes in chimney—type evaporative cooling tower. *Journal of Engineering Physics and Thermo physics., 66*(2), 126–132.
Goldstein, R. J., Sparrow, E. M., & Jones, D. C. (1973). Natural convection mass transfer adjacent to horizontal plates. *International Journal of Heat and Mass Transfer, 16*(5), 1025–1035.
Hayashi, Y., Hirai, E., & Ito, N. (1976). An analysis of crossflow cooling towers. *Journal of Chemical Engineering of Japan, 9*(6), 458–463.
Henderson-Sellers, B. (1983). The zone of flow establishment for plumes with significant buoyancy. *Applied Mathematical Modelling, 7*(6), 395–398.
Hilst, G. R. (1957). The dispersion of stack gases in stable atmospheres. *Journal of Air Pollution Control Association., 7,* 205–210.
Howell, J. R., Siegel, R., & Pinar Mengū¨, M. (2010). *Thermal radiation heat transfer* (5th edn). CRC Press, Taylor & Francis Group.
Hunt, G. R., & Kaye, N. B. (2006). Pollutant flushing with natural displacement ventilation. *Building and Environment, 41*(9), 1190–1197.
Ismail, K. N., Hamid, K. H. K., Kadir, S. A. S. A., Musa, M., & Savory, R. M. (2007). Woven stainless steel wire mesh supported catalyst for $NO_X$ reduction in municipal solid waste flue (MSW) gas: synthesis and characterization. *The Malaysian Journal of Analytical Sciences, 11*(1), 246–254.
Jiji, L. M. (2006). *Heat convection.* Netherlands: Springer.
Jörg, O., & Scorer, R. S. (1967). An experimental study of cold inflow into chimneys. *Atmospheric Environment, 1*(6), 645–646.
Kaye, N. B., & Hunt, G. R. (2009). An experimental study of large area source turbulent plumes. *International Journal of Heat Fluid Flow., 30*(6), 1099–1105.
Khan, A., Ashraf, M., Rashad, A. M., & Nabwey, H. A. (2020). Impact of heat generation on magneto-nanofluid free convection flow about sphere in the plume region. *Mathematics, 8*(11), 2010.
Kołodziej, A., & Łojewska, J. (2009). Mass transfer for woven and knitted wire gauze substrates: Experiments and modelling. *Catalysis Today, 147,* S120–S124.
Kondrashov, A., Sboev, I., & Dunaev, P. (2017). Heater shape effects on thermal plume formation. *International Journal of Thermal Sciences, 122,* 85–91.
Kratjig, W. B., Konke, C., Mancevski, D., & Gruber, K. (1998). Design for durability of natural draft cooling towers by life cycle simulation. *Engineering Structure, 20*(10), 899–908.
Kreith, F. (1999). *Fluid mechanics. Mechanical. Chemical engineering.* Taylor & Francis.
Laguerre, O., Amara, S. B., & Flick, D. (2005). Experimental study of heat transfer by natural convection in a closed cavity: application in a domestic refrigerator. *Journal of Food Engineering, 70*(4), 523–537.
Leu, J. S., & Jang, J. Y. (1994). The wall and free plumes above a horizontal line source in non-Darcian porous media. *International Journal of Heat and Mass Transfer, 37*(13), 1925–1933.
Lienhard, I. V., & John, H. (2000). *A heat transfer textbook*, published by John H. Lienhard IV, Cambridge, USA.
Lorenzini, G. (2006). Experimental analysis of the air flow field over a hot flat plate. *International Journal of Thermal Science., 45,* 774–781.
Lushi, E., & Stockie, J. M. (2010). An inverse Gaussian plume approach for estimating atmospheric pollutant emissions from multiple point sources. *Atmospheric Environment, 44*(8), 1097–1107.
Mehdizadeh, F., & Rifai, H. S. (2004). Modeling point source plumes at high altitudes using a modified Gaussian model. *Atmospheric Environment, 38,* 821–831.
Meyer, C. J., & Kröger, D. G. (1998). Plenum chamber flow losses in forced draught air-cooled heat exchangers. *Applied Thermal Engineering, 18*(9–10), 875–893.
Modi, V., & Torrance, K. E. (1987). Experimental and numerical studies of cold inflow at the exit of buoyant channel flows. *Journal of Heat Transfer, 109,* 692–699.
Nag, D., & Datta, A. (2007). *Variation of the recirculation length of Newtonian and non-Newtonian power-law fluids in laminar flow through a suddenly expanded axisymmetric geometry.*

Nemati, H., Moradaghay, M., Shekoohi, S. A., Moghimi, M. A., & Meyer, J. P. (2020). Natural convection heat transfer from horizontal annular finned tubes based on modified Rayleigh Number. *International Communications in Heat and Mass Transfer, 110,* 104370.

Nielsen, H. J., & Tao, L. N. (1965, January). The fire plume above a large free-burning fire. In *Symposium (International) on Combustion* (Vol. 10, No. 1, pp. 965–972). Elsevier.

Preez, A F Du, & Kroger, D. G. (1993). Effect of wind on performance of a dry-cooling tower. *Heat Recovery Systems and CHP, 13*(2), 139–146.

Rahman, M. M., & Chu, C. M. (2007). Design of air inlet duct to investigate the buoyancy effect of plume on the draft through laboratory scale forced draft air-cooled heat exchangers. In *Presented in the 21st Symposium of Malaysian Chemical Engineers.* University Putra Malaysia.

Ravi, M. R., Henkes, R. A. W. M., & Hoogendoorn, C. J. (1994). On the high-Rayleigh-number structure of steady laminar natural-convection flow in a square enclosure. *Journal of Fluid Mechanics, 262,* 325–351.

Saman, W. Y., & Alizadeh, S. (2001). Modelling and performance analysis of a cross-flow type plate heat exchanger for dehumidification/cooling. *Solar Energy, 70*(4), 361–372.

Schlichting, H., & Gersten, K. (2016). *Boundary-layer theory.* Springer.

Schreüder, W. A., & Du Plessis, J. P. (1989). Simulation of air flow about a directly air cooled heat exchanger. *Building and Environment, 24*(1), 23–32.

Scorer, R. S. (1954). Theory of airflow over mountains: III. Airstream characteristics. *Quarterly Journal of the Royal Meteorological Society, 80,* 417.

Shiozaki, T., Maruyama, S., Mohri, T., & Hozumi, Y. (2005). Fluid flow characteristics through a wire mesh at low Reynolds number for high temperature air combustion furnace. *Nihon Kikai Gakkai Ronbunshu, B Hen/Transactions of the Japan Society of Mechanical Engineers, Part B, 71*(709), 2375–2378.

Smrekar, J., Oman, J., & Širok, B. (2006). Improving the efficiency of natural draft cooling towers. *Energy Conversion and Management, 47*(9–10), 1086–1100.

Streeter, V. L. (1971). *Fluid Mechanics* (5th edn). International Student Edition, McGraw hill Kogakusha, Ltd. Singapore.

Vlasov, A. V., Dashkov, G. V., Solodukhin, A. D., & Fisenko, S. P. (2002). Investigation of the internal aerodynamics of the chimney type evaporative cooling tower. *Journal of Engineering Physics and Thermophysics, Springer, New York, 75*(5), 1086–1091.

Wang, I. T. (1996). Determination of transport wind speed in the gaussian plume difusion equation for low lying point sources. *Atmospheric Environment, 30*(4), 661–665.

Wei, Q. D., Zhang, B. Y., Liu, K. Q., Du, X. D., & Meng, X. Z. (1995). A study of the unfavorable effects of wind on the cooling efficiency of dry cooling towers. *Journal of Wind Engineering and Industrial Aerodynamics, 1*(54), 633–643.

Wong, S. C., & Chu, S. H. (2017). Revisit on natural convection from vertical isothermal plate arrays–effects of extra plume buoyancy. *International Journal of Thermal Sciences, 120,* 263–272.

Zhai, Z., & Fu, S. (2006). Improving cooling efficiency of dry-cooling towers under cross-wind conditions by using wind-break methods. *Applied Thermal Engineering, 26*(10), 1008–1017.

Zhou, X., Yang, J., Ochieng, R. M., Li, X., & Xiao, B. (2009). Numerical investigation of a plume from a power generating solar chimney in an atmospheric cross flow. *Atmospheric Research, 91*(1), 26–35.

Zinoubi, J., Maad, B. R., & Belghith, A. (2005). Experimental study of the resulting flow of plume-thermosiphon interaction: Application to chimney problems. *Applied Thermal Engineering, 25*(4), 533–544.

# Chapter 3
# Introduction to Solar Chimney and Its Applications

**Rumana Tasnim, Rezwan us Saleheen, Md. Tarek Ur Rahman Erin, and Farhan Mahbub**

## Introduction

Solar chimney as a dependable renewable energy system has successfully gained the interest of researchers over the past decades. Severe environment issues and energy crisis can be seen all over the world for continuous and excessive use of fossil energy as buildings can consume up to 42% energy usage of the entire world annually, mostly for cooling, heating, giving electricity and air-conditioning purpose (Shi & Chew, 2012). Traditional heating and cooling systems have a noticeable impact on greenhouse gas emissions. One of the best strategies for a building to diminish energy consumption is to increase the natural ventilation inside the surrounding area using solar chimney. The overall temperature inside a room can also be diminished using a solar chimney. According to another study, the requirement of daily fan shaft in a house in Tokyo can be reduced by 50% annually due to the implementation of natural ventilation (Miyazaki et al., 2006). Solar chimney contributes a great deal in efficient residential space heating as well as cooling. With the advancement of technology, researchers have been working relentlessly to develop solar chimney which can save both energy and ensure fire safety.

## Basics of Solar Chimney with Operation

Solar chimney is a distinct illustration of reasonable utilization of thermal energy. The elementary proposal implies the application of solar energy of sun to increase ventilation in a building for controlling the building temperature and offer ventilation.

---

R. Tasnim (✉) · R. Saleheen · Md. T. U. R. Erin · F. Mahbub
Department of Mechatronics Engineering, World University of Bangladesh, Uttara, Dhaka, Bangladesh
e-mail: rumana@mte.wub.edu.bd


Md. M. Rahman and C.-M. Chu (eds.), *Cold Inflow-Free Solar Chimney*,
https://doi.org/10.1007/978-981-33-6831-6_3

Solar chimneys are the ultimate solution for efficient building design. Solar chimneys are hollow containers that establish a connection between inside and outside part of the building. A solar chimney functions on the same principle as a fireplace in home. For instance, in a fireplace, the heat from the fire makes the warm air go up the chimney and out of the home, generally addressed as draft. This draft brings out cool fresh air inside home. Meanwhile, warm air rises taking the smoke from the fire along with it. In lieu of fire generating the heat, sun heats the chimney making the air inside go upward. As the warm air inside the chimney rises, it makes the similar draft effect that brings cool fresh air into the building, causing passive ventilation. Passive ventilation means for ventilating the building where no mechanical device is used. It not only ventilates a building, but also ensures a cost-effective method to cool a building devoid of using any mechanical energy (Maerefat & Haghighi, 2010a, b).

Some developed systems use geothermal designs to cool the incoming air to store the heat of the sun for use after sunset. Solar chimneys are simple and cost-effective means to heat and ventilate a building. An essential issue to consider when building a solar chimney is its location. The solar chimney must be positioned on the roof of a structure in a region that is hit by the sun's rays. The usual scenario is that it is located on a south-facing wall as long as the home is placed in the Northern hemisphere. Secondly, a chimney is constructed and colored in dark or black material with tinted glass and insulated glazing. It is usually painted in black for highest efficiency as this reduces the sunlight which is reflected off of the chimney and absorbs heat as well as ensures that maximum heat is passed to the air inside the building. Also, the size of it is another consideration. The length of the chimney is directly proportionate with the efficiency. The chimney allows the air inside to go up creating a draft which brings fresh air in from outside that provides a comfortable ventilation throughout the outhouse. It is an optimum design target to make the best use of ventilation effect by inducing enough temperature increase in the chimney with the target of solar irradiation (Khanal & Lei, 2011). There are many design considerations which makes the chimney more effective.

## Types of Solar Chimney

Based on a study, there are mainly three types of solar chimneys which can be installed on wall, roof, and window. They are (1) Trombe wall, (2) roof solar chimney, and (3) combining both 1 and 2. Figure 3.1a illustrates a schematic of Trombe wall (Type 1) which is used for the purpose of winter heating.

A Trombe wall is a huge wall that is painted in dark color to absorb thermal energy from sunlight. It has a glass as the outside glazing and storage wall inside. The outside glazing lets the solar radiation enter into the cavity of chimney for heating. Air in the cavity then goes up and passes in the room via top opening. Trombe wall is also used for summer cooling (Fig. 3.1c). In this condition, warm air inside the room can go all the way to outside atmosphere through chimney. Another composite Trombe–Michel

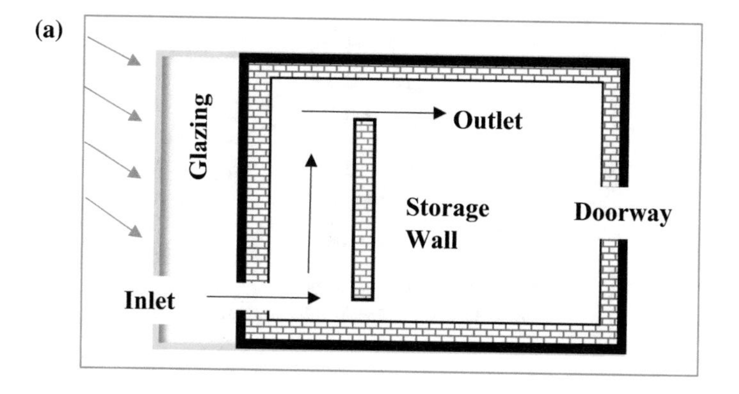


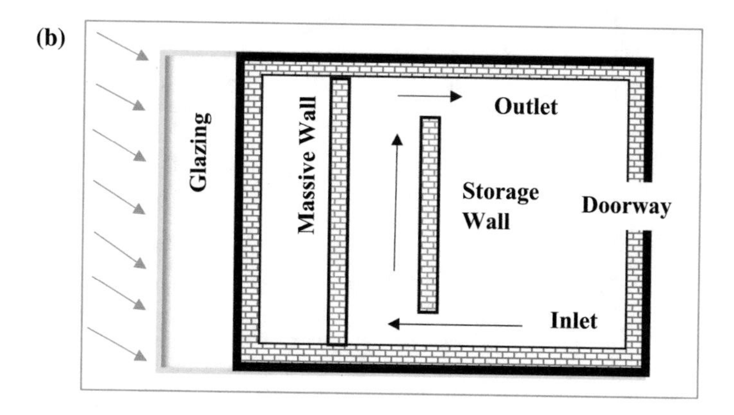


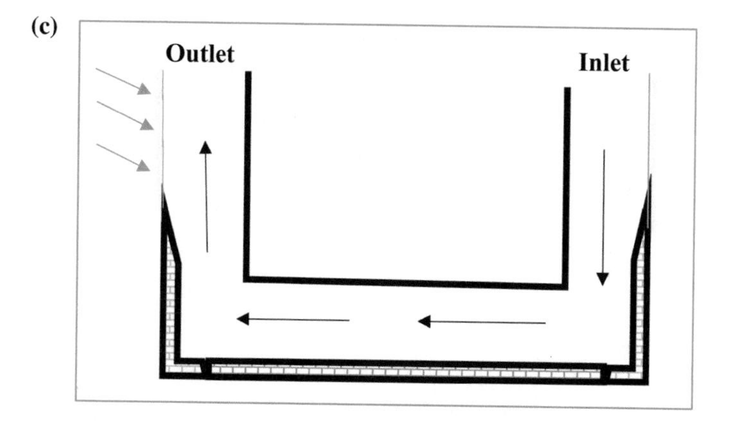


**Fig. 3.1** **a** Schematic of Trombe wall which is used for the purpose of winter heating. **b** Schematic of Trombe wall which is used for the purpose of winter cooling. **c** Schematic of a glazed solar chimney wall for tropical weather. **d** Schematic of an inclined implementation of roof solar chimney. **e** Schematic of a vertical implementation of roof solar chimney. **f** Schematic of a combined roof solar chimney

(d)

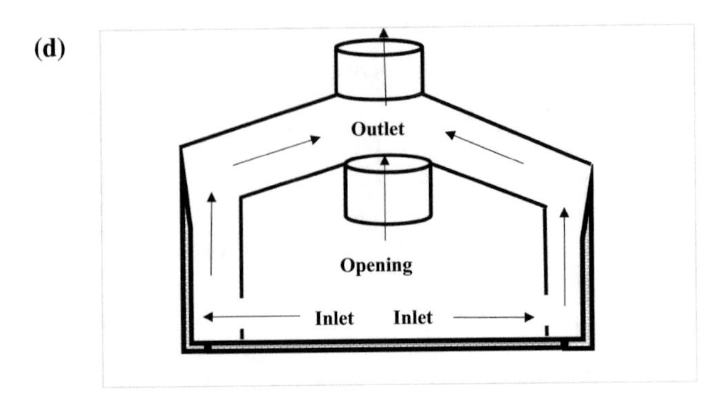


(e)

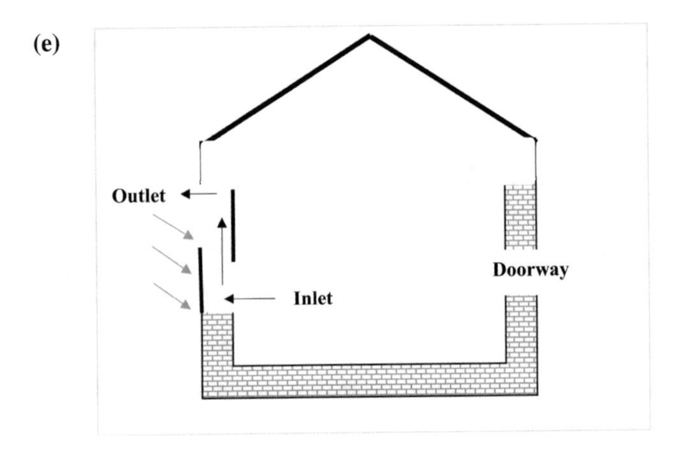


(f)

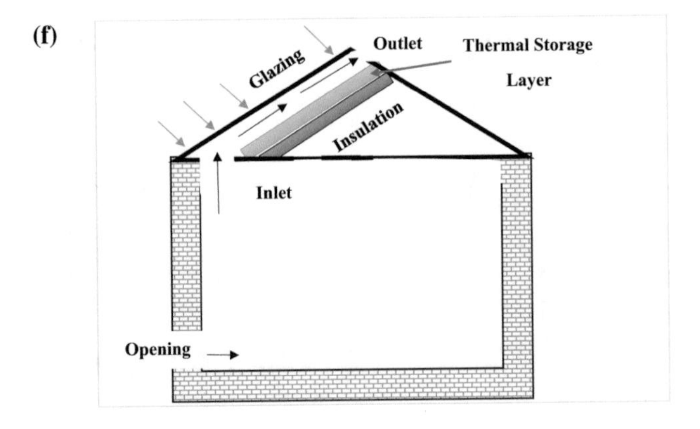


**Fig. 3.1** (continued)

wall structure is illustrated in Fig. 3.1b which can be implied for winter cooling. This structure helps to overcome heat losses from the inside room.

A glazed solar chimney wall for tropical weather is illustrated in Fig. 3.1c. It contains double glass panels with an air layer and openings positioned at the bottom (room side glass panel) and at the top (ambient side glass panel). The basic operation of the glazed solar chimney wall is similar to Trombe wall. Its function under other weather condition is troubled because of weak performance for limited size. Figure 3.1d shows a typical roof solar chimney (Type 2). As the performance of a solar chimney is based on the difference of temperature, a solar air heater (collector) at the roof is used for enhancing the temperature difference. A glazing is used outside for warming the air inside the cavity through absorption of solar radiation. To prolong heating period for late usage, a thermal storage layer is placed under the chimney cavity. Figure 3.1d shows inclined implementation of roof solar chimney, while Fig. 3.1e shows the vertical implementation. In the vertical implementation, an additional vertical chimney is used as an inlet. Both inlet and outlet are connected through two chimneys and linked with roof. One gathers solar radiation, and the other is a conventional chimney. Figure 3.1f shows a combined typed solar chimney (Type 3) consisting of both vertical and roof chimneys. Vertical solar collector is placed on roof, and ducts are gathered with the wall and roof. Air in the room can exhaust to outside via the top vertical solar collector, or via the ducts and then finally to the solar collector. In the middle of Fig. 3.1f, there is an opening to supply fresh air from outside which is on one side of the wall. Apart from the solar chimney stated above, another type of solar chimney is there which is implemented in solar chimney power plant.

## Solar Chimney for Space Cooling

The use of a solar chimney for cooling space is not like cooling using a Trombe wall. As a roof overhead cannot be placed along with a solar chimney, two extra vents are there. The first vent is placed above the chimney, whereas the second one is located at the opposite end of the building allowing ventilation. As soon as the side of chimney is heated by solar radiation, the air column inside the chimney becomes reheated. The vent above the chimney is kept open, so that the heated air does not get trapped. This heated air is drawn up to outside of the chimney, bringing fresh air inside as well as making a "draft" which offers cool and new air in the building. Figure 3.2 shows a solar chimney used for cooling activity.

Researchers have been working on enhancing the performance of chimney night ventilation (Aboul Naga & Abdrabboh, 2000) by combining a wall and a roof solar chimney. The combined system resulted in much higher air flow rate than roof solar chimney alone. In Thailand, performance of a solar chimney was analyzed in a one-room house (Khedari et al., 2003). The solar chimney was able to decrease noticeable electrical power consumption of an air conditioner on average daily. The result revealed about 10–20% decrease in average electrical power consumption

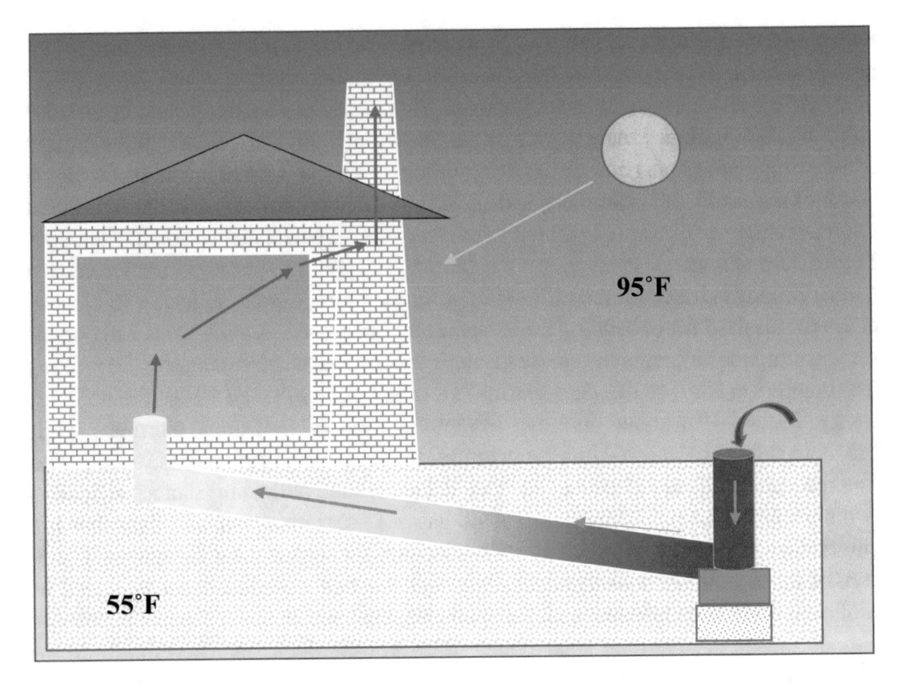


**Fig. 3.2** Solar chimney used for cooling activity

in Thailand. Solar chimney opening size was used for controlling the ventilation rate. To increase passive cooling and natural ventilation in a solar house, a system comprising of a solar chimney (SC) and an evaporative cooling cavity was used by some researchers (Maerefat & Haghighi, 2010a, b). The technique proved to be less power consuming. Furthermore, another group of researchers investigated the potential of a small-scale solar chimney making two small-scale models in a three-storied building in Thailand (Punyasompun et al., 2009). Solar chimneys were combined into the south-facing walls of a unit. The other unit operated as a reference. Room temperature data was recorded which revealed that the temperature in the room with solar chimney model was apparently 4–5 °C lower than the one with no solar chimney. After evaluating several setups, it was found that a solar chimney with an inlet opening at every floor and an outlet opening at the third floor was the most proficient.

## Solar Chimney for Space Heating

In recent world ecofriendly devices are widely appreciated. A solar chimney being an ecofriendly device has proven to be promising in moving air without the help of mechanical motor for space heating too. The use of a solar chimney for heating a space is quite simple. As soon as the side of chimney is hit by solar radiation, the air column inside the chimney gets reheated. If the vents on top exterior of the chimney are not open, the heated air rushes back to the living space. As soon as the air cools in the room, it is brought back into the solar chimney by reheating. When solar chimneys are used for heating, they function like Trombe walls.

Another key factor is determining the method of funneling the cool air to the chimney's base. The aim is to get the suction effect as well as enough ventilation. Normally, any of the two methods is used for this procedure. The first method utilizes windows on the bottommost level of the building with the purpose of capturing cooler air and funneling it through the dwelling to the chimney's base. This method is best for structures. The second method uses a pipe which is placed underground. The air inside pipe cools underground and is ultimately brought through the building to the base of the chimneys where using suction the air passes into the solar chimney. This air is later heated as well as released.

Many studies prove that the researchers have been concentrating more on space cooling. Nevertheless, during winter, the need for heating a space in some areas is much more than cooling. Very few works have been carried out by researchers in the case of space heating. A group of researchers analyzed the potential of solar chimney for this purpose (Haghighi & Maerefat, 2014). They carried out mathematical analysis of heat transfer using natural convection for a 2D vented room in contact with a cool ambience. Further studies on air gap, size of openings, outside air temperature, and solar radiation were conducted to find out the optimized operating criteria. The results reveal that the system is able to offer good indoor air condition in the room during daylight even with lessened solar intensity and low temperature.

## Components of a Solar Chimney System

A variety of solar chimneys exist nowadays. The fundamental components of a solar chimney are given below:

- The solar collector area: This is typically placed in the upper portion of the chimney. The placement, glazing type, insulation, and thermal properties of this element are essential for harnessing, retaining, and utilizing solar gains.
- Ventilation shaft: There are some essential components of this structure like position, height, cross section, and the thermal properties.
- Inlet and outlet air apertures: The location, size, and aerodynamic characteristics of these elements are also important.

## Review on Solar Chimney Power Plant Systems

A solar chimney system is built with a solar collector, a solar chimney, and turbine. The solar collector is kept on a certain height above ground, and the turbine is placed at the chimney base and is linked to an electric generator. Radiation from the sun enters collector and touches the ground surface underneath, which becomes heated and as a result heats the air, making it to go up into the central chimney. It causes the air beneath the collector to be drawn into the chimney. The air passing into the chimney drives the turbine which consequently generates electricity (Najmi et al., 2012). In 1903, the idea of a solar tower came to light for the first time by a Spaniard, Isidoro Cabanyes. A chimney was used for heating air of a house. In this house, a type of wind blade was used for generating electricity. The very first solar chimney power plant (SCPP) prototype was built by German structural engineering company, Schlaich Bergermann, in Spain during 1981 and 1982 (Schlaich, 1995). This power plant had a designed 50 kW peak power output. The solar chimney of this power plant was 194.6 m tall with 5.08 m diameter and 0.00125 m thickness. The radius of the collector was 122 m and the turbine was single rotor based with four blades. It operated from 1982 to 1989. The produced electricity was distributed in local power grid.

After the operation of this plant, many researchers worldwide started designing and developing solar chimney. A few researchers in West Hartford constructed a 10 W power capacity, while some other researchers built a 0.14 W microscale power plant in 1983 and 1985, respectively (Zhou et al., 2010a, b). Later in 1997, a group of researchers designed and developed a solar chimney model in Florida. In their design, they extended the collector base and added an absorber on the collector to enhance the power output (Pasumarthi & Sherif, 1998). In 2002, a 5 W pilot solar chimney power was made on a building roof in China (Zhou et al., 2007). In 2005, another pilot solar chimney system with glass-reinforced polyester was built (Ketlogetswe et al., 2008) in Botswana.

Later in 2011, a pilot of solar chimney (Kasaeian et al., 2010; 2011) was constructed with polycarbonate sheets at a university campus in Iran. The solar chimney had polyethylene pipe of about 12 m. In the same year, some other researchers in Iran designed a small-sized solar chimney. They carried out a thorough research on enhancing the performance of this chimney (Najmi et al., 2011). Based on their findings, to enhance the power output, asphalt or rubber can be used at the lowest end of the collector, and double glazing glasses can be used on the collector roof. Also, conical shape can be used at the chimney entrance. With the aim of enhancing functionality of solar chimney, some researchers conducted a study on the effects of diameter of collector and chimney and height on power generation capacity in solar insolation of 800 W/m$^2$ (Gholamalizadeh & Mansouri, 2013).

Also, in 2011, a pilot solar updraft power plant was constructed in Jordan (Al-Dabbas, 2012). Experimental study on the model was carried out where air velocity, temperature, solar radiation, and voltage difference were measured. Results revealed that the maximum height of solar updraft plant slowly increased with the effect of solar irradiation. In the same year, a SCPP was constructed in Turkey to determine the effect of the diameter of collector on air flow rate as well as chimney's temperature. As the researchers increased the collector area, the ground temperature increased and the temperature of air as well as air flow rate at chimney's base increased as well. In 2015, an experimental and mathematical analysis was conducted on a small-scale model of solar chimney (Shahreza & Imani, 2015). Air flow, heat transfer, and flow characteristics were computed and compared with the experimental findings. Intensifiers were used for intensifying the heat flux radiated by the sun on solar chimney. The findings revealed that use of intensifiers resulted in an overall increase in velocity in the chimney and more power generation.

Some researchers in Malaysia used six ground materials namely sand, black stone, dark green painted wood, sawdust, and pebble in order to convert solar radiation into kinetic energy. The experimental results proved the function of black stone and ceramic to be better than other materials. Ceramic with good storage capacity black stone with its abundance were suggested to as good absorbing material in solar chimneys. The usage of phase change material (PCM) enhances the thermal energy storage capacity of solar chimney. It can also prolong the usage of solar chimney to night. A study has been carried out to evaluate the performance of solar chimney with the usage of PCM in three heat fluxes: 500, 600 and 700 W/m$^2$ (Li & Liu, 2014). For all different heat fluxes under investigation, phase change period surpassed 13 h and 50 min. The air flow rates change according to the surface temperature of absorber.

Wind has an impact on the performance of solar chimney power plant. Some researchers conducted experiment to find the effects of wind speed and wind direction on inclined solar chimneys (Aja et al., 2013). The results showed that as the wind direction is from south moving north performance of system was good. However, performance failed as wind moves from east or west and performance degraded to some extent as wind moves from the north. It was also revealed that the wind speed had a considerable impact on the convective heat loss through the walls and the cover to ambient.

In 2014, some researchers in Iran built a 2 m tall solar chimney with 3 m collector radius (Kasaeian et al., 2014). They worked on the geometrical parameters of solar chimney and carried out mathematical analysis which was confirmed by the experimental data. Based on the result analysis, it was confirmed that the best substitute for the constructed solar chimney would be a chimney height of 3 m, collector inlet of 6 cm, and the chimney diameter of 10 cm. Also, according to the researchers, the diameter and height of the solar chimney are the most significant physical parameters for designing a solar chimney. In the next year, the same researchers further experimented with a solar chimney of 2 m height and 3 m collector diameter (Ghalamchi et al., 2015). The result was obtained for a number of collector inlet heights analyzing the air velocity and temperature distributions. It was revealed that the solar chimney performed better with small inlet size. To evaluate the effects of the air velocity and the internal heat load on the thermal atmosphere of the solar chimney ducts, another experiment was carried out (Tan & Wong, 2014). According to experiment findings, high ambient air speed above 2.00 m/s increases the air speed in the solar chimney ducts. Nevertheless, the ambient air speed lowers as solar irradiance becomes above 700 W/m$^2$. It was suggested that solar chimney is to be used under zero ambient air speed for better cross-ventilation. With the aim of getting new experimental data and optimizing solar chimney power plant (SCPP), a pilot solar chimney was built (Ghalamchi et al., 2016). The geometric dimension was analyzed which reveals that the best performance is achieved keeping chimney height same and making chimney diameter 10 cm and collector entrance distance 6 cm. Absorber material aluminum was proven to have better heat transfer rate than iron.

There have been some studies on large-scale solar chimney. In 2007, some researchers conducted a thorough research on optimizing large-scale solar chimney (dos Santos Bernardes et al., 1999). The use of double-roof collector enhanced the chimney operation at night. The results also presented that vegetation under the collector roof may survive with enough water but eventually will result in noticeable decrease in the plant performance. A number of studies proposed different geometries for large-scale solar chimney with multiple turbines. The performance of counter-rotating turbines was designed by some researchers for solar chimney power plants (Denantes & Bilgen, 2006). A comparative analysis was carried out between a counter-rotating turbine with inlet guide vanes and another with no inlet guide vanes. The performance analysis showed that the counter-rotating turbines without guide vanes have lower efficiency than a single-runner turbine. For different turbine layouts, better functional condition with counter-rotating turbines was established.

Another work analyzed the performance of large-scale solar chimney power plant (SCPP) in the area of Ber'Alganam (Azzawia-Libya) for an entire year (Ibrahim et al., 2019). Solar radiation was measured both experimentally and mathematically. The thermo-hydraulic behavior of the air in the solar collector and chimney was examined as well. There were in total four configurations of large-scale solar chimney power plant (SCPP) with nominal powers (5, 30, 100, and 200 MW). For an entire year, average monthly power curve and amount of annual electrical power for all the configurations were computed with capacity factors which demonstrated that the test area is a good location for this type of power plants.

Nizetic et al. are strong proponents of SCPP, after extensive analysis based on Mediterranean conditions, saying that while it is true that SCPPs operate on a very low overall efficiency (R0.1%) for electricity production, the energy source is free and renewable, so that the capital cost of construction and installation is offset by the benefits.

Ground covered by the solar collector acts as greenhouse that can be used for cultivating vegetables or fruits. This provides an opportunity to optimize the disadvantage of large installation cost to a source of possible additional revenues. But in order to cultivate foods using the ground covered by the solar collector, it requires proper irrigation with fresh water. Substantially, the prospective sites of SCPPs are selected in deserts, where land is reasonable and sunlight is ample. Hence, for maintaining the consistency in agricultural activities along with the power generation, land selection becomes an optimum challenge. South African researchers demonstrated a combined project of a SCPP and a large profitable greenhouse. During 1998, the project titled as "Greentowers" was initiated proposing a remedy for evaporation principle. Researchers implemented some black "shadowing nets" involving multiple purposes (Bonnelle, 2008). During daylight, solar radiation will be subsumed by those black shadowing nets which preserve the focal origin source of sensible heat to the moving air. Hence, no convection will occur, and a hot environment than the agricultural greenhouse air will be maintained. Usually, a slow photosynthesis phenomenon may occur due to shadowing a greenhouse. But, the light ambiences in bright sunlight regions can be accommodated using those black shadowing nets, where preservations of temperature and humidity are an additional advantage.

Usually, sun drying is an optimum process for postharvest handling process of several organisms having significant applications in processing and preservation of foods and drinks. This is an important process to dry fresh seaweed for shipping which can be accomplished spreading on a platform or by hanging with rope at coastal areas. However, attaining sun drying in an open space requires larger area and susceptible to contamination during raining seasons due to re-moistening. Solar drying is an optimum substitute for sun drying. Solar drying proposal avoids product quality loss rather shade drying and is particularly dependent on the design and operation. Several investigations in solar dryer were carried out during last decade. Significant among those demonstrations are passive dryer (Phang et al., 2015), active dryer (Djaeni & Sari, 2015) and hybrid dryer (Ali et al., 2014; Fudholi et al., 2013; Othman et al., 2012). However, the illustrations are in primary phase and require optimization for obtaining a remarkable outcome for practical implementation.

## Conclusion

Traditionally, the application of the solar chimneys is limited to use in agriculture sector for air movement in barns, silos, greenhouses, etc. It is also used to create sufficient draft in the solar dryer to dry crops, grains, fruits, and woods. The significance of air ventilation of greenhouses or gardens led toward producing healthy agricultural crops. Solar chimneys have the potential to be applied for avoiding pest distribution in agricultural products by ventilating the garden (Ghanbari & Rezazadeh, 2019). Solar-chimney-assisted passive cooling for building is another popular application of solar chimney (Hamdan, 2011). Therefore, for the past few decades, the solar chimney has been being applied to both commercial and residential buildings for decreasing heat transfer through walls as well as roofs. Solar chimney ensures natural ventilation as well.

According to the Mokheimer et al. (2017), the areas like Saudi Arabia, during summer where the demand of electricity is very high, the excess electricity is used to provide power to the cooling system and to ensure comfortable room during summer. Traditionally to meet the high demand of electricity, fossil fuels are used to be burnt that enhance its environmental footprints. Solar chimney power generation system will be one of the valid options for the area like Saudi Arabia (Mokheimer et al., 2017). Some researchers have studied the potential of solar chimney for use in rural areas of the developing countries for power generation, passive cooling, and drying agricultural products (Onyango & Ochieng, 2006; Dhahri & Omri, 2013; Chungloo & Limmeechokchai, 2009). The important application of solar chimney technology is its use for power generation. A number of energy sources such as coal, natural gas, and nuclear are used to generate large amount.

Guo et al. (2019) conducted a review of the state of the art of the solar chimney power plant SCPP, by posing seven questions still unresolved to date. They observed that to date no commercial SCPP has been constructed even though money has been proposed. Nevertheless, they are hopeful that new breakthrough will occur in future in this technology.

## References

Aboul Naga, M., & Abdrabboh, S. (2000). Improving night ventilation into low-rise buildings in hot-arid climates exploring a combined wall–roof solar chimney. *Renew Energy, 19,* 47–54.

Abuashe, I. A., Shuia, E. M., & Mariamy, A. M. (2019). Investigation of performance and production's potential of large-scale solar chimney power plant in the area of Ber'Alganam (Azzawia-Libya).

Aja, O. C., Al-Kayiem, H. H., & Abdul Karim, Z. A. (2013). Experimental investigation of the effect of wind speed and wind direction on a solar chimney power plant. In: *Proceedings of the 8th International Conference on Urban Regeneration and Sustainability*. Putrajaya, Malaysia: SC.

Al-Dabbas, M. A. (2012). The first pilot demonstration: Solar updraft tower power Plant in Jordan. *International Journal of Sustainable Energy, 31,* 399–410.

Ali, M. K. M., Wong, J. V. H., Ruslan, M. H., Sulaiman, J., & Yasir, S. M. (2014). *Internaitonal Journal of Modern Mathermatical Sciences, 10*(2), 125–136.

Bonnelle, D. (2008). Private communication.

Chungloo, S., & Limmeechokchai, B. (2009). Utilization of cool ceiling with roof solar chimney in Thailand: The experimental and numerical analysis. *Renewable Energy, 34*(3), 623–633.

Denantes, F., & Bilgen, E. (2006). Counter-rotating turbines for solar chimney power plants. *Renew Energy, 31,* 1873–1891.

Dhahri, A., & Omri, A. (2013). A review of solar chimney power generation technology. *International Journal of Engineering and Advanced Technology, 2*(3), 1–17.

Djaeni, M., & Sari, D. A. (2015). Low temperature seaweed drying using dehumidified air. *Procedia Environmental Sciences, 23,* 2–10.

dos Santos Bernardes, M. A., Molina Valle, R., & Cortez, M. F.-B. (1999). Numerical analysis of natural laminar convection in a radial solar heater. *International Journal Thermal Sciences, 38,* 42–50.

Fudholi, A., Sopian, K., Othman, M. Y., Ruslan, M. H., & Bakhtyar, B. (2013). Energy analysis and improvement potential of finned double-pass solar collector. *Energy Conversion and Management, 75,* 234–240.

Ghalamchi, M., Kasaeian, A., & Ghalamchi, M. (2015). Experimental study of geometrical and climate effects on the performance of a small solar chimney. *Renewable and Sustainable Energy Reviews, 43,* 425–431.

Ghalamchi, M., Kasaeian, A., Ghalamchi, M., & Mirza hosseini, A. H. (2016). An experimental study on the thermal performance of solar chimney with different dimensional parameters. *Renewable Energy, 91*, 477–483.

Ghanbari, M., & Rezazadeh, G. (2019). Application of solar chimney for pest control in agricultural crops. *Journal of Biosystems Engineering, 44*(4), 269–275.

Gholamalizadeh, E., & Mansouri, S. H. (2013). A comprehensive approach to design and improve a solar chimney power plant: A special Case-Kerman Project. *Applied Energy, 102,* 975–982.

Greentower. http://www.greentower.net.

Guo, P. H., Lia, T. T., Xu, B., Xu, X. H., & Lia, J. Y. (2019). Questions and current understanding about solar chimney power plant. *A Review, Energy Conversion and Management, 182*, 21–33.

Haghighi, A., & Maerefat, M. (2014). Solar ventilation and heating of buildings in sunny winter days using solar chimney. *Sustainable Cities and Society, 10,* 72–79.

Hamdan, M. O. (2011). Analysis of a solar chimney power plant in the Arabian Gulf region. *Renewable Energy, 36*(10), 2593–2598.

Kasaeian, A., Ghalamchi, M., & Ghalamchi, M. (2014). Simulation and optimization of geometric parameters of a solar chimney in Tehran. *Energy Conversion and Management, 83,* 28–34.

Kasaeian, A., Heidari, E., & NasiriVatan, S. (2010). Modeling of the temperature changes in a solar chimney. In *The Conference on Energy Management and Optimization, Tehran, Iran.*

Kasaeian, A. B., Heidari, E., & Vatan, S. N. (2011). Experimental investigation of climatic effects on the efficiency of a solar chimney power plant. *Renewable and Sustainable Energy Reviews, 15,* 5202–5206.

Ketlogetswe, C., Fiszdon, J. K., & Seabe, O. O. (2008). RETRACTED: Solar chimney power generation project—The case for Botswana. *Renewable and Sustainable Energy Reviews, 12,* 2005–2012.

Khanal, R., & Lei, C. (2011). Solar chimney—A passive strategy for natural ventilation. *Energy and Buildings, 43,* 1811–1819.

Khedari, J., Rachapradit, N., & Hirunlabh, J. (2003). Field study of performance of solar chimney with air conditioned building. *Energy, 28,* 1099–1114.

Li, Y., & Liu, S. (2014). Experimental study on thermal performance of a solar chimney combined with PCM. *Applied Energy, 114,* 172–178.

Maerefat, M., & Haghighi, A. P. (2010a). Natural cooling of stand-alone houses using solar chimney and evaporative cooling cavity. *Renewable Energy, 35*(9), 2040–2052.

Maerefat, M., & Haghighi, A. P. (2010b). Natural cooling of stand-alone houses using solar chimney and evaporative cooling cavity. *Renewable Energy, 35,* 2040–2052.

Miyazaki, T., Akisawa, A., & Kashiwagi, T. (2006). The effects of solar chimneys on thermal load mitigation of office buildings under the Japanese climate. *Renewable Energy, 31,* 987–1010.

Mokheimer, E. M., Shakeel, M. R., & Al-Sadah, J. (2017). A novel design of solar chimney for cooling load reduction and other applications in buildings. *Energy and Buildings, 153,* 219–230.

Najmi, M., Nazari, A., Mansouri, H., & Zahedi, G. (2011). Feasibility study on optimization of a typical solar chimney power plant. *Heat and Mass Transfer, 48,* 475–485.

Najmi, M., Nazari, A., Mansouri, H., & Zahedi, G. (2012). Feasibility study on optimization of a typical solar chimney power plant. *Heat and Mass Transfer, 48*(3), 475–485.

Nizetic, S., Ninic, N., & Klarin, B. (2008). Analysis and feasibility of implementing solar chimney power plants in the Mediterranean region. *Energy, 33,* 1680–1690.

Onyango, F. N., & Ochieng, R. M. (2006). The potential of solar chimney for application in rural areas of developing countries. *Fuel, 85*(17–18), 2561–2566.

Othman, M. Y., Fudholi, A., Sopian, K., Ruslan, M. H., & Yahya, M. (2012). Drying kinetics analysis of seaweed Gracilaria chnagii using solar drying system. *Sains Malaysiana, 41,* 245–252.

Pasumarthi, N., & Sherif, S. (1998). Experimental and theoretical performance of a demonstration solar chimney model—Part II: Experimental and theoretical results and economic analysis. *International Journal of Energy Research, 22,* 443–461.

Phang, H. K., Chu, C. M., Kumaresan, S., Rahman, M. M., & Yasir, S. M. (2015). Preliminary study of seaweed drying under a shade and in a natural draft solar dryer. *International Journal of Science and Engineering, 8*(1), 10–14.

Punyasompun, S., Hirunlabh, J., Khedari, J., & Zeghmati, B. (2009). Investigation on the application of solar chimney for multistory building. *Renew Energy, 34,* 2345–2361.

Schlaich, J. (1995). The solar chimney: Electricity from the sun: Edition Axel Menges.

Shahreza, A. R., & Imani, H. (2015). Experimental and numerical investigation on an innovative solar chimney. *Energy Conversion and Management, 95,* 446–452.

Shi, L., & Chew, M. Y. L. (2012). A review on sustainable design of renewable energy systems. *Renewable Sustainable Energy Reviews, 16,* 192–207.

Shi, L., Zhang, G., Yang, W., Huang, D., Cheng, X., & Setunge, S. (2018). Determining the influencing factors on the performance of solar chimney in buildings. *Renewable and Sustainable Energy Reviews.*

Tan, A. Y. K., & Wong, N. H. (2014). Influences of ambient air speed and internal heat load on the performance of solar chimney in the tropics. *Solar Energy, 102,* 116–125.

Zhou, X., Wang, F., & Ochieng, R. M. (2010a). A review of solar chimney power technology. *Renewable and Sustainable Energy Reviews, 14,* 2315–2338.

Zhou, X., Wang, F., & Ochieng, R. M. (2010b). A review of solar chimney power technology. *Renewable and Sustainable Energy Reviews, 14*(8), 2315–2338.

Zhou, X., Yang, J., Xiao, B., & Hou, G. (2007). Experimental study of temperature field in a solar chimney power setup. *Applied Thermal Engineering, 27,* 2044–2050.

# Chapter 4
# Effects of Physical Geometry on Solar Chimney Performance

**Ahmed Jawad, Mohd. Suffian bin Misran, Abu Salman Shaikat, Md. Tarek Ur Rahman Erin, and Md. Mizanur Rahman**

The global power demand increases due to the growth of population, change of lifestyle for modernization, and development of technology. All these factors highly depend on electricity which is known as one of the pure energy received from natural resources. Therefore, the power generation from renewable energy sources in a cheap, sustainable, and environmentally friendly method is becoming a hot topic in the world. In general, conventional electricity power is generated from fossil fuels which are non-renewable and responsible for pollution, global warming, and climate change. The traditional power generation systems are also responsible for other issues such as water and thermal pollution (Phan et al., 2016). These pollutants have a very high impact on the environments that cause disturbances in the natural ecosystem. According to the definition of renewable energy, it is infinite, abundant, sustainable, and environment-friendly. The most stable and accessible renewable energy sources are solar and wind energy that contributes significantly to power generation. It is estimated that wind and solar energy are expected to contribute 50% of the world's total electricity production. This is due to decreasing the cost, encouraging the expansion of renewable energy technologies, and reducing the dependencies on fossil fuel (Kempener et al., 2015). One of the well-established systems to harvest electricity from solar energy is the solar PV system. There are some more technologies such as solar chimney and solar concentrator can also be used for power generation. Solar chimney power generation system can be a suitable valid option to generate electricity. Although this technology is designed and developed for a long time, this power generation system is far behind the development due to a lack of research and

A. Jawad (✉) · Mohd. Suffian bin Misran
Faculty of Engineering, Universiti Malaysia Sabah, Kota Kinabalu, Sabah, Malaysia

Mohd. Suffian bin Misran
e-mail: suffian@ums.edy.my

A. S. Shaikat · Md. Tarek Ur Rahman Erin · Md. M. Rahman
Department of Mechatronics Engineering, World University of Bangladesh, 151/8, Green Road, Dhaka 1205, Bangladesh


Md. M. Rahman and C.-M. Chu (eds.), *Cold Inflow-Free Solar Chimney*,
https://doi.org/10.1007/978-981-33-6831-6_4

financing. The size of the solar chimney and capital cost are identified as the main barriers to the development of this technology. The first working prototype of the solar chimney is designed and developed in Manzanares, Spain, but it had less power generating capacity due to technical difficulties and complex adaptation (Brigitte, 2007). However, there is no combustion of fossil fuels present in the solar chimney power generation system, thus it is considered as low emission, pollution-free, and environmentally friendly technology (Chen, 2014). The solar chimney power generation system is considered a promising solution for future electricity generation since it is robust and can be installed in a wide variety of climate systems. The maintenance cost of the solar chimney power generation system is considered very low compared to other energy conversion technology. Besides, the infrastructure of a solar chimney is reliable as it can work for more than 80 years continually generating electricity efficiently (Grose, 2014).

The working principle of the solar chimney power generation system is very simple to explain. It has three main parts named solar collector, power generator, and chimney or draft. Solar collectors receive energy from solar radiation and increase the air temperature. The hot air from the solar collector rises upwards in the chimney due to the buoyancy effects. The chimney is equipped with a turbine or a power generation system that uses the kinetic energy of hot air movement and converts it into mechanical energy. This mechanical energy is used to generate electrical energy by a generator that is coupled with the mechanical turbine. The hot and less dense air exhausts into the surroundings through the chimney; this phenomenon is also known as the stack effect. The performance of the solar chimney power generator so far was estimated from numerical studies, mathematical models, and small-scale prototypes. The results indicate that this technology can be used for power generation where solar energy is abundant like deserts. The concept of a solar chimney power plant is very simple, and its application can also be used in other fields such as agriculture or thermal comfort system. The solar chimney can be used in the paddy and vegetable dryer as well as a passive cooling or ventilator for buildings (Chungloo & Limmeechokchai, 2007). Also, the chimney can be used for the distillation of wastewater as well as to produce drinking water efficiently from seawater (Zuo et al., 2011). The performance of the solar chimney depends on the diameter, pattern, and material quality of the solar collectors. Several studies have been carried out on solar collector qualities, design, and heat storage capacity. Some of the studies are focused on the enhancement of solar collectors performance at night as well as in winter when there is no sunlight or solar radiation is low compared to hot and sunny days (Bernardes, 2013; Kayiem, 2006; Choi et al., 2016).

Studies showed the thermal efficiency of the solar chimney power plant depends on the heat input in the solar collector and the depth of the solar collector is not considered as influencing factor of thermal performance. So the thermal performance of the solar chimney can be defined as the function of heat input only, whereas the mass flow rate in the solar chimney is highly influenced by the heat input and inlet area of the solar collector. The turbine efficiency is a function of mass flow rate; therefore, the turbine efficiency depends on heat input as well as inlet area of the solar collector. Study also showed that height of the chimney is responsible for the stack effects and also considers as one of the important performance dominant factors of solar chimney. The ambient air temperature at higher altitude is low, compared to ground level which is responsible for higher density difference between inlet and outlet of the chimney as a result producing more power. Although long chimney generates more power, but in terms of maintenance and operation point of view, the gigantic height of the chimney is considered one of the major drawbacks of a solar chimney. In addition, the cold air from the surrounding also penetrates the chimney, which is known as cold inflow or flow reversal, resulting in loss of draft in the chimney that leads to a reduction of power generation as well (Toghraie et al., 2018; Chu et al., 2012a, b; Burek & Habeb, 2007; Dai et al., 2003).

## Cold Inflow and Solar Chimney

The solar collector dimensions and the chimney height plays an important role in the efficiency of a solar chimney power plant. Therefore, many numerical and prototype models have been analyzed with different solar collector parameters and solar chimney's height. The height of the chimney is a major drawback since the infrastructure will need a large area, higher capital, and maintenance cost. Taller and sophisticated structure needs more technologically advanced equipment and more skilled labor to operate and maintain the system when needed. A good example would be solar chimney power plant projects with a chimney height of 1.5 km which were canceled or redesigned due to higher capital cost by the Namibian government (Cloete, 2008). Also, another issue with a tall solar chimney is the effect of cold inflow phenomena. In this phenomenon, the cold air enters the chimney exit from the surrounding area especially at high altitude where the temperature is generally lower (Moore & Garde, 1983). The cold inflow may decrease 20–30% of the thermal power plant efficiency. It was reported that natural draft cooling towers experienced cold inflow problems when the exit air velocity inside the chimney is small leading to a significant reduction in performance (Kloppers & Kröger, 2004). Pretrus investigated the effect of cold inflow phenomenon on solar chimney performance and observed the cold air from the surroundings enters the exit point of the chimney causing the hot air inside the chimney to cool down, thereby decreasing the velocity of air significantly and reducing the efficiency of the solar chimney (Petrus, 2007). He also reported that cold inflow or flow reversal exists in the chimney if any of the following incidents occur

- The exit air temperature drops, and it has become less than the surrounding air temperature.
- The chimney diameter is big enough so that cold air can penetrate from the side.
- Presence of strong crosswind may cause cold inflow in the chimney.

Three different techniques can be used to solve the problem of cold inflow. The first method is known as mechanical or forced ventilation system. A fan is introduced within the chimney that creates sufficient draft to create flow convection inside the chimney, but this will consume power and increase the operational cost. The forced draft ventilation system is risky during operation, and additional safety measures need to be taken. The second option is to increase the solar collector areas in the solar chimney power plant. The bigger the diameter, the solar collector receives more heat, and the temperature of the air inside the solar collector is higher; the high-temperature hot air creates sufficient draft so that the hot air flows from the chimney easily (Petrus, 2007). The third and most reliable solution is by installing a wire mesh screen at the chimney exit. The wire mesh screen at the chimney exit helps to enhance the exit air temperature and significantly increases the efficiency of the chimney by reducing the effects of cold inflow or flow reversal. According to Chu et al. (2012a, b), wire mesh screen on the top of a circular natural draft chimney is successfully able to stop the effect of cold inflow and enhanced the draft of the chimney (Chu et al., 2012a, b).

## Solar Chimney Power Plants

The uses of renewable energy are increasing day by day all over the world due to the reserve of fossil fuels which are now under threat. The consumption of fossil fuels is increasing significantly due to change of lifestyles and modernization. The reserves of fossil fuel are going to become insufficient if the consumption of fossil fuel is in current trends. (Chen, 2011; Erias & Grajetzki, 2016). Besides, decreasing reserve of fossil fuels is not the only concern, environmental pollution due to the combustion of fossil fuels is a major issue needed to be addressed. Environmental pollution may lead to respiratory and other health problems for human beings as well as other habitats (Bozkurt, 2010; Hausfather, 2014). According to the international renewable energy agency by the year 2030, the electricity generation from the renewable energy sources will be estimated about 40% for the total electricity generation (Saygin et al., 2015). In the beginning, wind and solar energies are used for mechanical work and drying mainly in the agricultural sector. Today, they are used mainly for electric power generation, which is subsequently used in power industry, homes, electric automobiles, and many more. Furthermore, wind energy can also be used for grinding and sailing ships. However, not all renewable energy is environmentally friendly. Other renewable energies such as hydro and thermal power are causing harm to the living habitat due to change of local environment significantly (McCully, 2001).

The first functional solar chimney power plant prototype rated at 50 kW was built in the 1980s whose chimney height and diameter were 195 m and 10 m, respectively. The height of the solar collector was maintained 1 m from the ground, and the diameter was 250 m. The plant was in operation for eight years before it was decommissioned due to quality of the construction materials. At the time, iron was used as the main construction material, and it was severely affected by rough water and rust. The solar chimney power plant was demolished in 1989. The prototype model was developed for research purpose where measurements of operating parameters such as temperature, velocity, and humidity were conducted. A total of 180 sensors were used in the experiments (Grose, 2014).

The experimental results indicated that if the height and effect of cold inflow of solar chimney can be eliminated, then solar chimney is a viable option for renewable energy generation since it provides clean energy and without excessively harming local habitats. The study also showed that cultivation of vegetables with enhanced greenhouse effect within the collector is also possible in the solar chimney power plant. The hot air from the collector can also be used for drying vegetables but this concept is still under research stage. National Geographic report on the solar chimney is inspiring for establishing solar chimney power plant since it supports ecological reasons. The solar chimney can be constructed at the wasteland, where clear land is in abundances, such as desert and remote areas. Solar chimney systems have better performance than a solar PV system compared, since operation performance of the PV system drops significantly in the desert due to high-temperature and cloudy condition. Besides, in the solar chimney power plant, the solar collector required less maintenance and cleaning compared to a solar PV system. The study also shows that wind turbine efficiency in an enclosed condition showed eight times better performance compared to an open space condition (Grose, 2014). The application of solar chimney is not limited to only power generation system, but it can be used for water distillation, which makes the solar chimney more desirable (Zuo et al., 2011; Ming et al., 2017).

China constructed the experimental prototype solar chimney model at Mongolia. The height of the chimney was 50 m and able to generate electricity about 200 kw. The electricity generation from solar chimneys saves the environment by reducing emission equivalent to burning of 100 tons of coal. Also, a 200 kW solar chimney power plant saves about 900 tons of water in comparison with thermal power plants. This optimistic result is a good indication to look forward to a bigger capacity solar chimney power plant to obtain a sustainable energy source (Chen, 2014). The US government also planned to establish a solar chimney power plant with a capacity of 200 MW at Arizona, Texas. The height of the proposed solar chimney will be about 1000 m which is considered the most challenging part of the project. At the same time, China and Australia governments also proposed almost similar projects, with a height of 1000 m and to generate about 200 MW electricity (Bansod, 2014). The Namibian government has also shown a great interest in the solar chimney power plant project and proposed solar chimney Green Tower in the year of 2008. The chimney is constructed under a considerable height of 1500 m with a diameter of about 280 m. The size of the solar collector is estimated to be about 27 km$^2$ whose electricity generation capacity is 400 MW (Brigitte, 2007).

## Working Principles of Solar Chimney Power Plant

Solar chimney power plant is also called a solar updraft power plant which is considered as a heat engine since the solar power plant heat or thermal energy is converted into mechanical engineering as shown in Fig. 4.1. From the figure, it is found that during each cycle operation of a solar chimney $Q_H$ is the amount of heat energy taken from the solar collector by the air and $Q_C$ is the heat energy provided to the environment which is considered as a cold sink. If the work done by the hot air to the turbine is denoted by $W$, then

$$W = Q_H - Q_c \tag{4.1}$$

Therefore, the efficiency of the solar chimney can be defined as

$$\eta = \frac{\text{net work done}}{\text{energy taken from solar collector}}$$
$$\eta = \frac{Q_H - Q_C}{Q_H} = 1 - \frac{Q_C}{Q_H} \tag{4.2}$$

In the solar chimney power plant, the $Q_C$ and $Q_H$ are the functions of the ambient temperature and solar collector temperature, respectively, then the theoretical maximum efficiency of the solar chimney can be written as

$$\eta = \frac{T_{\text{Collector}} - T_a}{T_{\text{Collector}}} = 1 - \frac{T_a}{T_{\text{Collector}}} \tag{4.3}$$

From Eqs. 4.2 and 4.3, for solar chimney operating at maximum efficiency

$$\frac{Q_C}{T_a} = \frac{Q_H}{T_{\text{Collector}}} \tag{4.4}$$

Therefore, the maximum theoretical efficiency of the solar chimney depends on the quantity of heat received by the air from the solar collector and the quantity rejected to the environment. In the solar chimney power plant, the solar radiation

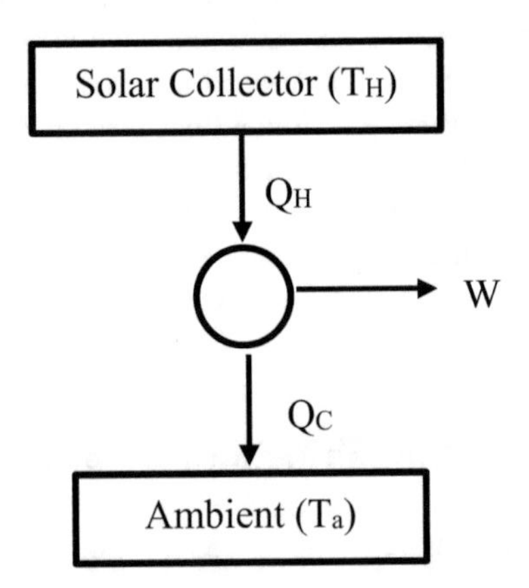


**Fig. 1** Solar chimney as heat engine

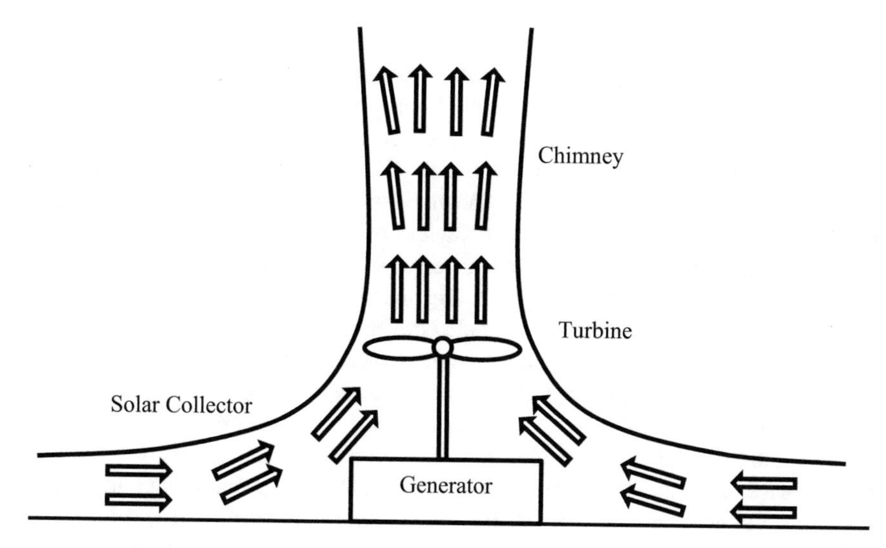


**Fig. 2** Solar chimney power plant

acts as a heat source that creates sufficient airflow to run a turbine situated in the solar chimney and generates electricity as shown in Fig. 2.

Solar chimney power plant consists of the three major components, namely solar collector, turbine and chimney. The solar collector is made with a transparent material which is either glass or plastic depending on the required demand of transitivity. The solar radiation is used to heat the air and change the internal energy of air molecules causing the air to flow. A turbine is placed either at the exit of the solar collector or at the beginning of the chimney. Electricity is generated by the turbine. The application of solar chimney is not limited to power generation only; it has received the highest appreciation as passive cooling devices as well as ventilators. This is also used for the wastewater and saline water distillation to achieve the fresh drinkable water.

## Solar Collector and Heat Sources

Solar collector or heat source is considered as the heart of the solar chimney power plant. The main component of solar collectors is the framework with the implanted transparent roof that can be cheap transparent plastic or glass. It works like a greenhouse, trapping heat from the solar radiation; therefore, the size (area) of the solar collector depends on the capacity of the solar chimney power plant. It can be from a few square meters to even thousands of square meters. The solar chimney power plant efficiency depends on the performance of the solar collector, solar collector design as well as solar irradiations (Al-kayiem et al., 2019). In the year 2004, Bonnelle proposed a new concept design of the solar collector with ribs and large inlet area which reduce the friction losses compared to the conventional circular disk-shaped solar collector (Bonnelle, 2004). Another important factor suggested by Ali in the year 2017 is the solar collector and solar chimney material (Ali, 2017).

Li et al. in the year 2016 developed a mathematical solution on the circular and square-shaped solar collector. The outcome of the model indicates that the solar collector has a circular shape and has shown better performance than square-shaped solar collector (Guo et al., 2016). Al-Kayiem et al. (2017) have reported experimental and numerical results for four different solar collectors. The solar chimney height is maintained about 6 m. The investigation results indicated that a wider solar collector enhances air velocity. This collector has shown efficient performance at night time, and it has more tendency to collect solar radiation for heat storage (Al-Azawiey et al., 2017). Fathi et al. (2016) have presented a report on the performance of solar chimney by using numerical analysis on the Manzanares prototype. The study aims to determine the effect of solar radiation on solar collector and solar chimney performance. The numerical investigation results show that that higher solar radiation tends to generate more power than the lower (Fathi et al., 2016). In the year 2006, Al-Kayiem has mentioned a numerical study on convergence, straight conventional and divergent solar collectors. The numerical investigation results indicated that convergent solar collectors harm the efficiency of solar collectors, whereas solar collectors with a divergence angle of 0.5° have shown efficiencies. The performance of a solar collector has been enhanced by 10%. The divergence angle has increased from 0.5 to 1°. The change of divergence angle has boosted up the performance of the solar collector up to 14%, but the construction of a solar collector with a divergence angle 1° is challenging (Kayiem, 2006). Klimenta and Peuteman, in the year 2014, have stated a design of pyramid shape solar chimney and conducted a mathematical analysis to estimate the performance of it. The results showed that the performance of this model was much similar to the prototype used in Spain. Due to the large sloped angle of the collector, it will receive more solar radiation and perform with higher efficiency (Klimenta & Peuteman, 2014). In the year 2016, Ilinca has reported a simulation study on the conventional solar collectors with angles of 0, 2.5, and 5°. The solar chimney was kept constant with solar collector radius 495 m under the solar radiations of 200 W/ to 1100 W/. The simulation results indicated that a solar collector with an inclination angle of 5° is capable of generating more electric power than other two types of models. It was estimated about 4 times higher power generated by the 5° inclination solar collector-assisted solar chimney model than the conventional solar collector chimney model (Ilinca, 2016). Study also showed that the inclined solar chimney has shown more efficiency than the cylindrical chimney. Since the inclined configuration allows it to receive more solar irradiation compared to others. In addition, in the inclined solar chimney, vegetables and paddy can also be dried at the exit of the draft. In the year 2013, Cao and Koonsrisuk reported about the solar chimney performance after mathematically analyzing a cylindrical-shaped inclined solar chimney. The analysis shows more satisfactory results to support the innovative concept of sloped designed solar chimney model. (Koonsrisuk & Chitsomboon, 2013; Cao et al., 2017). Furthermore, in the year 2011, Panse reported simulation results of 1 MW theoretical capacity solar chimney performance. The simulations results encourage the inclined solar chimney compared to conventional solar chimney because of inclination, and the solar collector receives maximum solar radiations. Also, inclined solar chimneys are very easy to construct on a mountain; it may reduce a significant amount of construction cost (Panse et al., 2011).

Lebbi and Daimallah (2020) reported about a numerical study on a sloped collector associated with conventional solar chimney power generation systems. The simulation results indicate that the modified power generation system can generate up to 16 times more power than a conventional design solar power plants. In this study, the best performance configuration is shown for the sloping distance of 0.8 and inclination angle of 9.1° (Lebbi & Daimallah, 2020). Nasraoui and Kchaou (2020) presented a numerical study using CFD software and the results presented in a paper published in the year 2020. Three different collector models such as conventional circular, parallel flow, and counter flow solar collectors as shown in Fig. 3 are designed for this study. The CFD simulation results showed that the maximum air velocity was observed in the counter flow solar collector model than the other two models (Nasraoui et al., 2020). Pretorius and Kro have published numerical simulation results of solar chimney power plant output in the year 2006 for different inlet areas of solar collectors. The simulation results indicate that a solar chimney with lower air inlet area has shown more efficiency than others. The lower inlet area has experience of less heat loss during operation (Pretorius & Kro, 2006). In the year 2016, Al-Azawiey has printed experimental results of four different solar collectors with different heights. The outcomes from the experiments indicate that the solar collector which has a lower height is performed more efficiently than the higher height solar collector. The study showed that the higher solar collector has a higher face area that has experienced cold inflow and crosswind. The effects of these are lowering the solar collector air temperature and significantly drop the performance of the solar chimney (Al-Azawiey et al., 2016).

In the year 2011, Kasaeian et al. published experimental results of the 5 and 15 cm high solar collector. The experimental results indicate that lower inlet size of the collector with higher collector area and chimney height tends to generate more power compared to others (Kasaeian et al., 2011). Many researchers have conducted numerical, mathematical, and experiment analysis on various physical

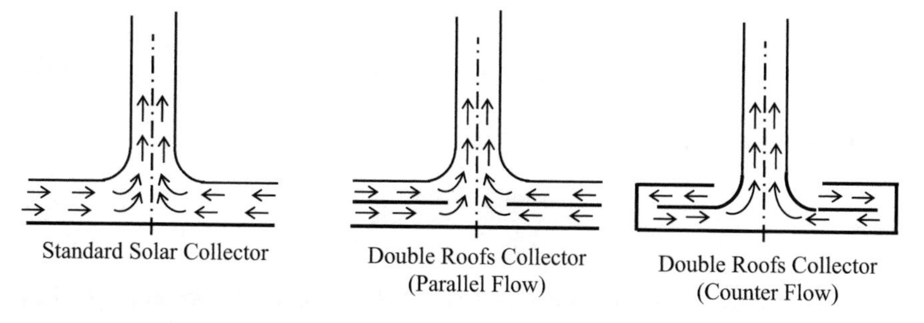


**Fig. 3** Solar collector configuration. *Source* Nasraoui et al. (2020)

geometric parameters such as solar collector height, air inlet area, solar collector area, chimney height, and chimney radius. The outcome of the study showed that the efficiency increases with increment chimney height, solar collector area, and solar collector inclination angle (Ghorbani et al., 2015; Zou & He, 2015; Bansod et al., 2016). Also, some of the researchers have analyzed the combined effect of physical parameters on the solar chimney performances. Their findings indicate that higher solar chimney, lower solar collector air gap, and bigger solar collector area have shown better performance than others (Ghalamchi, et al., 2013; Milani, et al., 2017).

## Design of Solar Collector

Solar collector is one of the main parts of a solar chimney power plant where solar radiation both direct and diffuse strikes at the collector glass or plastic roof. The air in solar collector received a fraction of energy from the reflection of solar radiation. The quantity of energy depends on the amount of solar radiation, solar incident angle as well as transmissivity performance of roof materials. The roof materials also haves reflective index, thickness, and extinction coefficient. An absorber plate is placed in the solar collector that diverts part of the energy and eventually transmitted to the air in the solar collector.

The amount of heat energy is gained by the air in the solar chimney collector can be expressed as

$$Q_{\text{coll}} = \dot{m}C_p(T_o - T_i) \tag{4.5}$$

where $\dot{m}$ is the mass flow rate; $C_p$ is the specific enthalpy of air in the solar collector; $T_o$ and $T_i$ are the solar collector outlet and inlet temperature, respectively.

The amount of energy calculated by Eq. 4.5 is considered as the output energy of the solar collector. The input energy in the solar collector depends on the solar radiation ($G$) which is measured in the standard unit W/m$^2$ and the total area of the collector ($A_{\text{coll}}$). Therefore, the total input energy of the solar collector will be the product of solar collector area ($A_{\text{coll}}$) and solar radiation ($G$) dropped in the collator area.

$$Q_{\text{in,coll}} = GA_{\text{coll}} \tag{4.6}$$

So the efficiency of the solar collector can be calculated as

$$\eta_{\text{coll}} = \frac{\text{Output Energy}}{\text{Input Energy}} = \frac{Q_{\text{coll}}}{GA_{\text{coll}}} \tag{4.7}$$

To estimate the energy flow rate, heat balance equation can be developed under a steady-state condition. The energy balance equation can be expressed as

$$\dot{Q}_{\text{coll}} = \dot{m}C_P(T_o - T_i) = \tau\alpha A_{\text{coll}}G - \beta\big(T_p - T_a\big)(A_{\text{coll}}) \tag{4.8}$$

where $\alpha$ and $\tau$ are the coefficients of transmittance and absorbance, respectively; $\beta$ is the heat loss coefficient when the heat is transferred from the absorber to the air (W/m$^2$.K); $T_p$ and $T_a$ are the plates and ambient temperature, respectively.

If the efficiency of the solar collector is considered, then the energy balance equation can be rewritten as

$$\dot{Q}_{\text{coll}} = GA_{\text{coll}}\eta_{\text{coll}} \tag{4.9}$$

From Eqs. 4.8 and 4.9

$$\eta_{\text{coll}} = \tau\,\alpha - \frac{\beta(T_P - T_a)}{G} \tag{4.10}$$

The hot air from the solar collector is passing through the chimney. So the mass flow rate from the chimney can be estimated the following equation

$$\dot{m} = \rho_{\text{air}}A_{\text{chim}}V_{\text{chim}} \tag{4.11}$$

The energy transfer rate Eq. 4.9 can be rewritten as

$$\eta_{\text{coll}} = \frac{\rho_{\text{air}}A_{\text{chim}}V_{\text{chim}}C_p(T_o - T_i)}{GA_{\text{coll}}} \tag{4.12}$$

From Eqs. 4.10 and 4.12, the velocity of air in the chimney can be estimated which is

$$V_{\text{chim}} = \frac{\tau\alpha A_{\text{coll}}G - \beta(T_P - T_a)A_{\text{coll}}}{\rho_{\text{air}}A_{\text{chim}}C_P(T_o - T_i)} \tag{4.13}$$

In this equation, the pressure losses due to friction and energy losses in the ground storage are not considered; therefore, Eq. 4.13 is independent on the solar chimney collector height.

According to Amin 2016, the performance of the solar chimney collector depends on the average air and hot plate temperature and that can be estimated from the following equation

$$T_{\text{avg,air}} = T_{\text{in,coll}} + \frac{\dot{Q}_{\text{coll}}}{A_{\text{coll}}\beta F_R}\left(1 - F''\right) \tag{4.14}$$

$$T_{\text{avg},P} = T_{\text{in,coll}} + \frac{\dot{Q}_{\text{coll}}}{A_{\text{coll}}\beta F_R}(1 - F_R) \tag{4.15}$$

where $F_R$ is the heat removal coefficient that can be estimated from the equation below

$$F_R = \frac{\dot{m}C_p}{\beta A_{\text{coll}}}\left(1 - \exp\left(\frac{\beta A_{\text{coll}} F'}{\dot{m}C_P}\right)\right) \tag{4.16}$$

In this equation, the factor $F'$ is the coefficient of efficiency of the solar collector and $F''$ is the airflow factor in the solar collector. The relations between all these factors can be express as

$$F'' = \frac{F_R}{F'} \tag{4.17}$$

The average airflow temperature in the solar collector can be solved by calculating the arithmetic mean from the inlet air temperature and the hot plate temperature. The arithmetic mean equation can be written as

$$T_{\text{avg,air}} = \frac{T_i + T_p}{2} \tag{4.18}$$

## Chimney

A chimney is a part of a plant that is used to remove unwanted heat or hot gases or dust particles from the combustion place to the surrounding environment. The movement of the hot gases depends on the stack effect due to chimney height and the temperature difference between the process side and the ambient.

$$\Delta P_{\text{stack}} = g H_{\text{chim}} \Delta\rho \tag{4.19}$$

This equation can be rewritten as

$$\Delta P_{\text{stack}} = g H_{\text{chim}} \frac{(\Delta T)}{T_{i,\text{chim}}} \tag{4.20}$$

The height of the chimney can be determined by balancing the stack effect with Bernoulli energy equations.

$$\rho g H_{\text{chim}} \frac{(\Delta T)}{T_{o,\text{coll}}} = \frac{1}{2}\rho v^2 \sum K_{\text{lossess}} \tag{4.21}$$

The velocity of moving gas can be estimated from this equation, and the volumetric flow rate of the gas in the chimney is

$$q = C_d.A_{\text{chim}}.\sqrt{2g H_{\text{chim}} \frac{(T_{o,\text{coll}} - T_a)}{T_{o,\text{coll}}}} \tag{4.22}$$

where the coefficient of discharge is $C_d$ and depends on the draft loss in the chimney; the cross-sectional area of the chimney is $A_{\text{Chim}}$; volumetric flow rate $q$ and $T_{o,\text{coll}}$ and $T_a$ are the chimney collector outlet and ambient temperature, respectively.

The application of the chimney is not only limited in the industries or in the household sector; it can be used for power generation which is known as solar chimney power plants. In the solar chimney power plant, large size chimney plays an important role to create sufficient pressure drop as well as create buoyancy effects or stack effect. The combined effect of solar collectors and chimneys creates sufficient airflow inside the chimney that can be used to run a turbine for generating electricity. The geometry such as height, cross-sectional area, and shape of the chimney plays an important role to keep a sufficient differential pressure and free from external environmental effects. According to Rahman et al. 2012, if the natural draft chimney height is more than the effective plume chimney height (EPCH), then the cold air can penetrate at the exit point of the chimney, resulting in chimney losing draught as well as loss of performance. Also, the chimney efficiency can be estimated from the amount of heat converted into kinetic energy and heat output from the collector of the solar chimney power plant. Therefore, according to El-Ghonemy 2016, the solar chimney efficiency equation can be written as

$$\eta_{\text{sc}} = \frac{\text{KE}}{\dot{Q}_{\text{coll}}} = \frac{g H_{\text{chim}}}{C_p T_{\text{amb}}} \tag{4.23}$$

This equation is also indicated that the efficiency of the chimney is influenced by the height and ambient temperature only. The amount of kinetic energy can be projected by using the following equation

$$\text{KE} = \eta_{\text{sc}} \times \dot{Q}_{\text{coll}} = \frac{g H_{\text{chim}}}{C_p T_{\text{amb}}} \times \rho_{\text{air}} \times V_{\text{chim}} \times A_{\text{chim}} \times (\Delta T_{\text{chim}}) \tag{4.24}$$

The differential pressure in the chimney is produced due to the stack effect and can be expressed as

$$\Delta P = \rho_{\text{air}} \times g \times H_{\text{chim}} \times \frac{\Delta T}{T_a} \tag{4.25}$$

Therefore, to get maximum efficiency from the solar chimney power plant, the geometry (size and shape) of the chimney is needed to be studied (El-Ghonemy 2016; Ahmed and Hussain 2018; Sakir et al. 2014; Ortega, 2011; Ekechukwu and Norton, 1997). Studies showed that the efficiency of the solar chimney depends on its height, diameter, and shape. Materials selection for the chimney is usually based on the size and shape; therefore, the chimney materials can be stainless steel, concrete reinforcement, and fabric as well. A cylindrical shape prototype model was designed and developed in Germany whose height was 195 m and the diameter of the chimney was 10 m. The success of the model indicates that solar chimney power generation systems can be considered as an economic and environmentally friendly technology. Large-scale solar chimney power plants can be designed to generate power up to 400 MW (Haaf et al., 1983; Haaf, 1984). Sakir et al. 2014 published experiment results of 3.05 m high solar chimney model. The chimney model plant covered approximately 16.4 $m^2$ area. The output of the plant is varied from 3 to 20 W depending on the solar radiation intensity. The maximum efficiency of the model plant was recorded 0.11% because of insufficient solar radiation, and the data was taken during winter. The results of the experiments suggested that solar chimneys can be constructed on top of a building for ventilation or passive cooling (Sakir et al. 2014).

In the year 2009, Zhou et al. presented theoretical analysis results of different solar chimney heights. The study aimed to maximize the height of the chimney and avoid negative buoyancy or flow reversal or cold inflow. From the theoretical analysis results, it can be concluded that the optimal chimney height is needed 615 m to generate electric power 102.2 kW. The solar chimney power generation efficiencies are also calculated from the theoretical data and found that the maximum efficiency is increased from 0.18 to 0.59% at an optimum chimney (Zhou et al. 2009). Maia et al. (2009) conducted an analytical and numerical study to determine the effect of geometric parameters on the airflow of solar chimney. The heights of the solar chimney were varied from 10 m to 50 m, and the radius is maintained at 1 m. The radius and air gap of the solar collector were 20 m and 0.5 m, respectively. The mass flow rate has a significant relation with chimney height and diameter of the solar collector. The study also showed that larger mass flow rate and velocity have a significant effect on the temperature where it decreases at larger mass flow and velocity (Maia et al., 2009). According to the Koonsrisuk and Chitsomboon, 2013, chimney height of 400 m with solar collector 200 m can generate sufficient amount of electricity to fulfill the demand for the typical village in Thailand, and investment cost is also affordable by the local government (Koonsrisuk & Chitsomboon, 2013).

El-ghonemy (2016) develops mathematical models to estimate the power output, pressure drop across the turbine or electricity generator, maximize the chimney height, airflow rate and temperature as well as the overall efficiency. The diameter and height of the chimney were 10 m and 200 m, respectively. A 500 m diameter solar collector is used to generate sufficient heat that can generate a monthly average power of 118 kW to 224 kW (El-ghonemy, 2016). Fasel et al. (2013) published CFD analysis data on solar chimney power plants. The focus of the investigation was the effect of geometric dimension on fluid dynamics and heat transfer in the solar chimney. The scaling factors that are used in this study are 1:250, 1:30, 1:10, 1:5, 1:2, 1:1, and 5:1. The study showed that the mass flow influences that performance of the solar chimney and it depends on the shape of the chimney and temperature of the collector. Fasel et al. (2013) defined the maximum available power as integrating the product of dynamics pressure and velocity at 10% of the chimney height. The simulation results show that power generation increases with the increment of chimney height and solar collector size of solar chimney power plants.

Furthermore, the tall cylindrical-shaped towers, buildings, and drafts mostly face problems like oscillations caused by the crosswinds. The crosswind effects cause flow separation on cylindrical shapes that generates vortices and circulation phenomena at the downstream. This problem can cause the tower to collapse due to vibration and oscillations under natural frequency. Wind tunnel tests on various chimney configuration were conducted and achieved the said conclusion. The crosswind creates excessive vibration due to natural frequency and causes failure on the structure (Lupi et al., 2017).

Koonsrisuk et al. (2010) discussed a theory to analyze the geometry of the solar chimney. The research aimed to maximize the operation parameters such as flow rate and temperature under a fixed area and volume. The study presented that the efficiency of the solar chimney power plant increased with the increment of the size of the plant. Koonsrisuk et al. (2010) suggested that the pressure losses at the collector entrance and the transition section between collector and chimney can be neglected. The author also suggested that the friction losses in the collector can also be neglected. Panse et al. (2011) suggested an inclined solar chimney that is cost-effective, less technical and easier to maintain compared to a vertical solar chimney. The author also claimed that an inclined solar chimney is more feasible than vertical chimney for power generation. The inclined chimney can be constructed along the inclined face of a hill that helps to reduce the construction cost significantly. Also, an inclined solar chimney allows energy to gather from both solar and wind energy (Panse et al., 2011).

In 1995, Schlaigh introduced an innovative design of solar chimney to save the infrastructure cost. The design consisted of a frame made of iron with fabric coating introduced as a floating solar chimney. Floating solar chimney has the advantage of low construction and material cost along with excellent stability in the event of an earthquake (Schlaich, 1995). In the year 2004, Papageorgiou demonstrated the concept of the self-floating solar chimney (FSC) which used its net uplift forces for stability. The configuration of FSC consists of the rings and heavy metal assembly to support the chimney during the crosswinds (Papageorgiou, 2004). In the year 2016, Beneke proposed an octagon-shaped solar chimney. Numerical analysis has been done to evaluate the performance of the new type of solar chimney and compare with conventional cylindrical shape solar chimney to find out the optimum configuration. It is found that an octagon shape chimney shows better results since the airflow increases insignificantly (Beneke et al., 2016). The divergent-shaped solar chimney tremendously increases the fluid velocity at the throat, which turns the turbine faster than usual speed, producing more energy than a conventional cylindrical chimney. Hu and Leung (2017) conducted several experiments and numerical studies on divergent chimneys. A mathematical model is developed to understand the hydrodynamics feature of the divergent chimney. The outcomes of mathematical models are compared with the simulation results to validate the results of the divergent chimneys. It is found that the optimal area ratio (entrance over exit) of the divergent chimney was changed when the solar chimney height is varied from 100 m to 300 m. It is also found that a 300 m tall divergent chimney can generate 4 times more power than a 100 m divergent solar chimney (Hu & Leung, 2017). Furthermore, the authors performed numerical solution on various designs of chimney such as cylindrical, fully divergent and three different types of diffuser-shaped draft, and the results show that divergent-shaped draft is 13 times more efficient than a cylindrical solar chimney. This is making a promising approach for the future of solar chimney application, because of its reduced capital cost if the height of the chimney can reduce (Hu et al., 2017). Ahmed and patel (2017) reported the dimensions of the solar chimney model which is optimized by using CFD simulation study. A four meter height solar chimney model was designed and developed for the experiment. The experimental and simulation results indicate that the maximum air velocity is observed at chimney throat. The maximum recorded velocity was observed 4.67 m/s for inlet opening 0.02 m. Therefore, it can be concluded that a small opening of a solar collector shows better performance than larger opening and the divergent angle $2^{o}$ is suitable options for solar chimney (Ahmed & Patel, 2017). (Zhou & Xu, 2018) analyze solar chimney power plants to determine the pressure losses in different segments. The backflow phenomena in the divergent-shaped solar chimney are also found during analysis of solar chimney. It is also found that the pressure loss is directly related to the mass flow rate (Zhou & Xu, 2018). Lebbi et al. (2015) study the effect of outlet and inlet to the solar chimney air flow rate. Therefore, numerical analysis was done for the solar chimney with different exit and inlet diameters. It is concluded that the outlet and inlet ratio is considered as an important parameter of the understanding of thermo-hydrodynamic flow behavior of solar chimney (Lebbi et al., 2015).

Studies using numerical simulation and experimental work have shown that divergent-shaped chimney is more efficient compared to the convergent-shaped solar chimney. The result analysis shows that the height and area ratio of the solar collector and divergent angle considered an important parameter that influences the flow in the solar chimney. Also, the divergent-shaped chimney has higher kinetic energy at the base of the chimney that can extract more power potential from the turbine. One of the significant drawbacks observed in the convergent shape chimney is the static pressure head reduced significantly at the exit point of the chimney due to the effect of air velocity increase. A high-pressure drop can cause the chimney to experience backflow or reverse flow (Ubhale et al., 2016; Lebbi et al., 2015; Pattanashetti & Madhukeshwara, 2014; Chu, et al. 2012, b; Jörg & Scorer, 1967).

Mehla et al. (2019) have completed an experimental study on convergent-shaped solar updraft tower with four (water jackets, black polyethylene, sand and small stones) different heat storage materials. The energy storage capacity of each material was calculated based on energy balance analysis. The maximum effect of the system is observed when small stones are used as heat storage material with the convergent-shaped solar updraft tower (Mehla et al., 2019). Cao et al. (2017) have performed simulation analysis on a solar chimney with six heat storage materials such as granite, limestone, sand, soil, water bags, and sandstone. The analysis results suggest that sand has better heat storage performance than others (Cao et al., 2017). Al-Azawiey and Hassan (2016) have performed experiments on a solar chimney with six different heat storage materials such as ceramic, sand, sad dust, green painted word, pebbles, and black stone. The experimental results suggested that black stones act as black bodies and ideal for the absorption and emitting radiation that makes black stone better heat storage material compared to others (Al-Azawiey & Hassan, 2016). Some researchers are using water jackets and water bags as a heat storage material (Choi et al., 2016; Bernardes et al., 2013). According to Chikere et al. (2013), the installation of inlet guide vanes can minimize the heat losses due to the crosswind. It also enhanced the efficiency of solar chimney (Chikere et al., 2013). There are some other issues that help to change the performance of the solar collector. Among them, crosswind has significant effect on collector performance. It harms solar collectors and enhances losses of heat. The positive impact of crosswind is found at the exit of the chimney. The strong crosswind spills the exit gas faster so the flow rate increases (Shen et al., 2014).

## Turbine

Solar chimney uses wind turbines to generate electric power. In the solar chimney power plant, a turbine is placed at the inlet of the chimney or the base of the chimney. Turbine used kinetic energy from the airflow, which is converted into mechanical energy and followed by electrical energy. Single shaft wind turbines are used to generate electric power in the solar chimney power plant. So the mechanical power developed by the wind turbine in the solar chimney power plant is

$$P_{t,\mathrm{Max}} = \frac{2}{3} \times V_{\mathrm{chim}} \times A_{\mathrm{chim}} \times \Delta P_{\mathrm{Stack}} \tag{4.26}$$

If the efficiency of solar collector, the height of the chimney and collector area is considered, and then the equation of maximum power can be rewritten as

$$P_{t,\mathrm{Max}} = \frac{2}{3} \times \eta_{\mathrm{coll}} \times \frac{g}{C_P T_{o,\mathrm{coll}}} \times H_{\mathrm{chim}} \times A_{\mathrm{coll}} \times G \tag{4.27}$$

If the efficiency of the turbine ($\eta_{\mathrm{turbine}}$) is considered, then the actual electric power ($P_e$) generated by the wind turbine in the solar chimney is

$$P_e = \frac{2}{3} \times \eta_{\mathrm{coll}} \times \eta_{\mathrm{turbine}} \times \frac{g}{C_P T_{\mathrm{amb}}} \times H_{\mathrm{chim}} \times A_{\mathrm{coll}} \times G \tag{4.28}$$

From Eq. 4.28, it can be concluded that the solar chimney electrical power output is a function of chimney height ($H_{\mathrm{chim}}$) and solar collector area ($A_{\mathrm{coll}}$). Namely solar chimney electrical power output is directly proportional to the chimney height and collector area, $H_{\mathrm{chim}}$ and $A_{\mathrm{coll}}$. Thus, the height of the chimney can be reduced by increasing the area of the solar collector optimizing the dimension of the solar chimney. Other factors such as cost of collector, chimney, and mechanical components are needed to be considered as well. Recent studies also showed that size, placement, and design of the turbine are important factors to achieve maximum efficiency of the plant. Schlaich (1995) suggested three different turbine configurations: horizontal turbines, multi-vertical turbines, and single huge vertical turbine. He suggests that the turbine in solar chimney power plant is placed at the base of the plant. Alternatively, Pasumarthi and Sherif (1998) introduced a concept where the turbine is placed at the top of the chimney. The installation and maintenance of turbines at the chimney top, however, will be a challenging issue due to solar chimney height (Schlaich, 1995; Pasumarthi & Sherif, 1998). From the experimental and numerical study, it is found that the single huge vertical turbine has minimum aerodynamic losses with high torque, but assembly of a single vertical turbine for a large-scaled chimney is challenging due to design constraints at the gear train (Fluri & Von Backström, 2008; Zuo et al., 2020). Effect of numbers of turbine blades on solar chimney performance is investigated by (Tingzhen et al., 2008). Five blade turbines in a 400 m height solar chimney associated with a 1500 m radius solar collector can generate 10 MW electric powers. The power conversion efficiency reached up to 50%. Another study showed that 6 blades wind turbine efficiency can reach up to 57%. From these studies, it can be concluded that a higher number of blades increased the rotational speed, as a result, enhanced the performance of the wind turbine power generation. (Tingzhen et al., 2008; Hanna et al., 2016; Kasaeian et al., 2017).

## References


Ahmed, O. K., & Hussein, A. S. (2018). New design of solar chimney (case study). *Case Studies in Thermal Engineering, 11,* 105–112.

Ahmed, M. R., & Patel, S. K. (2017). Computational and experimental studies on solar chimney power plants for power generation in Pacific Island countries. *Energy Conversion and Management, 149,* 61–78.

Ali, B. (2017). Techno-economic optimization for the design of solar chimney power plants. *Energy Conversion and Management, 138*, 461–473. https://doi.org/10.1016/j.enconman.2017.02.023.

Al-Azawiey, S. S., & Hassan, S. B. (2016). Heat absorption properties of ground material for solar chimney power plants. *International Journal of Energy Production and Management, 1*(4), 403–418.

Al-Azawiey, S. S., Al-Kayiem, H. H., Hassan, S. B. (2016). Investigation on the influnce of collector height on the performance of solar chimney power plant. *ARPN Journal of Engineering and Applied Sciences, 11*, 12197–12201.

Al-Azawiey, S. S., Al-Kayiem, H. H., & Hassan, S. B. (2017). On the influence of collector size on the solar chimneys performance. *MATEC Web of Conferences* 131. https://doi.org/10.1051/matecconf/201713102011.

Al-Kayiem, H., Mohammad, S. (2019). Potential of renewable energy resources with an emphasis on solar power in Iraq: An outlook. *Resources, 8*, 42. https://doi.org/10.3390/resources8010042.

Bansod, P, J., Thakre, S. B., Wankhade, N. A. (2016). Study of influence of size parameters on power output in chimney operated solar power plant. *13*, 64–68. https://doi.org/10.9790/1684-1302036468.

Beneke, L. W., Fourie, C. J. S., & Huan, Z. (2016). Investigation of an octagon-shaped chimney solar power plant. *Journal of Energy in Southern Africa, 27*(4), 38–52.

Bernardes, M. A. dos S. (2013). On the heat storage in solar updraft tower collectors - Influence of soil thermal properties. *Solar Energy, 98*, 49–57. https://doi.org/10.1016/j.solener.2013.07.014.

Bernardes, S., Zhou, X., Aure, M. (2013). On the heat storage in solar updraft tower collectors – Water bags. *91*, 22–31. https://doi.org/10.1016/j.solener.2012.11.025.
Bonnelle, D. (2004). Solar chimney, water spraying energy tower, and linked renewable energy conversion devices: Presentation, criticism and proposals.
Bozkurt, Đ. (2010). Energy resources and their effects on environment vocational school of technical studies electrical programme 2 energy resources 3 fossil fuels 4 energy resources. *WSEAS Transactions on Environment and Development, 6*, 327–334.
Brigitte, W. (2007). Green tower not all hot air. 02-20-2007.
Burek, S. A. M., & Habeb, A. (2007). Air flow and thermal efficiency characteristics in solar chimneys and Trombe Walls. *Energy and Buildings, 39*(2), 128–135.
Cao, F., Mao, Y., Zhu, T., & Zhao, L. (2017). TRNSYS simulation of solar chimney power plants with a heat storage layer. *Turkish Journal of Electrical Engineering & Computer Sciences, 25*(4), 2719–2726.
Chen, F. F. (2011). An indispensable truth.
Chen, S. (2014). Solar chimneys' may help solve China's energy woes. South China Morning Post.
Chikere Aja, O., Al-Kayiem, H. H., & Ambri Abdul Karim, Z. (2013). Analytical investigation of collector optimum tilt angle at low latitude. *Journal of Renewable and Sustainable Energy, 5*(6), 063112.
Choi, Y. J., Kam, D. H., Park, Y. W., & Jeong, Y. H. (2016). Development of analytical model for solar chimney power plant with and without water storage system. *Energy, 112*, 200–207.
Chungloo, S., Limmeechokchai, B. (2007). Application of passive cooling systems in the hot and humid climate: The case study of solar chimney and wetted roof in Thailand. *Building and Environment, 42*, 3341–3351. https://doi.org/10.1016/j.buildenv.2006.08.030.
Chu, C. C. M., Chu, R. K. H., & Rahman, M. M. (2012a). Experimental study of cold inflow and its effect on draft of a chimney. *Advanced Computational Methods and Experiments in Heat Transfer XII, WIT Transactions on Engineering Sciences, 75*, 73–82.
Chu, C. M., Rahman, M. M., & Kumaresan, S. (2012b). Effect of cold inflow on chimney height of natural draft cooling towers. *Nuclear Engineering and Design, 249*, 125–131.
Cloete, R. (2008). Namibia backing 400 MW solar tower-agri concept. http://www.engineeringnews.co.za/article/namibia-backing-400-mw-solar-toweragri-concept-2008-07-24/rep_id:4136.
Dai, Y. J., Huang, H. B., & Wang, R. Z. (2003). Case study of solar chimney power plants in Northwestern regions of China. *Renewable Energy, 28*(8), 1295–1304.
El-Ghonemy, A. (2016). Solar chimney power plant with collector. *IOSR Journal of Electronics and Communication Engineering (IOSR-JECE), 2*, 28–35.
Ekechukwu, O. V., & Norton, B. (1997). Design and measured performance of a solar chimney for natural-circulation solar-energy dryers. *Renewable Energy, 10*(1), 81–90.
Erias, A., Grajetzki, C. (2016). World energy resources 2016. *World Energy Council, 2016*, 6–46. https://doi.org//www.worldenergy.org/wp-content/uploads/2013/09/Complete_WER_2013_Survey.pdf.
Fasel, H. F., Meng, F., Shams, E., & Gross, A. (2013). CFD analysis for solar chimney power plants. *Solar Energy, 98*, 12–22.
Fathi, N., Aleyasin, S. S., & Vorobieff, P. (2016). Numerical-analytical assessment on Manzanares prototype. *Applied Thermal Engineering, 102*, 243–250. https://doi.org/10.1016/j.applthermaleng.2016.03.133.
Fluri, T. P., & Von Backström, T. W. (2008). Performance analysis of the power conversion unit of a solar chimney power plant. *Solar Energy, 82*(11), 999–1008.
Ghalamchi, M., Ghalamchi, M., Ahanj, T. (2013). Numerical simulation for achieving optimum dimensions of a solar chimney power plant. *1*, 26–31. https://doi.org/10.12691/rse-1-2-3.
Ghorbani, B., Ghashami, M., Ashjaee, M. (2015). Electricity production with low grade heat in thermal power plants by design improvement of a hybrid dry cooling tower and a solar chimney concept. *Energy Conversion and Management, 94*, 1–11. https://doi.org/10.1016/j.enconman.2015.01.044.

Grose, T. K. (2014). Solar chimneys can convert hot air to energy, but is funding a mirage? In *(2014) Nationa lGeographic*. Retrieved from http://news.nationalgeographic.com/news/energy/2014/04/140416-solar-updraft-towers-convert-hot-air-to-energy/.

Guo, P., Li, J., Wang, Y., & Wang, Y. (2016). Evaluation of the optimal turbine pressure drop ratio for a solar chimney power plant. *Energy Conversion and Management, 108,* 14–22. https://doi.org/10.1016/j.enconman.2015.10.076.

Haaf, W. (1984). Solar chimneys: part II: Preliminary test results from the Manzanares pilot plant. *International Journal of Sustainable Energy,* 2(2), 141–161.

Haaf, W., Friedrich, K., Mayr, G., & Schlaich, J. (1983). Solar chimneys part I: Principle and construction of the pilot plant in Manzanares. *International Journal of Solar Energy,* 2(1), 3–20.

Hanna, M. B., Mekhail, T. A. M., Dahab, O. M., Esmail, M. F. C., & Abdel-Rahman, A. R. (2016). Experimental and Numerical Investigation of the solar chimney power plant's Turbine. *Open Journal of Fluid Dynamics, 6*(04), 332.

Hausfather, Z. (2014). Climate impacts of coal and natural gas. *Berkerley Earth,* 1–21.

Hu, S., & Leung, D. Y. (2017). Mathematical modelling of the performance of a solar chimney power plant with divergent chimneys. *Energy Procedia, 110,* 440–445.

Hu, S., Leung, D. Y., & Chan, J. C. (2017). Impact of the geometry of divergent chimneys on the power output of a solar chimney power plant. *Energy, 120,* 1–11.

Ilinca, A. (2016). Conventional and sloped solar chimney. 1–6.

Jörg, O., & Scorer, R. S. (1967). An experimental study of cold inflow into chimneys. *Atmospheric Environment, 1*(6), 645–654.

Kasaeian, A. B., Heidari, E., Vatan, S. N. (2011). Experimental investigation of climatic effects on the efficiency of a solar chimney pilot power plant. *Renewable and Sustainable Energy Reviews, 15*, 5202–5206. https://doi.org/10.1016/j.rser.2011.04.019.

Kasaeian, A., Mahmoudi, A. R., Astaraei, F. R., & Hejab, A. (2017). 3D simulation of solar chimney power plant considering turbine blades. *Energy Conversion and Management, 147,* 55–65.

Kayiem, H. H. A. NQAA (2006). Geometry alteration effect on the performance of a solar-wind power system. *International Conference on Energy and Environment,* 50–55.

Kebabsa, H., Said, M., Lebbi, M., Daimallah, A. (2020). Thermo-hydrodynamic behavior of an innovative solar chimney. *Renewable Energy, 145*, 2074–2090. https://doi.org/10.1016/j.renene.2019.07.121.

Kempener, R., Lavagne, O., Saygin, D., Skeer, J., Vinci, S., & Gielen, D. (2015). Off-grid renewable energy systems: status and methodological issues. *The International Renewable Energy Agency (IRENA)*.

Klimenta, D., Peuteman, J. (2014). A solar chimney power plant with a square-based pyramidal shape: Theoretical considerations. *13*, 331–336..

Kloppers, J. C., Kröger, D. G. (2004). Cost optimization of cooling tower geometry. *Engineering Optimization, 36*, 575–584. https://doi.org/10.1080/03052150410001696179.

Koonsrisuk, A., & Chitsomboon, T. (2013). Mathematical modeling of solar chimney power plants. *Energy, 51,* 314–322.

Koonsrisuk, A., Lorente, S., & Bejan, A. (2010). Constructal solar chimney configuration. *International Journal of Heat and Mass Transfer, 53*(1–3), 327–333.

Lebbi, M., Boualit, H., Chergui, T., Boutina, L., Bouabdallah, A., & Oualli, H. (2015, March). Tower outlet/inlet radii ratio effects on the turbulent flow control in a solar chimney. In *IREC2015 The Sixth International Renewable Energy Congress*, pp. 1–6. IEEE.

Lupi, F., Niemann, H. J., & Höffer, R. (2017). A novel spectral method for cross-wind vibrations: application to 27 full-scale chimneys. *Journal of Wind Engineering and Industrial Aerodynamics, 171,* 353–365.

Maia, C. B., Ferreira, A. G., Valle, R. M., & Cortez, M. F. (2009). Theoretical evaluation of the influence of geometric parameters and materials on the behavior of the airflow in a solar chimney. *Computers & Fluids, 38*(3), 625–636.

McCully, P. (2001). Rivers no more: The environmental effects of large dams. *Silenced Rivers: The Ecology and Politics of Large Dams.*

Mehla, N., Kumar, K., & Kumar, M. (2019). Thermal analysis of solar updraft tower by using different absorbers with convergent chimney. *Environment, Development and Sustainability, 21*(3), 1251–1269.

Milani Shirvan, K., Mirzakhanlari, S., Mamourian, M., Kalogirou, S. A. (2017). Optimization of effective parameters on solar updraft tower to achieve potential maximum power output: A sensitivity analysis and numerical simulation. *Applied Energy 195*, 725–737. https://doi.org/10.1016/j.apenergy.2017.03.057.

Ming, T., Gong, T., de Richter, R. K., Liu, W., & Koonsrisuk, A. (2016). Freshwater generation from a solar chimney power plant. *Energy Conversion and Management, 113,* 189–200.

Ming, T., Gong, T., de Richter, R. K., Cai, C., & Sherif, S. A. (2017). Numerical analysis of seawater desalination based on a solar chimney power plant. *Applied Energy, 208*(June), 1258–1273. https://doi.org/10.1016/j.apenergy.2017.09.028.

Moore, F. K., Garde, M. A. (1983). Aerodynamic losses of highly flared natural draft cooling towers. In: Third Waste Heat Management and Utilization Conference, pp. 221–223.

Nasraoui, H., Driss, Z., Kchaou, H. (2020). Novel collector design for enhancing the performance of solar chimney power plant. *Renewable Energy, 145*, 1658–1671. https://doi.org/10.1016/j.renene.2019.07.062.

Ortega, E. P. (2011). *Analyzes of Solar Chimney Design* (*Master's Thesis*). Department of Energy and Process Engineering, Norwegian University of Science and Technology, Oslo, Norway.

Panse, S. V., Jadhav, A. S., Gudekar, A. S., & Joshi, J. B. (2011). Inclined solar chimney for power production. *Energy Conversion and Management, 52*(10), 3096–3102.

Papageorgiou, C. D. (2004). *External wind effects on floating solar chimney* (pp. 159–163). EuroPES: IASTED proceedings of power and energy systems.

Pattanashetti J. S., & Madhukeshwara, N. (2014, Jan). Numerical Investigation and Optimization of Solar Tower Power Plant. *International Journal of Research in Aeronautical and Mechanical Engineering, 2*(1), 92–104.

Pasumarthi, N., & Sherif, S. A. (1998). Experimental and theoretical performance of a demonstration solar chimney model—Part I: mathematical model development. *International Journal of Energy Research, 22*(3), 277–288.

Petrus, J. (2007). Optimization and control of a large-scale solar chimney power plant by. *Mechanical Engineering.*

Phan, L., Singh, N., Jesneck, J., Falkowski, J., Hausfather, E., Morris, W., & Scaramellino, T. (2016). U.S. Patent Application No. 15/174,073.

Pretorius, J. P., Kro, D. G. (2006). Critical evaluation.

Sakir, M. T., Piash, M. B. K., & Akhter, M. S. (2014). Design, construction and performance test of a small solar chimney power plant. *Global Journal of Research in Engineering, 14*(1) Version 1.0.

Schlaich, J. (1995). The solar chimney: electricity from the sun. Edition Axel Menges.

Shen, W., Ming, T., Ding, Y., & Wu, Y. (2014). Numerical analysis on an industrial-scaled solar updraft power plant system with ambient crosswind. *Renewable Energy, 68,* 662–676.

Tingzhen, M., Wei, L., Guoling, X., Yanbin, X., Xuhu, G., & Yuan, P. (2008). Numerical simulation of the solar chimney power plant systems coupled with turbine. *Renewable Energy, 33*(5), 897–905.

Toghraie, D., Karami, A., Afrand, M., & Karimipour, A. (2018). Effects of geometric parameters on the performance of solar chimney power plants. *Energy, 162,* 1052–1061.

Ubhale, N. N., Mallah, S. R., & Bothra, L. S. (2016). A review: numerical simulation for solar chimney by changing its radius and height. *International Journal on Recent Technologies in Mechanical and Electrical Engineering, 3*(6), 05–08.

Zhou, X., Yang, J., Xiao, B., Hou, G., & Xing, F. (2009). Analysis of chimney height for solar chimney power plant. *Applied Thermal Engineering, 29*(1), 178–185.

Zhou, X., & Xu, Y. (2018). Pressure losses in solar chimney power plant. *Journal of Solar Energy Engineering, 140*(2).
Zuo, L., Zheng, Y., Li, Z., & Sha, Y. (2011). Solar chimneys integrated with sea water desalination. *Desalination, 276*(1–3), 207–213. https://doi.org/10.1016/j.desal.2011.03.052.

# Chapter 5
# Lazy Plume Stack Effect Above Chimneys

**Chi-Ming Chu**

**Abstract** A natural draft chimney is defined by the existence of buoyancy and a solid wall barrier between two regions of fluids that differ in density. This familiar concept is approximately true for chimneys of relatively small flow area-to-height ratios. When its plume source parameter exceeds 1.0, the usual chimney does not fully define the region of buoyancy difference, but the height is extended by a plume-chimney, the magnitude has to date not been experimentally measured. Plumes are flows of free boundary layer in nature, making it virtually impossible to measure the pressure drop. The approach taken here was to employ a CFD software to perform simulations of heated chimney systems at four source Richardson numbers ranging between 0.044 and 0.53 under normal natural convection mode, and then simulated again effectively as jets, but matching the plumes' Reynolds numbers and temperature changes, achieving partial dynamic similarity. The values of effective plume-chimney height (EPCH) agreed with existing empirical formulae by 2–75%. A good correlation was found between the EPCH and the inverse square of the maximum entrainment coefficient, signifying that the degree of stack effect depends on hindering the entrainment process.

**Keywords** Lazy plume · Chimney · CFD · Zero-gravity · Large source area · Richardson number

## Nomenclature

| | |
|---|---|
| $F(z)$ | Buoyancy flux at height $z$ ($m^4s^{-3}$) |
| $g$ | Gravitational acceleration constant at sea level ($ms^{-2}$) |
| $h_o$ | Effective plume-chimney height (m) |
| $h_{SW}$ | Solid-walled chimney height (m) |
| $K_e$ | Velocity head entrance loss coefficient (–) |

C.-M. Chu (✉)
Faculty of Engineering, Universiti Malaysia Sabah, Jalan UMS, Kota Kinabalu 88300, Sabah, Malaysia
e-mail: chrischu@ums.edu.my



Md. M. Rahman and C.-M. Chu (eds.), *Cold Inflow-Free Solar Chimney*,
https://doi.org/10.1007/978-981-33-6831-6_5

| | |
|---|---|
| $K_{\text{ex}}$ | Velocity head exit loss coefficient (–) |
| $L$ | Characteristic dimension of plume source (m) |
| $M(z)$ | Momentum flux ($m^4s^{-2}$) |
| $Q(z)$ | Mass flux at height $z$ ($kgm^{-2}s^{-1}$) |
| $\text{Ri}_o$ | Richardson number at source (–) |
| $r$ | Plume radius (m) |
| $r_o$ | Plume source radius (m) |
| $T_a$ | Ambient temperature (°C) |
| $T_o$ | Hot air temperature in collector and tower (°C) |
| $u$ | Angular velocity ($ms^{-1}$) |
| $v_e$ | Entrainment velocity (radial) ($ms^{-1}$) |
| $v_{\text{max}}$ | Maximum entrainment velocity at a given height ($ms^{-1}$) |
| $w_c$ | Plume vertical centreline velocity ($ms^{-1}$) |
| $w_{co}$ | Plume vertical centreline velocity at source ($ms^{-1}$) |
| $w_o$ | Plume source mean velocity ($ms^{-1}$) |
| $z$ | Vertical height from plume source (m) |
| $\alpha$ | Entrainment coefficient (–) |
| $\Gamma_o$ | Plume source parameter (–) |
| $\Delta p_{\text{total}}$ | Total pressure drop (Pa) |
| $\Delta p_{\text{pipe}}$ | Pipe wall frictional pressure drop (Pa) |
| $\Delta p_{\text{Inlet}}$ | Pipe inlet entrance pressure loss (Pa) |
| $\Delta p_{\text{Outlet}}$ | Pipe outlet pressure loss (Pa) |
| $\Delta p_{\text{Compressor}}$ | Available compressor pressure head (Pa) |
| $\Delta T$ | Temperature rise in the collector (K) |
| $\rho_a$ | Ambient air density ($kgm^{-3}$) |
| $\rho_{av}$ | Mean heated air density in the cylinder ($kgm^{-3}$) |
| $\rho_o$ | Plume source mean density ($kgm^{-3}$) |

## Introduction

Plumes are ubiquitous on a wide range of scales in both the natural and the manmade environments. Examples that immediately come to mind are the vapour plumes above industrial smoke stacks or the ash plumes forming particle-laden clouds above an erupting volcano. However, plumes also occur where they are less visually apparent, such as the rising stream of warm air above a domestic radiator, of oil from a subsea blowout or, at a larger scale, of air above the so-called urban heat island. In many instances, not only the plume itself is of interest but also its influence on the environment as a whole through the process of entrainment (Hunt & van den Bremer, 2010).

## *History of Effective Plume-Chimney Height*

For over 40 years, heat transfer engineers have employed plume-chimney to close the solution loop of the natural convection performance of air-cooled heat exchangers, which have relatively large flow area-to-height ratios, or large Richardson numbers. Doyle and Benkly (1973) pioneered the application of "effective plume height" in the modelling of natural convection performance of forced draft air-cooled heat exchangers, when balancing the buoyant driving force against the frictional pressure drop. The "effective plume height" (EPH), an equivalent chimney height of a lazy plume, was at first defined by Doyle and Benkly (1973) as

$$h_o = \frac{\int_{h=0}^{h=h\max} \rho dh}{\rho_o} \tag{5.1}$$

The basis of the definition in Eq. (5.1), however, hypothesises that the entire plume column up to the maximum height is driving the draft through the source, which is not believed to be the case and will be shown by the determination of the value of the plume-chimney height in this work.

The plume in such a situation is a lazy plume where the centreline velocity at the plume source, at the top of a forced draft air-cooled heat exchanger, rises to reach a peak at some point above, with the necking of the plume radius.

Conventional operation of the air-cooled heat exchanger with the fans off has two worst possible cases and their consequences are identified as

1. tube blockage due to freezing and
2. vapour transmission due to inability to condense.

Reliable performance predictions for air-cooled heat exchangers operating without fans (as percentage of normal duty) are therefore important to designers and operators. Moreover, fanless cooling is a possible design option to increase operational safety and may lead to potential savings in fuel cost. The simulation of the natural convection performance of air-cooled heat exchangers requires the solution of the buoyancy and the heat transfer equations (Chu, 2002). The "effective plume-chimney height" is denoted by $h_o$ which is the height of the plume rising above the heat exchanger that contributes to the natural convection flow through the tube bundle.

The effective plume-chimney height (EPCH) was first estimated by Doyle and Benkly (1973) who produced an empirical formula derived from Morton et al. (1956) atmospheric plume rise model. As far as the author is aware, his was the first attempt to determine experimentally the value of EPCH in a project related to prototype fast breeder reactor project at Dounreay under the scenario as described above (Chu, 1986). Chu et al. (1988) reported that the experimental results yielded a negligible near-zero value of EPCH for a laboratory-scale natural draft air-cooled heat exchanger of 0.457 m × 0.457 m exit face dimensions in comparison with the chimney portion's overall height. Later, improved equations were developed using

Nusselt number correlations for heated horizontal square plates (Chu, 2002, 2006) and which have been found to predict reasonably an industrial-scale forced draft air-cooled heat exchanger natural convection performance with an error band of + 25/−40%. The procedure of applying it as part of the heat transfer and pressure drop balance iterations is described in Chu (2005) and has been suggested for the design of natural convection air-cooled heat exchangers (Sinnott & Towler, 2009). The EPH was renamed as "effective plume-chimney height" (EPCH) to signify the stack effect, and that it was a plume. It was also found by simulating the industrial test rig using CFD that the additional draft through the air-cooled heat exchanger could not be accounted for unless an EPCH was included in the pressure drop balance (Chu et al., 2017). The main issues with experimental measurements are the unstable flows and that the magnitudes of plume velocity and frictional loss are beyond the normal range of current instruments, with velocities going down to below the threshold 0.4 $ms^{-1}$ of vane anemometers and typically at 0.1 $ms^{-1}$, which though within reach of hot wire anemometers can only measure a local point velocity and the differential pressure is less than 5 Pa (Hunt & van den Bremer, 2010; Rahman et al., 2017).

## *The Concept of Plume-Chimney*

A clear-cut means to measure EPCH would benefit the field of lazy plume studies. A lazy plume is defined by a plume function at source ($z = 0$) with a value of greater than 1.0, based on Morton (1959) work on forced plumes in his attempt to correct far-field plume models with virtual origins. A lazy plume is depicted by Fig. 5.1 with the characteristic necking in the near-field region where there is acceleration, and the centreline temperature is maintained.

The principal subject of investigation here was the plume-chimney and with it, the entrainment in the near-field. The approach chosen was to divide the phenomenon into near-field and far-field as popularised by Chen and Nikitopolous (1979) rather than attempting to unify the plumes that exhibit varying entrainment coefficients by the approach of virtual origin which was first applied by Morton (1959).

For there to be an EPCH, a defining concept of a plume-chimney needs to be established. While the term chimney plume is used to describe the commonly observed plumes discharging from chimneys, plume-chimney describes the chimney or stack

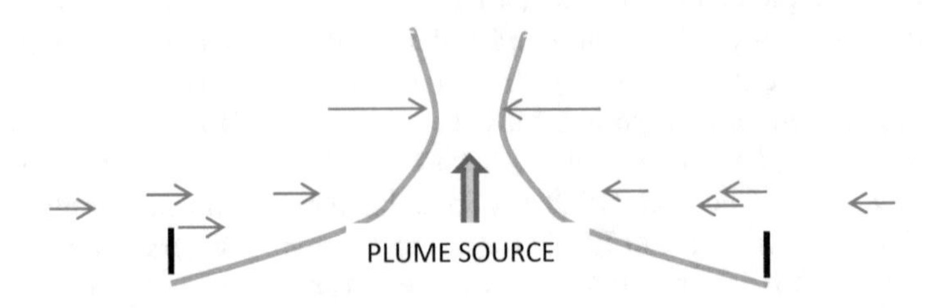


**Fig. 5.1** Lazy plume rising from a large source area

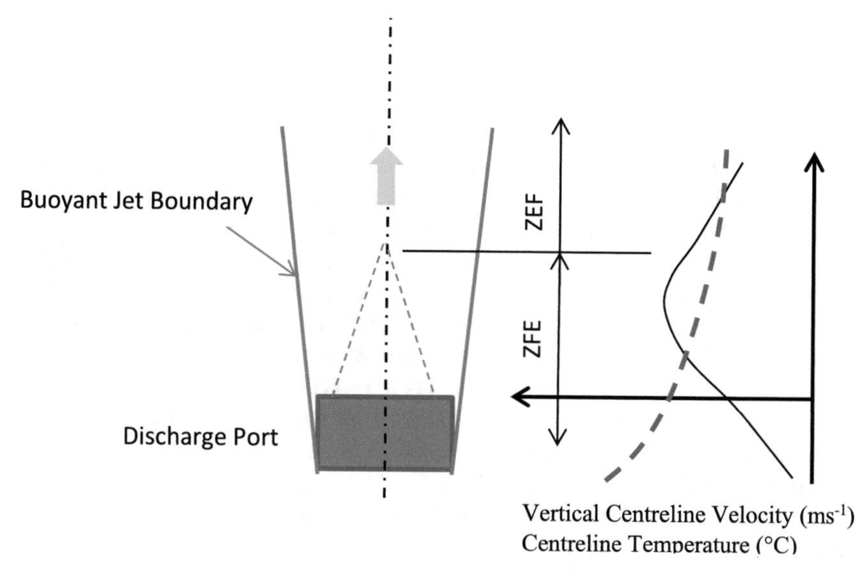


**Fig. 5.2** Zone of flow establishment with the velocity peaking at the neck and zone of established flow with both temperature and velocity in decay for a vertical buoyant jet

effect of a plume, which by definition is buoyant. If plume-chimney behaves like a chimney, then by definition, it has both buoyancy and height of a surrounding "wall", effectively extending the solid-walled chimney if it is being discharged from one. The effect is actually implied in the buoyant jet as the zone of flow establishment (ZFE) (Fig. 5.2), which Chen and Nikitopouls (1979) had distinguished from the zone of established flow (ZEF). The ZFE is defined as the region where the plume or buoyant jet accelerates from the source and reaches a peak. This approach in distinguishing two regions of plumes as "near-field" and "far-field" was also used in fire plume research (Quintiere & Grove, 1998). The existence of EPCH is believed to be due to the inefficient entrainment of a lazy plume (Chu, 2002). The stack effect if designed can enhance existing large diameter natural draft chimneys such as solar chimneys. While cold inflow impairs a chimney's performance (Chu et al., 2016), the plume stack effect enhances it; and these two are mutually exclusive.

## *Entrainment Coefficient and Plume Function*

Closely related to EPCH is the entrainment coefficient, defined by Morton et al (1956) as $\alpha = v_e/w_c$, where $v_e$ is the radial velocity (Y-component) at the radius where the vertical velocity (Z-component) has reduced to 1/e-th value of centreline vertical velocity at the same height. The value that they had recommended was 0.0833 for far-field axisymmetric plumes with self-similar radial profiles, and 0.116 for "top hat" profiles. Many researchers have concluded that Morton et al. (1956) classical assumption of a constant entrainment coefficient does not apply universally to plumes and especially near-field and lazy plumes (Fox, 1970; List & Imberger, 1973; Chen & Nikitopoulos, 1979; Henderson-Sellers, 1983; Malin, 1986; Carazzo et al., 2008; Kaye & Hunt, 2009; Marjanovic et al., 2017; Carlotti & Hunt, 2017). Quintiere and Grove (1998) had used a near-field value of 0.22 for axisymmetric plumes in their modelling of fire plumes, and Chen and Nikitopoulos (1979) showed their simulated predictions of water discharge where at source Richardson number of 0.125 and at a normalised plume height $z/r$ of 2.0, the coefficient was approximately 0.048. Malin's (1986) values were 0.054 for experimental data and 0.051 in his numerical predictions of the plume region. Carazzo et al. (2008) compiled an extensive database from six sources including Morton et al. (1956), Sneck and Brown (1974), Fan (1967), Abraham and Eysink (1969), Fox (1970) and Crawford and Leonard (1962), and applied Briggs (1969) plume rise model to the data. The range of scatter of maximum plume height predictions was shown to be reasonably contained by the entrainment coefficient values of 0.05–0.16.

For self-similar plumes, Morton et al. (1956) suggested the formula to estimate $\alpha$ from the plume radius $r$ as shown in Eq. (5.2):

$$r = \frac{6\alpha}{5} z \tag{5.2}$$

By this equation, the plume radius can be conveniently determined at the radial distance where the vertical velocity reaches a set percentage of the centreline velocity, e.g., 5%.

Marjanovic et al. (2017) results included an entrainment coefficient defined by the ratio of the rate of increase of integrated volume flux with height over the square root of integrated momentum flux across an infinitely long radius. Their method of solution was by fixing the plume inlet velocity (source) and solved as though it were a forced flow, i.e., the solution of heat transfer and the solution of fluid flow were not mutually dependent.

Based on Morton (1959) work on virtual origins of forced plumes, the plume function for any height is defined as (Hunt & Kaye, 2005; Caulfield, 1991)

$$\Gamma(z) = \frac{5F(z)Q^2(z)}{4\alpha M^{5/2}(z)} \tag{5.3}$$

where $Q(z)$ is the mass flux, $F(z)$ is the buoyancy flux and $M(z)$ is the momentum flux, and $z$ is the plume height.

The plume source parameter, the plume function at $z = 0$, is related to the source Richardson number $\mathrm{Ri}_o$ by

$$\Gamma_o = \frac{5\pi^{1/2}}{4\alpha} \frac{\rho_a^{3/2}}{\rho_o} \mathrm{Ri}_o \tag{5.4}$$

where $\mathrm{Ri}_o = \frac{r_o g(\rho_a - \rho_o)}{\rho_o w_o^2}$ and $\alpha$ is based on far-field value of 0.116 with "top hat" profile.

The plume function should converge to 1.0 as the plume progresses to the far-field.

## *CFD Simulation*

Simulation has been very useful where it is too expensive or dangerous or virtually impossible to carry out experiments. In this paper, I hope to present a CFD approach to determine this EPCH in a non-intrusive manner by measuring the pneuma-static buoyancy head in a zero-gravity condition at the same mass flowrate as a chimney under natural convection to determine the net buoyancy head which is the effective plume-chimney height (EPCH). This would be a first step in a determination of this parameter. In the data analysis, the parameter would be related to the entrainment coefficient, the zone of flow establishment, the zone of constant temperature, source Richardson number and the plume source parameter.

## Methodology

Simulations were conducted for four vertical cylinders at a fixed diameter and different lengths of 3 m, 10 m, 20 m and 50 m to produce lazy plumes at source Richardson numbers of 0.53, 0.201, 0.099 and 0.044, respectively. Data were then extracted and analysed for plume characteristics and how they relate to the EPCH. After making the comparisons with literature results primarily in terms of the far-field entrainment coefficient and plume source parameter to demonstrate the validity of the simulation, the plume-chimney will be shown to be essential for accurate modelling of the near-field entrainment process.

## *Centreline Temperature and Velocity Profiles*

The vertical velocity which was the Z-component of the plume velocity would be extracted from the simulation results and similarly the centreline temperature for all cases. These would be plotted to determine the ZFE and the zone of constant temperature, ZCT, the region of height above the source where the source temperature was maintained or raised slightly. These were analysed with reference to Richardson number to identify characteristics of the plume-chimney and the magnitude of the EPCH.

Morton et al. (1956) had suggested that the determination of entrainment coefficient could be carried out at an arbitrary plume radius, set a certain percentage of the

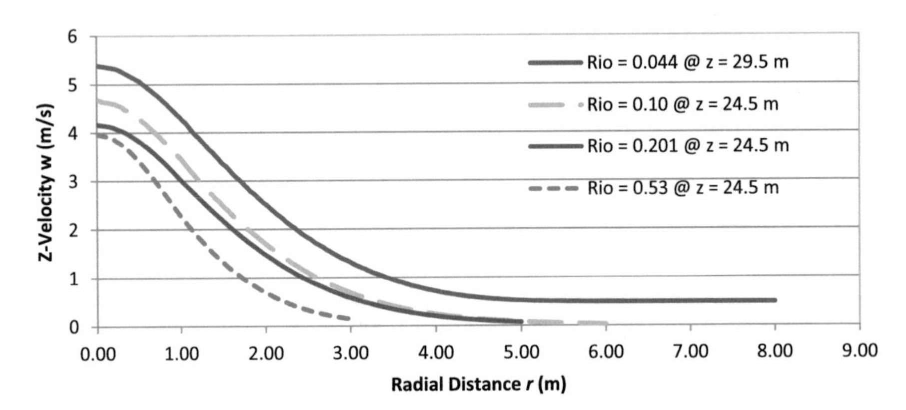


**Fig. 5.3** Far-field plume vertical velocity radial profiles where Rio = 0.53, 0.201, 0.10 and 0.044 correspond to cylinder heights of 3 m, 10 m, 20 m and 50 m respectively

centreline vertical velocity $w_c$. As the averaging of the vertical velocity is sensitive to plume radius, plume radius is determined arbitrarily and the radial profile shape changes dramatically in the source near-field, it was decided that for the purpose of observing the trends in the near-field the entrainment coefficient be defined as the ratio of the maximum radial velocity along the horizontal $Y$-axis $v_{max}$ to the centreline vertical velocity at any given height $w_c$. As will be seen later in Fig. 5.12, the maximum radial velocity at any given height appeared to be at the radius where the entrainment flow impacted on the plume boundary. The maximum radial velocity was chosen in defining the entrainment coefficient because it was representative of the laziness of a plume, as the higher a maximum radial velocity was at any given centreline vertical velocity, the greater would be the laziness or the curve bend of a near-field plume.

It is generally expected for the entrainment coefficient to asymptotically reach the value of 0.116 for "top hat" or 0.0833 for Gaussian profile in the far-field plume where the plume function would also monotonically decrease to a value of 1.0 as axial distance increases. To validate the simulation for plume flows, the far-field plume radius was estimated at the location of 5% of the axial velocity for three plume heights, and Eq. (5.2) was used to estimate the Gaussian entrainment coefficient of the plume. Typical far-field plume radial profiles of the $z$-velocity are Gaussian in nature as shown in Fig. 5.3.

## *EPCH Determination*

Vertical cylinders in one diameter size and four different heights were chosen as the simulated models with heated air flowing through. The software has been validated for cylindrical flow via two means: firstly by entering the calculated natural convection flowrate and densities through a vertical cylinder 25 m in length and 0.1 m in diameter to balance the total pressure drop calculated from published values of sharp entrance ($K_e = 0.5$) and sudden expansion ($K_{ex} = 1.0$) loss coefficients (ASHRAE, 2017) and pipe friction factor (Haaland, 1983) with the available pressure drop from the buoyancy. The agreement was within ±0.1% for heat load and ±7% for air mass flowrate, which is reasonable. Secondly, a simple experiment with a vertical pipe of 1580 mm × 0.101 m dia. that was heated at 202 W showed a steady-state inlet velocity of 0.88 ms$^{-1}$, which was within ±6% of the simulated value of 0.93 ms$^{-1}$. Near-exit plume temperature experimental data were also reported to be in good agreement with the software (Tan, 2019).

A normal means to calculate the performance of a natural draft chimney system is to iteratively balance the buoyancy head of a solid-walled column of height $h_{sw}$ against the sum of all the resistance forces (Li et al, 2017; Zhou et al, 2009). The proposed plume-chimney stack effect here $h_o$ is in addition to the buoyancy of the solid-walled column. Since the total frictional loss is balanced by the total buoyancy head, the unknown EPCH can be determined by deducting the amount of solid column buoyancy head from the amount of total resistance head, as shown in Eq. (5.3):

$$[h_o(\rho_a - \rho_o) + h_{SW}(\rho_a - \rho_{av})]g = \Delta p_{\text{total}} = \Delta p_{\text{pipe}} + \Delta p_{\text{Inlet}} + \Delta p_{\text{Outlet}} \quad (5.5)$$

where $\Delta p_{\text{pipe}}$ was estimated by using Haaland (1983) equation for the Darcy friction factor, the inlet and outlet losses by ASHRAE (2017) *K*-coefficients for air pipes and ducts. The recommended *K*-coefficients for the sharp entrance effect were 0.5, and the outlet sudden expansion was 1.0.

However, the use of friction factor, entry and exit losses was not favoured as these are calculated from published coefficients of kinetic head and would add to the uncertainty in the determination. Errors within the CFD software, however, can cancel each other internally when comparing two cases. It was decided to determine the overall buoyancy head of the column operating under natural convection (LHS of Eq. 5.5) by applying a forced convection to an identical column at the same flowrate and heat load as the natural convection case, except it is operating under zero-gravity condition, effectively turning the plume into a jet. To do so, the LHS of Eq. (5.5) was replaced by $\Delta p_{\text{Compressor}}$ as shown in Eq. (5.6):

$$\Delta p_{\text{Compressor}} = \Delta p_{\text{pipe}} + \Delta p_{\text{Inlet}} + \Delta p_{\text{Outlet}} \quad (5.6)$$

The buoyancy head of the solid column, $h_{SW}(\rho_a\text{-}\rho_{av})\, g$ on the LHS of Eq. (5.5), was compared to that of the pressure head required to shift the air by forced convection, LHS of Eq. (5.6), and the residual difference would be the value of $h_o(\rho_a - \rho_o)g$. The pressure head to be generated by the compressor had to be adjusted by trial-and-error in matching the flowrate and cylinder exit temperature. The approach thus achieved partial similarity between the plume and the jet: thermal and dynamic similarity in the solid-walled chimney, and partial dynamic similarity between the plume and jet, since the Reynolds number only applies to the jet but the plume is characterised by both Reynolds and Richardson numbers. Like the heater, the compressor was set to occupy the entire inner column as it was found that any smaller in size would cause the air to form recirculation vortices within. The natural convection mode of the plume would be buoyancy-driven as per normal, while running in forced convection mode employing a compressor would make the plume purely momentum driven. This procedure created a full thermal, inertial and frictional loss equivalence of the near-field regions of the buoyancy-driven plume and the momentum-driven jet. The frictional loss the compressor had to work against in the jet case is reckoned to be on the lower side of the plume value, since the frictional loss due to entrainment by the momentum of the jet without the buoyancy effect at virtually the same velocity would be less than that of a plume with the buoyancy plus momentum-driven entrainment.

Ideally, the air within the solid cylindrical column should be set at a uniform temperature throughout to allow for easy density difference balance. However, by doing this in the simulation, a permanent offset of temperature convergence was found to be created, the convection was then switched to be controlled by heat flux, and density variation through the column would be averaged by using the trapezium-rule as well as by simple averaging. The two averaging methods have produced very close values. The heat loads were set to yield approximately 60–70 °C of averaged outlet temperature, similar to industrial chimneys such as cooling towers.

Figure 5.4 depicts the domain and the model in the simulation. The structured mesh grid was employed for the direct solution of the Reynolds-averaged Navier–Stokes equations and run on Phoenics 2019 software (CHAM). The mesh grid has been optimised as given in Table 5.1 using two densities, where $\times$ 1 was the default and $\times$ 2 was double the default density.

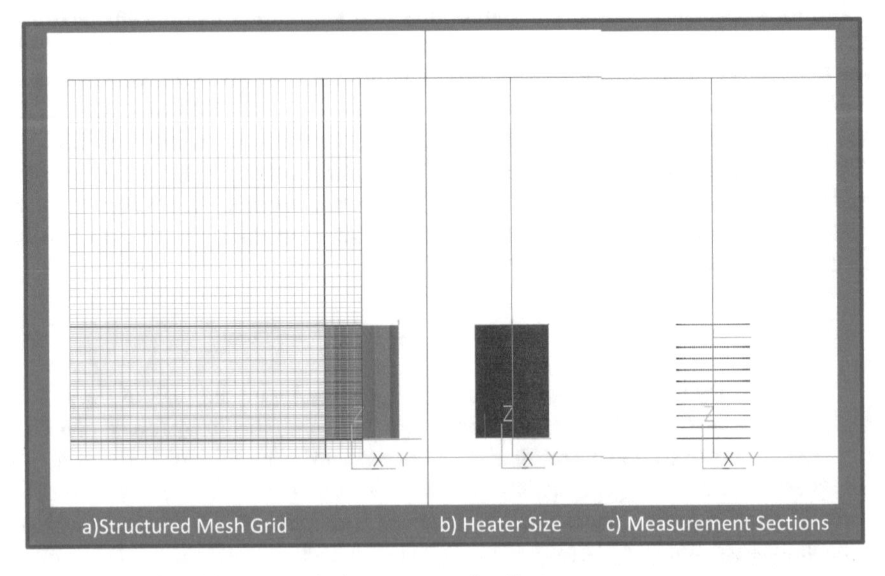


**Fig. 5.4** CFD simulation geometrical arrangement for all cases

**Table 5.1** Structured solution mesh grid optimisation for the simulation models

| Model geometry | $X$ = Angular dimension; $Y$ = Radial dimension and $Z$: Vertical dimension | |
|---|---|---|
| 6.28 rad × 3 m × 2 m dia. Domain size 6.28 rad × 8 m dia. × 25 m | Mesh grid density by number of cells in regions | Mass flowrate |
| 1 × Default mesh grid density | $X$: 1 Y: 3, 18<br>$Z$: 3, 3, 2, 3, 2, 3, 2, 2, 3, 2, 3, 19 | 5.04 |
| 2 × Default mesh grid density | $X$: 1 Y: 6, 36<br>$Z$: 6, 6, 4, 6, 4, 6, 4, 6, 4, 6, 4, 24, 14 | 5.02 |
| | Percentage Difference = | -0.40 |
| 6.28 rad × 10 m × 2 m dia. Domain size 6.28 rad × 8 m dia. × 40 m | Mesh grid density by number of cells in regions | |
| 1 × Default mesh grid density | $X$: 1 Y: 3, 18<br>$Z$: 2, 4, 4, 4, 4, 4, 4, 4, 3, 4, 4, 18 | 8.25 |
| 2 × Default mesh grid density | $X$: 1 Y: 6, 36<br>$Z$: 4, 8, 8, 8, 8, 6, 8, 8, 8, 6, 8, 19, 17 | 8.27 |
| | Percentage difference = | 0.24 |
| 6.28 rad × 20 m × 2 m dia. Domain size 6.28 rad × 8 m dia. × 50 m | Mesh grid density by number of cells in regions | Mass flowrate |
| 1 × Default mesh grid density | $X$: 1 $Y$: 3, 18<br>$Z$: 2, 7, 6, 6, 5, 6, 6, 6, 6, 5, 5, 17 | 11.20 |
| 2 × Default mesh grid density | $X$: 1 $Y$: 4, 34<br>$Z$: 4, 14, 12, 12, 10, 12, 12, 12, 12, 10, 17, 16 | 11.23 |
| | Percentage difference = | 0.27 |
| 6.28 rad × 50 m × 2 m dia. Domain size 6.28 rad × 8 m dia. × 100 m | Mesh grid density by number of cells in regions | Mass flowrate |
| 1 × Default mesh grid density | $X$: 1 $Y$: 2, 17<br>$Z$: 1, 6, 5, 6, 7, 8, 7, 6, 5, 6, 7, 16 | 16.69 |
| 2 × Default mesh grid density | $X$: 1 $Y$: 4, 34<br>$Z$: 2, 12, 10, 12, 14, 16, 14, 12, 10, 12, 14, 16, 16 | 16.95 |
| | Percentage difference = | 1.56 |

## *CFD Model*

The number of cells in the angular X-dimension (angular) was set at 1 throughout, thus assuming axisymmetric plume. The doubling of cells was effected in Y-radial and Z-vertical only in all regions as demarcated by magenta lines in Fig. 5.4a. An exception is in the last region of Z-dimension, which was developing into far-field plume, where a NULL object was inserted at halfway up to reduce the region's size while keeping the same number of cells, creating an additional region, 13 in total. The doubling showed no significant difference in the chosen criterion, mass flowrate, at 1.6% or less between the two mesh grid densities. The heat load also showed the same result.

## Navier–Stokes Equation and the Turbulence Model

The RANS-type governing equations were solved for steady-state, two-dimensional axisymmetric flow with essentially no tangential velocity, by setting one cell in the $\theta$ arc (X-), and uniform specific heat. The two-dimensional single-phase governing equations are as used in Chu et al. (2016). Chen and Kim (1987) $k$-$\varepsilon$ turbulence model was applied.

## Solution Domain and Boundary Conditions

The axisymmetric solution domain has a radial extent of 8 m and a longitudinal length of 10 m, 40 m, 50 m or 100 m for the hollow cylinder height of 3 m, 10 m, 20 m or 50 m, respectively. At the inlet of the chimney, the ambient air temperature was set to 30 °C, and the chimney heating source was located inside the chimney made to occupy its entire volume and consisted of the domain material causing no flow obstruction. Heat loads were fixed for each cylinder size so as to impart thermal energy to the air instantaneously upon contact in rising through the chimney. At the low and high boundaries of the solution domain, a fixed pressure condition is used in the simulations, so that air is entrained into the solution domain at the low boundary at ambient temperature.

At wall surfaces, hydraulically smooth equilibrium log-law wall functions are used for the momentum and turbulence transport equations (Launder & Spalding, 1974). The wall was assumed to be adiabatic and to have zero thickness.

## Solution Procedure and Convergence

As the chimney system that includes a significant plume-chimney is a natural convection system, my approach was not by fixing the plume inlet velocity, but by fixing the heat load and let the CFD software determine the velocity and chimney exit temperature through buoyancy-frictional balance. The upstream parametric values were therefore affected by their downstream values, since the temperature and velocity were both unknowns and needed to be iterated for solution.

Initial values at the monitoring probe were ambient pressure and temperature, $1.0 \times 10^{-10}$ for $u$ (angular), $v$ (radial velocity), $w$ (vertical velocity) and turbulent parameters. The RANS conservation equations were represented in finite volume discretised form and solved in steady state, using the SIMPLEST technique (CHAM), with a global convergence criterion of 0.01% for pressure, velocity and temperature at the monitoring probe, positioned above the top exit of all the cylinders. To approximate the location of the object of interest in an infinite domain, the walls of the domain are frictionless, and at both the top and bottom ends of the domain, the openings are set to be at fixed pressure and not at fixed mass flow. All the simulations were at steady state. The maximum number of sweeps at steady state was set at 6,000 to 10,000; most of the time the convergence was achieved in fewer iterations. When most of the parameters had converged to within the global criterion and often to below 0.01%, leaving only the vertical velocity $w$ unconverged at less than 0.5% due to the fluctuating flow, the simulation was deemed to have converged satisfactorily. The following table summarises the CFD model (Tables 5.2 and 5.3).

**Table 5.2** CFD simulation assumptions and settings

| Assumptions |
| --- |
| The domain material air was an ideal gas and was incompressible |
| No heat was lost by the chimney through the wall |
| Cylinder wall thickness was zero |
| Normal wall function was applied |
| Density variation is included in the solution |
| Heat source was uniform throughout the interior of solid chimney column |
| Transfer of heat by the heater to the air was instantaneous |
| The compressor operated to shift the fluid from the entry to the exit of the column |
| No friction was caused by the compressor in the zero-gravity case |
| Symmetry in the cylinder at all points of the circumference |
| Mesh grid was optimised by the criterion of mass flowrate |
| Settings |
| Diameter was set at 2 m, vertical cylinder lengths at 3, 10, 20 and 50 m |
| Ambient temperature was 30°C and outlet temperature set by trial-and-error at 60–70 °C |
| Heat loads were at 200, 300, 400 and 600 kW with increasing cylinder length |
| Navier–Stokes conservation equations were used as the mathematical fluid dynamics model |
| Turbulence model was that of Chen and Kim (1987) $k$-$\varepsilon$ |
| Global convergence criterion was set at ±0.01% |
| Software employed was Phoenics 2019 |

**Table 5.3** Chimney settings for identical flow and heating conditions in CFD simulation

| Chimney configuration | Settings | | | | | Computed simulation results | |
|---|---|---|---|---|---|---|---|
| | Conv. model | Grav. force | Driving pressure | Heat load (kW) | Ambient temp. (°C) | Ave. outlet temp. (°C) | Mass flowrate ($kgs^{-1}$) |
| 3 m × 2 m dia. | Natural (free) | Yes | Solid wall + Plume | 200 | 30 | 69.7 | 5.02 |
| 3 m × 2 m dia. | Forced | No | Compressor | 200 | 30 | 69.7 | 5.02 |
| 10 m × 2 m dia. | Natural (free) | Yes | Solid wall + Plume | 300 | 30 | 66.1 | 8.27 |
| 10 m × 2 m dia. | Forced | No | Compressor | 300 | 30 | 66.2 | 8.25 |
| 20 m × 2 m dia. | Natural (free) | Yes | Solid wall + Plume | 400 | 30 | 65.5 | 11.23 |
| 20 m × 2 m dia. | Forced | No | Compressor | 400 | 30 | 65.7 | 11.26 |
| 50 m × 2 m dia. | Natural (free) | Yes | Solid wall + Plume | 600 | 30 | 65.3 | 16.95 |
| 50 m × 2 m dia. | Forced | No | Compressor | 600 | 30 | 65.4 | 16.90 |

## Results and Discussion

### *Centreline Temperature and Velocity*

For temperature, it appears to persist after exiting the chimney as can be seen in the Z-dimension profile plots of the simulations that for both natural convection and compressor-driven flows it is the same (Fig. 5.5). For reference purpose in this paper, the region where this temperature maintains at or slightly above the source value is named as the zone of constant temperature or ZCT. From Fig. 5.6, the centreline vertical velocity was as expected increasing with height in the near-field region of a lazy plume (Chen & Nikitopoulos, 1979; Henderson-Sellers, 1983; Chu, 1986; Malin, 1986; Quintiere and Grove, 1998; Kaye, 2008). The centreline velocity profile with height however, differs between the buoyancy-driven and the compressor-driven flows, where the latter does not have a ZFE.

While the patterns of centreline vertical velocity with plume height appear very dissimilar in which the buoyancy-driven plume in Fig. 5.6 exhibits a maximum velocity at a certain height and then decays steadily, the momentum-driven jet commences its decay immediately upon exiting the cylinder. This would have posed a serious problem to the current zero-gravity method of determination of EPCH but

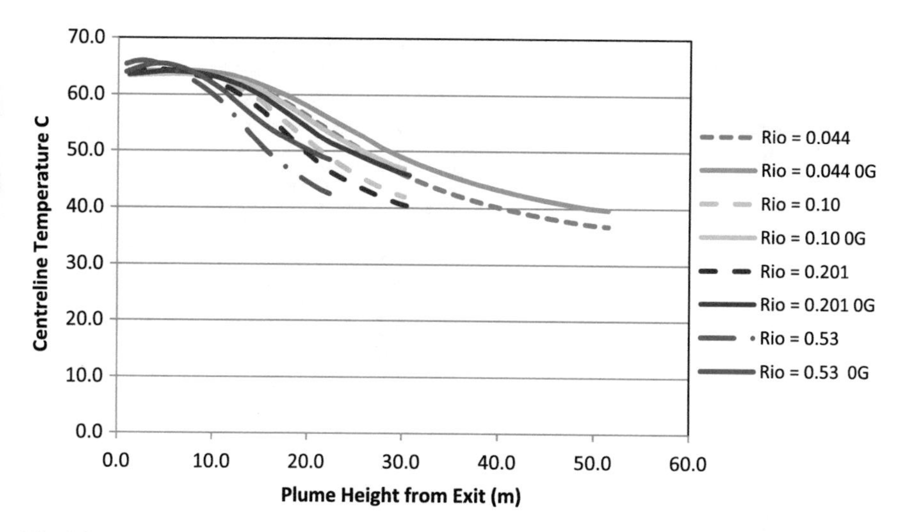


**Fig. 5.5** Centreline temperature decay profile with plume height for buoyancy- and momentum-driven mode

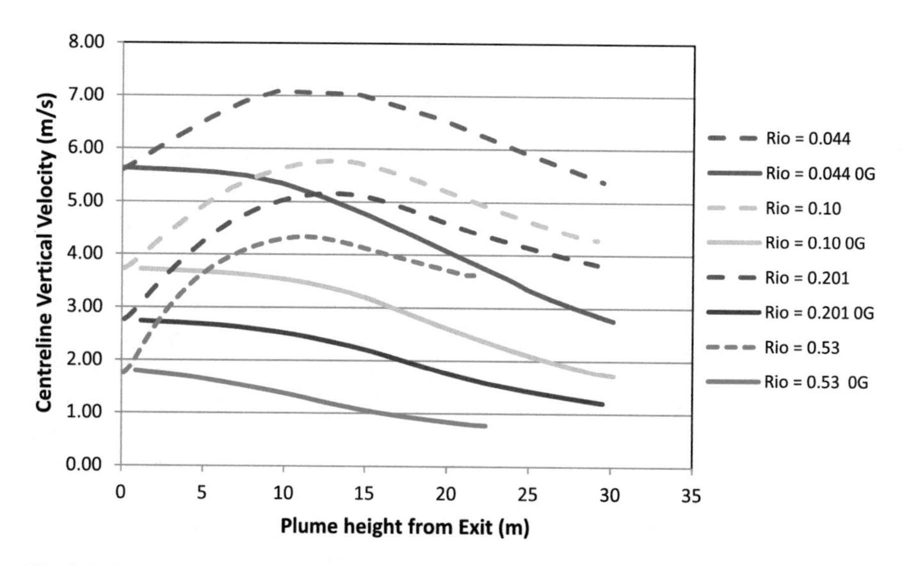


**Fig. 5.6** Centreline vertical velocity profile with plume height for buoyancy-driven mode

for the relative shortness of the region of interest, i.e., the ZCT of the near-field plume immediately upon exiting from the cylinder, as will be seen later in Table 5.7.

In Fig. 5.7, the ZCTs appear to be linearly related to the source Reynolds number. But the ZFE appears to be linear to the plume source parameter $\Gamma_o$ only when it is above 1.0 (Fig. 5.8), and peaked at around 2.0. The parallel ZCTs of the buoyancy-driven plumes and the momentum-driven jets suggest that they respond similarly to

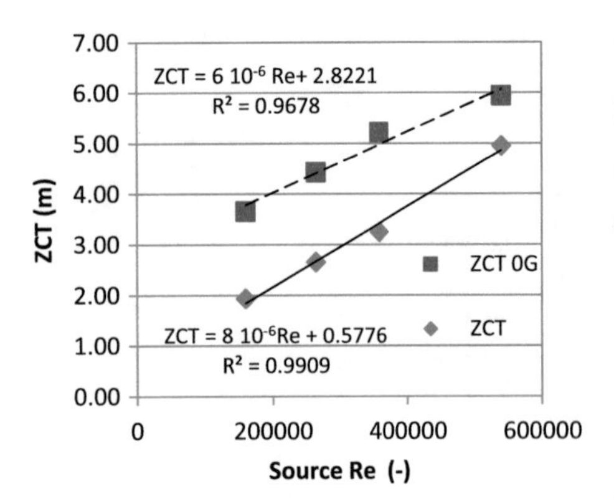


**Fig. 5.7** Zones of constant temperature for both modes have similar patterns of relationship with the source reynolds number

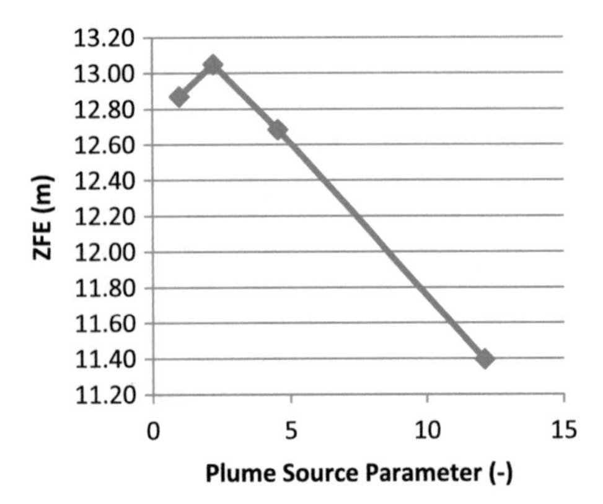


**Fig. 5.8** Pattern of relationship of the zone of flow establishment with the plume source parameter $\Gamma_o$

the Reynolds number, in a linearly proportional manner. The comparison shows the rate of mixing to be lower for the jets than for the plumes, but the relationship with Re was similar. The lower rate of mixing found in the jets means less frictional loss by the jets than the plumes, and the estimated EPCH value should be lower than the actual. This pattern supports the present approach of a zero-gravity synthetic plume driven by a compressor to study the amount of plume buoyancy convertible to kinetic energy upon discharge from a source.

Contour plots are only shown for the 20-m-long cylinder case as they are similar among the cases. The contour plot of temperature in Fig. 5.9 indicates the temperature continued unabated in the ZCT, while in Fig. 5.10 it is observed that the velocity vectors increase in lengths upon discharge and then reduce after some distance. Pressure contour is also shown in Fig. 5.11; however, it does not add more information than the temperature and velocity contours.

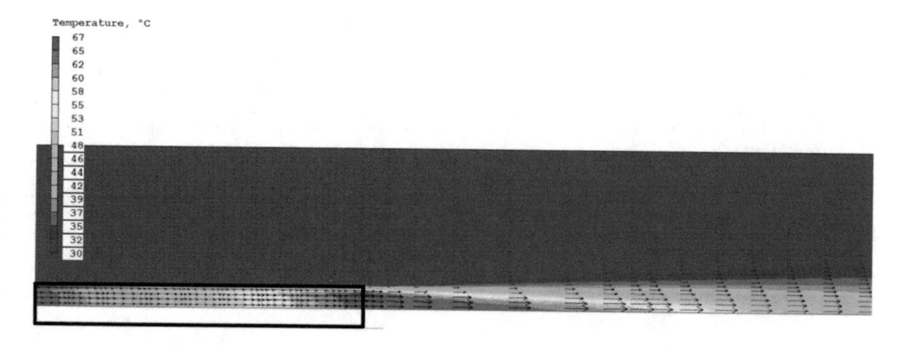


**Fig. 5.9** Plume discharge temperature profile of the 20 m × 2 m dia. vertical cylinder

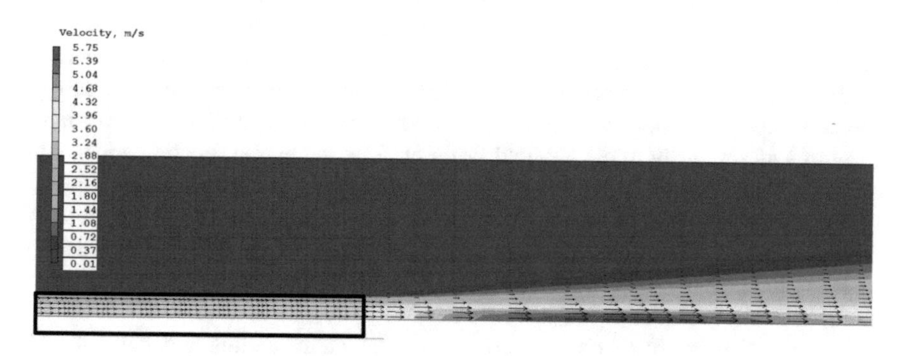


**Fig. 5.10** Plume discharge velocity profile of the 20 m × 2 m dia. vertical cylinder

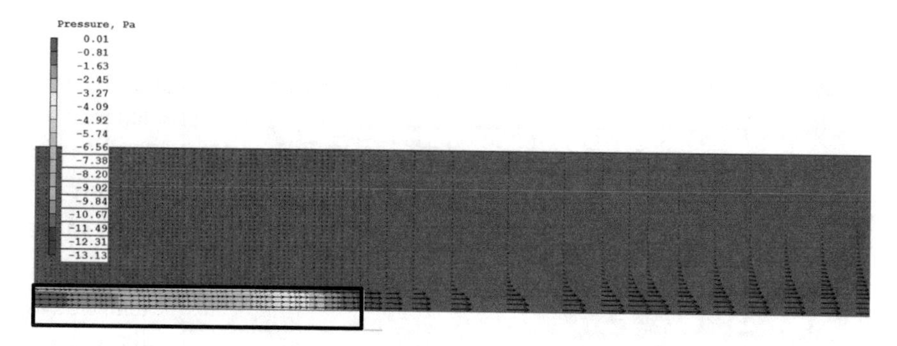


**Fig. 5.11** Plume discharge pressure profile of the 20 m × 2 m dia. vertical cylinder

## *Entrainment Coefficient*

The far-field plume entrainment coefficients and the plume function were determined by Eqs. (5.2) and (5.3), respectively, as given in Table 5.4.

**Table 5.4** Far-field plume entrainment coefficient and plume function values

| Source richardson number | 0.53 | 0.20 | 0.10 | 0.044 |
|---|---|---|---|---|
| Normalised plume height *z/r* | 16.5 | 24.5 | 24.5 | 29.5 |
| Entrainment coefficient | 0.113 | 0.103 | 0.093 | 0.091 |
| Plume function | 8.69 | 2.92 | 3.38 | 1.64 |
| Plume decay power $w_c = b(z/r)^n$ | −0.184 | −0.393 | −0.381 | −0.373 |
| Jet decay power $w_c = b(z/r)^n$ | −0.784 | −0.983 | −1.074 | −1.164 |

*Note* b and n are numerical coefficients

The entrainment coefficients appeared to diverge with increasing source Richardson numbers. The range of the far-field entrainment coefficients appears acceptable based on Carazzo et al. (2008) work. The plume function value was evaluated by utilising the entrainment coefficient estimated from Eq. (5.2) and generally tended towards 1.0 asymptotically and looked to require a longer distance to attain as source Richardson numbers became higher. While the radial profiles indicate attainment of self-similarity in the far-field (see Fig. 5.3), the plume function appears to maintain values significantly above 1.0 in this simulation. The simulated data appear to obey the power laws of velocity decay with height from discharge for plumes of −1/3 and jets of −1, with a significant departure only at $Ri_o = 0.53$, perhaps due to its lazy plume characteristics persisting as reflected by the local plume function of 8.69.

The radial velocity (Y-component) across the plume radius and beyond was obtained at several heights up to the maximum height of simulation for each $Ri_o$ as illustrated in the example of $Ri_o = 0.53$ (3 m cylinder) (Fig. 5.12), showing it

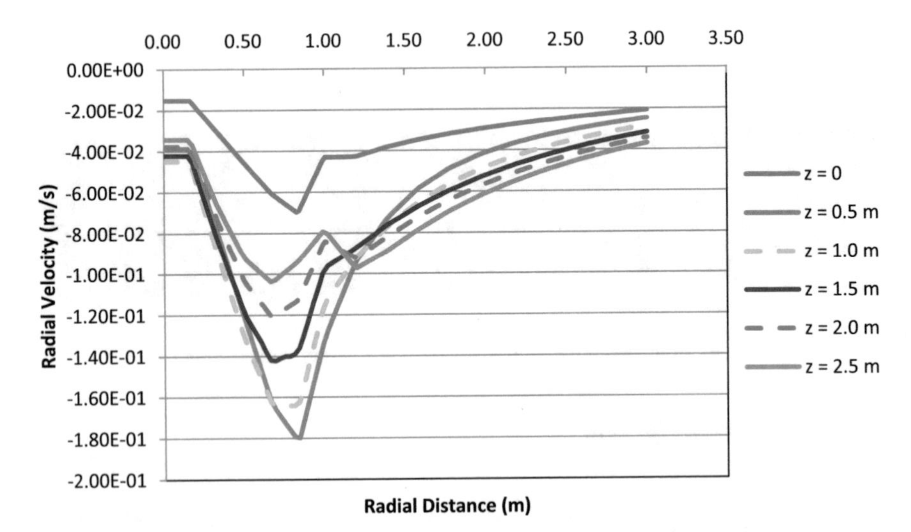


**Fig. 5.12** Locating the maximum magnitude of radial velocity in the entrainment coefficient

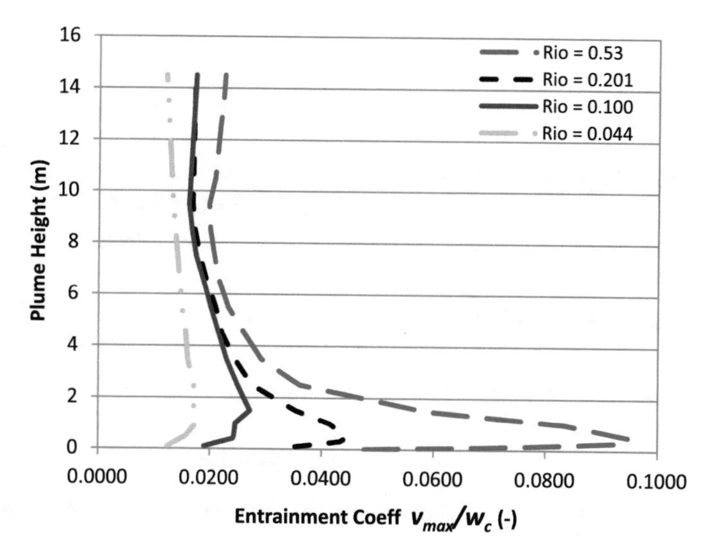


**Fig. 5.13** Entrainment coefficient variation with height for the four lazy plumes

reaches a peak at or within the radius. This peak did not abruptly dropped to zero as the case would be a flow impact on a solid wall, but dropped gradually towards the centre. Entrainment coefficients as the ratio of $v_{max}/w_c$ were determined for all the four geometries as shown in Fig. 5.13. The entrainment coefficient appears to sharply increase from a nonzero value at source and after peaking to as high as 0.095 decays to a far-field value, at approximately between 0.01 and 0.025. The maximum values ranged from 0.017 to 0.095. The trends of the coefficients in Fig. 5.13 show similar patterns as Marjanovic et al. (2017) results. On the basis of the definition of virtual origin by Hunt and Kaye (2005), Marjanovic et al. (2017) had applied a virtual origin shift to an otherwise non-monotonic decay of the plume function to develop an analytical model that appeared to fit the simulated results well.

A plot of the maximum entrainment coefficient against the source Richardson number shows a remarkably linear relationship as shown in Fig. 5.13. The first implication of this relationship is that the coefficient is not a universal constant, as already concluded by many workers (Marjanovic et al., 2017; Chen & Nikitopoulos, 1979; Kaye & Hunt, 2009; Hunt & van den Bremer, 2010). The second is that the maximum rate of entrianment is determined by the source Richardson number or the plume function. The maximum rate appears to be at or within the height over radius ratio $z/r$ of 1.0 for each case (Fig. 5.14).

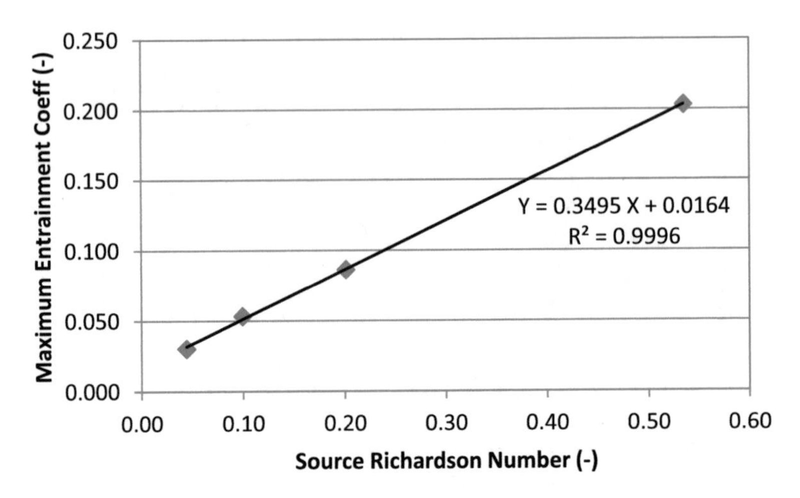


**Fig. 5.14** Relationship of maximum entrainment coefficient ($Y$) with source Richardson number ($X$)

## *EPCH*

Using the steps outlined in Sect. 2.2, the EPCH was determined by the difference between compressor pressure and solid-walled buoyancy pressure as given in Table 5.5.

The EPCH was also predicted using Chu (2002) correlation. The predicted values were 2–75% of the CFD derived values, acceptable accuracy but can do with the inclusion of a momentum term since the higher velocity resulted in lower deviation. In Table 5.5, the CFD derived values of EPCH show that there was some dependency on the vertical velocity which effect has not been included by the empirical correlation of Chu (2002). To verify the uncertainty involving the use of industrial coefficients for cylinder inlet and outlet losses and friction factor correlation, a comparison was made in Table 5.6 between the current method of determining the EPCH and the conventional means of determining the EPCH. It can be seen that the conventional way of determining the EPCH gave values that were way beyond the acceptable range in Table 5.5.

In the current method, the frictional loss incurred by the boundary layers of buoyancy-driven plume and the momentum-driven jet, corresponding to the sudden expansion loss in Table 5.6, was assumed to be approximately the same. Earlier, Fig. 5.6 shows the pattern of centreline velocity with plume height that differs between the two modes of flow, where the momentum-driven jet was made to have the same flowrate and temperature rise as the buoyancy-driven plume. As hypothesised earlier, the plume column is not likely to have a stack effect all the way to its maximum height in the high altitude, and the zone of established flow is when the plume detaches above the zone of constant temperature from the convection system

**Table 5.5** Determined values of EPCH for the four cylinders with heat input

| Chimney configuration | Ave. temp. within solid wall (°C) | Solid-walled buoyancy pressure (Pa) | Forced convection compressor pressure (Pa) | Ave. outlet temp. (°C) | Ave. outlet vertical velocity ($ms^{-1}$) |
|---|---|---|---|---|---|
| 3 m × 2 m dia. | 49.97 | 2.12 | 2.53 | 69.7 | 1.55 |
| 10 m × 2 m dia. | 48.09 | 6.45 | 6.78 | 69.7 | 2.53 |
| 20 m × 2 m dia. | 47.75 | 12.67 | 13.05 | 66.1 | 3.43 |
| 50 m × 2 m dia. | 47.64 | 31.47 | 32.15 | 66.2 | 5.17 |
| Chimney configuration | EPCH (m) | % Solid-walled | Predicted EPCH (m) Chu (2002) | Source Richardson number $Ri_O$ | Plume source parameter $\Gamma_o$ |
| 3 m × 2 m dia. | 0.26 | 9 | 0.46 | 0.535 | 12.103 |
| 10 m × 2 m dia. | 0.27 | 2.7 | 0.48 | 0.201 | 4.551 |
| 20 m × 2 m dia. | 0.35 | 1.7 | 0.48 | 0.099 | 2.203 |
| 50 m × 2 m dia. | 0.49 | 1.0 | 0.48 | 0.044 | 0.972 |

around the source. Instead, as the natural convection chimney system being considered here consisted of the entire vertical cylinder and the region between the plume inlet (source) up to the EPCH, which was short relative to the entire plume rise height, the largest change in vertical velocity $w$ was found to be a mere 4.2%; hence, the error in frictional loss calculation is considered minimal as shown in Table 5.7.

On this basis, the use of the compressor pressure is justified to achieve the same heat load and flowrate of a buoyancy-driven plume to determine by trial-and-error the total buoyancy pressure head required to drive the flow. The jet's frictional loss was in fact likely to be lower than that of the plume's, so that any determined value of EPCH would therefore be a conservative estimate, that is, if it is possible to determine the EPCH accurately, the significance of the accurate value will be greater.

The ZCT and the EPCH have been found to correlate strongly with the inverse source Richardson number as shown in Figs. 15 and 16.

This analysis demonstrates a significant stack effect that continues on with a plume at the exit of a buoyancy-driven discharge. Placing the EPCH values against the criterion of greater than 1.0 in the plume source parameter for lazy plume shows the plume formed from the 3-m, 10-m and 20-m-long cylinders qualified to be classified as lazy, and the plume rising from the 50 m length cylinder was a forced plume. The

**Table 5.6** Determination of EPCH by summation of all frictional losses compared with the current method (0G CFD)

| $h_{sw}$ (m) | Reynolds number (–) | Pipe friction factor (–) | Pipe pressure drop (Pa) | Sharp entrance loss (Pa) | Sudden expansion loss (Pa) | Solid-walled chimney buoyant pressure (Pa) | EPCH by sum $\Delta P$ (m) | EPCH by 0G CFD (m) | EPCH by Chu (2002) |
|---|---|---|---|---|---|---|---|---|---|
| 3 | 159,792 | 0.016 | 0.03 | 0.55 | 1.24 | 2.12 | −12 | 0.26 | 0.46 |
| 10 | 263,242 | 0.015 | 0.23 | 1.49 | 3.32 | 6.45 | −83 | 0.27 | 0.48 |
| 20 | 357,563 | 0.014 | 0.81 | 2.74 | 6.13 | 12.66 | −184 | 0.35 | 0.48 |
| 50 | 539,535 | 0.014 | 4.26 | 6.24 | 13.93 | 31.48 | −451 | 0.49 | 0.48 |

**Table 5.7** Velocity change from the plume source to the height of EPCH for both modes of simulation

| Cylinder height | $Ri_o$ | Natural convection | | | Compressor-driven | | |
|---|---|---|---|---|---|---|---|
| | | $w_{co}$ | $w_c$ (EPCH) | % Diff. | $w_{co}$ | $w_c$ (EPCH) | % Diff. |
| 3.0 | 0.535 | 1.72 | 1.79 | 4.2 | 1.79 | 1.79 | -0.1 |
| 10.0 | 0.201 | 2.68 | 2.75 | 2.6 | 2.74 | 2.74 | 0.0 |
| 20.0 | 0.099 | 3.63 | 3.70 | 2.1 | 3.72 | 3.72 | 0.0 |
| 50.0 | 0.044 | 5.60 | 5.68 | 1.5 | 5.63 | 5.63 | 0.0 |

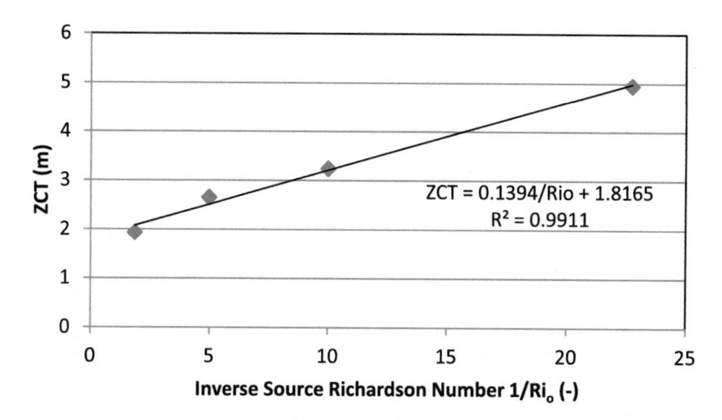


**Fig. 5.15** Strong correlation between the ZCT and the inverse of source $Ri_o$

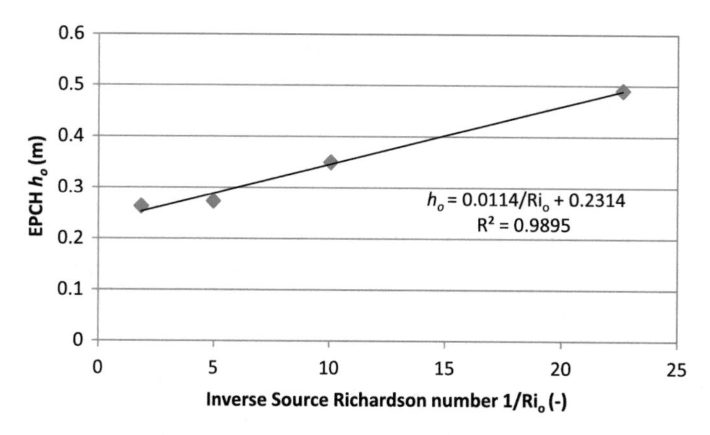


**Fig. 5.16** Strong correlation between the EPCH and the inverse of source Richardson number ($1/Ri_o$)

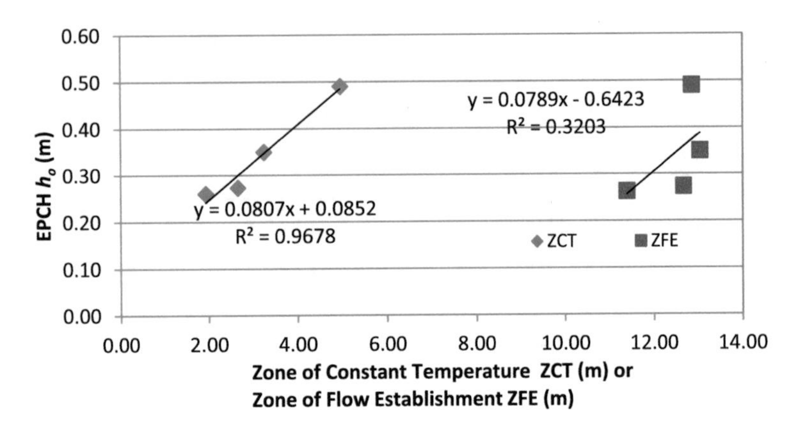


**Fig. 5.17** Correlation of EPCH with ZCT and ZFE

EPCH shows a very good linear correlation with the coefficient of determination $R^2 = 0.95$ versus ZCT, but with ZFE, the relationship appears to be more complex (Fig. 17).

The definition of a core in a conical shape, namely a zone of constant temperature (ZCT), closely identifies with a plume-chimney, and the associated EPCH should enable the development of an entrainment model that accounts for a chimney wall with characteristics more robust than the free boundary layer of self-similar plumes.

A primitive version of the concept had been used to derive an estimation method for the effective plume-chimney height above a forced draft air-cooled heat exchanger (Chu, 2002, 2006), where the plume source was modelled by using a square flat plate with only horizontal entrainment flow at the surface, i.e., the plume source parameter was infinite. There is no reason why a very high Richardson number plume source with zero axial centreline velocity at the source, as in forest fires and warm ocean surfaces, should not exhibit this chimney wall characteristic. The existence of the plume-chimney means that the numerical solution at any step would depend on the values downstream because it is behaving as a solid-walled natural convection chimney would. By modelling a near-field plume as a chimney with a boundary layer effectively acting as a wall, the entrainment coefficient, or the rate, can be more accurately modelled.

A plot of inverse square of the entrainment coefficient $v_{max}/w_c$ against EPCH shows that the higher the EPCH, the higher is the centreline vertical velocity $w_c$, and the lower is the maximum horizontal radial velocity $v_{max}$ (Fig. 5.18). This is consistent with the hypothesis that there should be a wall resisting the entrainment in the ZCT, as a consequence constituting a chimney wall around the lazy plume in the near-field. When the lazy plume has a significant initial momentum, which is the $Ri_o = 0.044$ case, the latter has the pre-existing resistance put up by the momentum resulting in a higher EPCH.

On the other hand, as $w_c/v_{max}$ tends to zero, such as a small $w_c$ in relation to the $v_{max}$, there is a residual EPCH, which means that much of the EPCH is directed

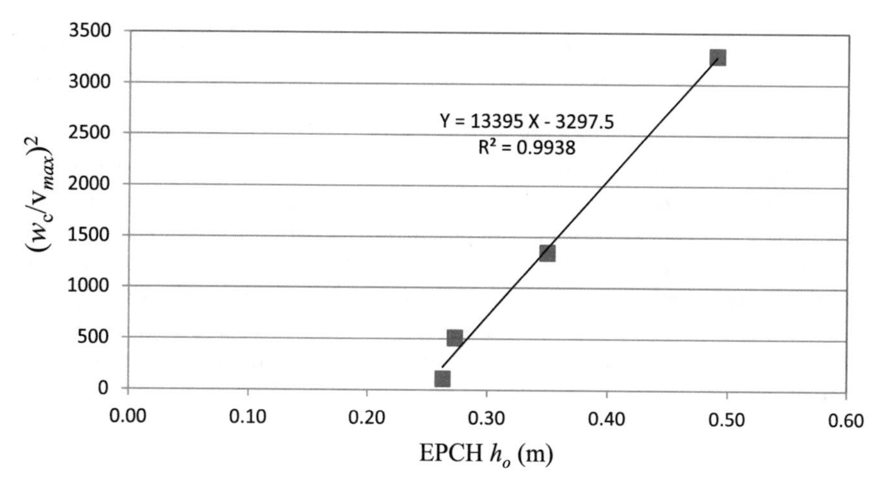


**Fig. 5.18** EPCH (X) and the entrainment coefficient $v_{max}/w_c$ (*Y*) has an inverse square relationship

towards entrainment. This may explain the large entrainment coefficient found in studies to date (Quintiere & Grove, 1998; Marjanovic et al., 2017). The buoyant pressure head of the EPCH being available regardless of whether it is directed to upflow or diverted to entrainment had been explored and shown to be effective (Chu, 2002, 2006), which related the EPCH to the horizontal entrainment along the flat plate, and then applied to air-cooled heat exchangers that had upflow as well as horizontal entrainment.

To sum up the analysis:

(i) Lazy plumes exhibit the characteristics of a chimney by the persistence of plume source temperature, signifying a near-field region unmixed by entrainment; and the peaking of the radial velocity as the entrainment impacts the plume core region is indicative of a wall-like quality.
(ii) The magnitude of the EPCH has been determined and corroborated by an empirical correlation;
(iii) The EPCH has a linearly proportional relationship to the inverse source Richardson number and an inverse square relationship with the entrainment rate; and
(iv) For engineering purposes, the approach of separating the plume into near-field and far-field entities appears to be more useful than using the method of virtual origins.

When this is played out at a geophysical scale, the wall-like characteristic can modify the understanding of the dynamics of the plume rise. Near-field of large source area natural convection phenomena such as forest fires, aircraft turbulence, hurricane or typhoon cyclogenesis and volcano eruptions would probably be better understood. This may lead to explaining the term "Hot Towers" in hurricane studies (Kelley & Stout, 2004), and the often reported description of a fire storm having "a life of its own" weather pattern. Experimental determination similar to the setup of the CFD should be carried out, where a mechanistic approach in defining the EPCH will lead to a much insight into many natural phenomena involving near-field lazy plumes.

While the simulated data look promising, the results of the current study are based on a limited range of Richardson number and plume source parameter, where in geometry the diameter was only of a single size and at an industrial chimney scale, with a single turbulence model and on a single software platform. Rigorous simulation studies are called for that might relate the EPCH with the distance of virtual origin to the actual plume source, and possibly construct the plume-chimney within the ZCT using vortex models and DNS and LES methods. Lastly, a most glaring need of all is to design an experimental facility that can validate the simulation over a diverse range of fields.

## Conclusions and Future Studies

- Effective plume-chimney height (EPCH) of natural convection flows has been determined in a non-intrusive manner by CFD simulation of jets, synthetic plumes with zero-gravity, by applying justifiable partial similarity between the plumes and the jets.
- The EPCH values are all positive and significant at high Richardson numbers, i.e., lazy plumes, and close to the order of magnitude as predicted by an empirical correlation to within 2–75%.
- The EPCH values are considered significant as they were conservatively determined where the frictional losses of a momentum-driven jet are less than that of a buoyancy-driven plume, all other conditions being equal.
- EPCH is linearly proportional to the zone of constant temperature, the ZCT is strongly proportional to the inverse of source Richardson number, and the maximum entrainment coefficient's linearly proportional relationship with the source Richardson number is remarkable.
- Near-field lazy plume entrainment rate cannot be estimated properly without a fluid- static balance that includes the plume-chimney where downstream values of parameters such as density affect the upstream values of the same parameters.
- The much larger entrainment rate near a lazy plume source than that of a self-similar plume may be related to the EPCH.

**Acknowledgements** The author would like to offer his sincere thanks to the Ministry of Higher Education, Malaysia, for the kind assistance provided through fundamental grant No. FRG0022-TK-1/2006, and the provision of funding by Heat Transfer and Fluid Flow Services (HTFS) towards plume studies above air-cooled heat exchanger project at National Engineering Laboratory (NEL-TÜV), U.K.

## References


Abraham, G., & Eysink, W. D. (1969). Jets issuing into fluids with a density gradient. *Delft Hydraulic Laboratory Publication, 66,* 145–175.

Ashrae, (2017). *ASHRAE Handbook- Fundamentals*, Fluid Flow, 3.8, S.I. Edition, ISBN 978-1-939200-58-7.

Briggs, G. A. (1969). Optimum formulas for buoyant plume rise. *Philosophical Transactions of the Royal Society of London, 265,* 197–203.

Caulfield, C. C. (1991). *Stratification and Buoyancy in Geophysical Flows.* PhD thesis, University of Cambridge, UK.

Carazzo, G., Kaminski, E., & Tait, S. (2008). *Journal of Geophysical Research, 113* (B09201), 1–19. https://doi.org/10.1029/2007jb005458.

Carlotti, P., & Hunt, G. R. (2017). An entrainment model for lazy turbulent plumes. *Journal of Fluid Mechanics, 811*, 682–700. Cambridge.

CHAM. http://www.cham.co.uk/phoenics/d_polis/d_lecs/general/maths.htm. Concentration, Heat and Momentum Limited, Bakery House, 40 High Street, Wimbledon Village, London, SW19 5AU.

Chen, Y. S. & Kim, S. W. (1987). Computation of turbulent flows using an extended k-e turbulence closure model, *NASA CR-179204.*

Chen, C. J., Nikitopoulos, C.P. (1979). On the near field characteristics of axisymmetric turbulent buoyant jets in uniform environment. *International Journal of Heat Mass and Transfer, 22*, 245–255. Elsevier.

Chu, C. M. (2006). Use of chilton-colburn analogy to estimate effective plume chimney height of a forced draft air-cooled heat exchanger. *Heat Transfer Engineering, 27* (9), 81–85. Taylor and Francis, Philadelphia, U.S.A.

Chu, C. M. (2005). Improved heat transfer predictions for air-cooled heat exchangers. *Chemical Engineering Progress, A.I.Ch.E. 101* (11), 46–48. November, New York, NY.

Chu, C. M. (2002). A preliminary method for estimating the effective plume chimney height above a forced-draft air-cooled heat exchanger operating under natural convection. *Heat Transfer Engineering, 23,* 3–13. Taylor and Francis, London.

Chu, C. C. M. (1986). *Studies of the Plumes above Air-Cooled Heat Exchangers Operating under Natural Convection*, Ph.D. thesis, Department of Chemical Engineering, University of Birmingham, United Kingdom.

Chu, C. M., Rahman, Hieng R. Y. T., & M. M. (2017). Simulation of effective plume-chimney above natural draft air-cooled heat exchangers, POWERENERGY2017-3435. In *Proceedings of the ASME 2017 Power and Energy Conference*, PowerEnergy2017, June 26–30, 2017, Charlotte, North Carolina, USA.

Chu, C. M., Rahman, M. M., Kumaresan, S. (2016). Improved thermal energy discharge rate from a temperature-controlled heating source in a natural draft chimney, *Applied Thermal Engineering, 98,* 991–1002. Elsevier.

Chu, C. M., Farrant, P. E., & Bott, T. R. (1988). Natural convection in air-cooled heat exchangers, In *2nd UK National Conference on Heat Transfer*, pp 1657–1688, IChemE/IMechE Publication.

Crawford, T. V., & Leonard, A. S. (1962). Observations of buoyant plumes in calm stably air. *Journal of Applied Meteorology, 1,* 251–256.

Doyle, P. T., & Benkly, G.J. (1973). Use fanless air coolers. *Hydrocarbon Processing, July,* 81–86.

Fan, L. (1967). *Turbulent buoyant jets into stratified or flowing ambient fluids, Report KH-R-15.* Pasadena, California, USA: California Inst. of Technology.

Fox, D. G. (1970). Forced plume in a stratified fluid. *Journal Geophysical Research, 75*(33), 6818–6835.

Haaland, S. E. (1983). Simple and explicit formulas for the friction factor in turbulent pipe flow. *Journal of Fluids Engineering, March,* 83–90.

Henderson-Sellers, (1983). The zone of flow establishment for plumes with significant buoyancy. *Applied Mathematical Modelling, 7*, 395–397. Butterworth.

Hunt, G. R., & van den Bremer, T. S. (2010). Classical plume theory: 1937–2010 and beyond, *IMA Journal of Applied Mathematics*, 76, 424 – 448. Oxford University.

Hunt, G. R., & Kaye, N. B. (2005). Lazy plumes. *Journal of Fluid Mechanics, 533*, 329–338. Cambridge.

Kaye, N. B. (2008), Turbulent plumes in stratified environments: a review of recent work. *Atmosphere-Ocean 46* (4), 433–441. Taylor and Francis.

Kaye, N. B. & Hunt, G. R. (2009). An experimental study of large area source turbulent plumes, *International Journal of Heat and Fluid Flow, 30*, 1099–1105. Elsevier.

Kelley, O. & Stout, J. (2004). A "Hot Tower" above the eye can make hurricanes stronger. https://www.nasa.gov/centers/goddard/news/topstory/2004/0112towerclouds.html.

Launder, B. E., & Spalding, D. B. (1974). The numerical computation of turbulent flow. *Computer Method in Applied Mechanics and Engineering, 3,* 269.

Li, X. X., Duniam, S., Gurgenci, H., Guan, Z. Q., & Veeraragavan, A. (2017). Full scale experimental study of a small natural draft dry cooling tower for concentrating solar thermal power plant. *Applied Energy, 193*, 15–27. Elsevier.

List, E. J. & Imberger, J. (1973). Turbulent entrainment in buoyant jets and plumes. *Journal of the Hydraulics Division Proceedings ASCE, 99*, 1461–1474.

Malin, M. R. (1986). The decay of mean and turbulent quantities in vertical forced plumes. *Applied Mathematical Modelling, 11*, 301–314, Butterworth.

Marjanovic, G., Taub, G. N., & Balachandar, S. (2017). On the evolution of the plume function and entrainment in the near-source region of lazy plumes. *Journals Fluid Mechanics, 830*, 736–759. Cambridge University Press.

Morton, B. R., Taylor, G. I., & Turner, J. S. (1956). Turbulant gravitational convection from maintained and instantaneous sources. In *Proceedings of the Royal Society of London*, Series A, 234, 6 March, pp. 1–22.

Morton, B. R. (1959). Forced plumes. *Journals Fluid Mechanics, 5*(1), 151–163. Cambridge.

Quintiere, J. G., & Grove, B.S. (1998). A unified analysis for fire plumes. In *Twenty-Seventh Symposium (International) on Combustion/The Combustion Institute*, pp. 2757–2766.

Rahman, M. M., Chu, C. M., Tahir, A. M., Misran, M. A., Ling, L. (2017). Experimentally identify the effective plume chimney over a natural draft chimney model. In *IOP Conference Series: Materials Science and Engineering 217 012002, International Conference on Materials Technology and Energy*, 20–21 April 2017, Curtin University, Malaysia.

Sinnott, R.K., & Towler, G. (2009). *Coulson and Richardson's Chemical Engineering: Chemical Engineering Design* (5th ed., Vol. 6). Oxford: Pergamon Press.

Sneck, H. J., & Brown, D.H. (1974). Plume rise from large thermal sources such as dry cooling towers, *ASME Journals of Heat Transfer*, 232–238.

Tan, K. J. Y. (2019). Plume-Chimney temperature profile simulation and data analysis using computational fluid dynamics (CFD), *Final Year Project thesis, Chemical Engineering Programme*, Universiti Malaysia Sabah.

Zhou, X. P., Yang, J. K., Xiao, B., Hou, G. X., & Xing, F. (2009). Analysis of Chimney Height for Solar Chimney Power Plant, *Applied Thermal Engineering, 29*, 178–185. Elsevier.

# Chapter 6
# Numerical Analysis of Solar Chimney Design for Power Generation

**Chi-Ming Chu, Mohd. Suffian bin Misran, Heng Jin Tham, and Md. Mizanur Rahman**

## Abbreviations

| | |
|---|---|
| $A_c$ | Tower inner diameter (m⁾ |
| $A_{coll}$ | Total collector area (m$^{2)}$ |
| $c_f$ | Blade, transmission and generator efficiency (–) |
| $c_p$ | Air-specific heat capacity (J kg$^{-1}$ K$^{-1}$) |
| $D$ | Tower inner diameter (m) |
| f | Darcy friction factor (–) |
| $f_t$ | Turbine efficiency or power ratio (–) |
| $G$ | Solar irradiation rate (Wm$^{-2}$) |
| $g$ | Gravitational acceleration constant at sea level (ms$^{-2}$) |
| $h_c$ | Solar chimney height (m) |
| $h_o$ | Effective plume-chimney height (m) |
| $h_p$ | Cold inflow penetration depth (m) |
| $L$ | Characteristic dimension of plume source (m) |
| $\dot{M}$ | Air mass flow rate (kg s$^{-1}$) |
| $\dot{Q}$ | Heat gained by air in the collector (m) |
| $T_a$ | Ambient temperature (°C) |
| $T_o$ | Hot air temperature in collector and tower(°C) |
| $\dot{V}$ | Volumetric flow rate (m$^3$s$^{-1}$) |
| $W$ | Shaft work (W) |
| $w_c$ | Tower inlet mean velocity (ms$^{-1}$) |
| $w_{copt}$ | Optimum tower inlet mean velocity (ms$^{-1}$) |
| $\alpha$ | Solar irradiation absorption factor of collector (–) |

C.-M. Chu (✉) · Mohd. Suffian bin Misran · H. J. Tham
Faculty of Engineering, Universiti Malaysia Sabah, Kota Kinabalu, Sabah, Malaysia
e-mail: chrischu@ums.edu.my

Md. M. Rahman
Department of Mechatronics Engineering, World University of Bangladesh, Dhaka, Bangladesh



Md. M. Rahman and C.-M. Chu (eds.), *Cold Inflow-Free Solar Chimney*,
https://doi.org/10.1007/978-981-33-6831-6_6

| | |
|---|---|
| $\alpha_H$ | Heat transfer coefficient ($Wm^{-2} K^{-1}$) |
| $\beta$ | = αH (Wm−2 $K^{-1}$) |
| $\Delta p_{tot}$ | Total pressure drop (Pa) |
| $\Delta p_t$ | Turbine pressure drop (Pa) |
| $\Delta p_w$ | Tower wall frictional pressure drop (Pa) |
| $\Delta p_{ex}$ | Tower exit pressure loss (Pa) |
| $\Delta p_{ten}$ | Tower entrance pressure loss (Pa) |
| $\Delta p_{cen}$ | Collector entrance pressure loss (Pa) |
| $\Delta T$ | Temperature rise in the collector (K) |
| $\eta_{SC}$ | Thermal efficiency of solar chimney (−) |
| $\eta_{SP}$ | Overall efficiency of solar chimney (−) |
| $\eta_{coll}$ | Collector efficiency (−) |
| $\eta_{wt}$ | Mechanical to electrical transmission efficiency (−) |
| $\rho_a$ | Ambient air density ($kg\ m^{-3}$) |
| $\rho_o$ | Hot air density ($kg\ m^{-3}$) |

Solar chimney operates on the principle of natural convection flow. The sizing of the solar chimney and the collector area begins with the natural convection pneuma-static balance of the solar chimney. Solar chimney for power generation will be designed in the following sections. The design approach is to enable different combinations of parameters to be assessed prior to doing rigorous designs requiring sophisticated tools like CFD and Multiphysics software.

## Thermal Pneumatic Balance

The solution of the balance is based on the following assumptions:

- the air is assumed to be dry and is an ideal gas. If there is considerable humidity this has to be included in the air properties.
- the air density stratification in the atmosphere is negligible.
- the ambient environment is a still surrounding.
- the solar irradiation rate is constant.
- the heat loss through the tower wall is negligible, and
- The flow is one dimensional and one directional through the tower.

The balance is straightforward:

$$[h_c(\rho_a - \rho_o) + h_o(\rho_a - \rho_o)]g = \Delta p_{tot} \tag{6.1}$$

where $\Delta p_{tot}$ is the total pressure drop available, $h_c$ is the chimney height, and $h_o$ is the effective plume-chimney height. The two unknowns, exit temperature and flow rate through the tower, will require the simultaneous solution of heat transfer in the collector area and pressure drop by iterative trial and error. The density $\rho_o$ is calculated from the heated air temperature at the collector exit to the tower inlet, and the ambient density $\rho_a$ is at the collector inlet, where ambient air enters. If a constant heat flux is used based on a constant solar irradiation rate, $G$, Wm$^{-2}$, an absorption coefficient $\alpha = 0.75$ (Schlaich, 1995) to account for the heat loss from the collector to supports by conduction and to air by convection (radiation is minimal at this temperature) then the temperature rise inside the tower can be estimated using Eq. (6.2):

$$\dot{Q} = GA_{\text{coll}} = \dot{M}c_p(T_o - T_a) + \alpha_H A_{\text{coll}}(T_o - T_a) \tag{6.2}$$

If simplification is intended in estimating $T_o$, $\alpha$ can be set to 1.0 and $\alpha_H$ to 0.0 for no loss at all. Equations 1 and 2 need to be solved together with two unknowns. A typical heat transfer coefficient $\alpha_H$ for air on a horizontal surface is estimated by Eq. 6.3 (Geankoplis, 2003):

$$\alpha_H = 1.52\Delta T^{(1/3)} \tag{6.3}$$

since the Gr.Pr value is likely to be greater than $2 \times 10^7$, which yields about 5 Wm$^{-2}$ K$^{-1}$.

$\Delta p_{\text{tot}}$ consists of the tower wall frictional loss $\Delta p_w$, the tower top exit sudden expansion loss $\Delta p_{\text{ex}}$, the tower inlet entrance loss $\Delta p_{\text{ten}}$, the collector entrance loss $\Delta p_{\text{cen}}$, and the turbine pressure drop as shown below:

$$\Delta p_{\text{tot}} = \Delta p_w + \Delta p_{\text{ex}} + \Delta p_{\text{ten}} + \Delta p_{\text{cen}} + \Delta p_t \tag{6.4}$$

While the first four types of pressure drop are readily calculated from published $K$-coefficients in velocity head formulae, the turbine pressure drop $\Delta p_t$ is given by manufacturers and the typical value is 0.25–0.45 (Windpower Engineering and Development, 2020).

## Chimney Efficiency

The theoretically maximum mechanical work that can be extracted from a solar chimney is $\dot{V}\Delta p_{\text{tot}}$, which is without any of the losses listed in Eq. (6.3). The Carnot thermal efficiency of a solar chimney (SC) has been provided by Schlaich (Schlaich, 1995), Michaud (1999) as shown in Eq. (6.5)

$$\eta_{\text{SC}} = \frac{\dot{V}\Delta p_{tot}}{\dot{M}c_p\Delta T} = \frac{h_c g}{c_p T_a} \tag{6.5}$$

This efficiency is obtained from the balanced Eqs. (6.1) and (6.2) and is approximated by

$$\eta_{SC} = 33 \times 10^{-6} h_c \tag{6.6}$$

Equations (6.4) and (6.5) provide a quick method to estimate the maximum power output of a solar chimney, by assuming that 100% of the solar irradiation rate $G$ over the collector area is converted into heat in the tower, for example, the total solar collector area of the 50 kWel prototype in Manzanares, Spain is $A_{\text{coll}} = 46{,}000\ \text{m}^2$. At the maximum solar irradiation rate of 1000 $\text{Wm}^{-2}$, the maximum theoretical power output is 294 kW.

## Overall Tower Efficiency

In order to estimate the power output from solar irradiation rate, it is necessary to analyze the overall efficiency of a SC power plant. Generally, the overall efficiency of a SC power plant $\eta_{\text{sp}}$ is defined as the product of the partial efficiencies of the SC components:

$$\eta_{\text{sp}} = \eta_{\text{SC}} \eta_{\text{coll}} f_t \eta_{\text{wt}} \tag{6.7}$$

where $\eta_{\text{sc}}$, $\eta_{\text{coll}}$, $f_{\text{t}}$ are the efficiencies of the chimney, collector, turbine and $\eta_{\text{wt}}$ of the transmission, blades and generator efficiency, respectively (Nizetic & Klarin, 2010). The alternative name of turbine efficiency is the power ratio (Eq. 6.9).

The collector efficiency $\eta_{\text{coll}}$ is defined as the ratio of the thermal energy absorbed from the irradiation rate $G$ over the projected collector area $A_{\text{coll}}$ and is obtained from the measured values as shown in Eq. (6.8a):

$$\eta_{\text{coll}} = \frac{\dot{Q}}{A_{\text{coll}} G} = \frac{\rho_o w_c A_C c_p \Delta T}{A_{\text{coll}} G} \tag{6.8a}$$

However, Eq. 6.8a is unable to estimate the convective heat loss of the collector without knowing the mass flow rate, and Eq. (6.7) should be used

$$\eta_{\text{coll}} = \alpha - \frac{\beta \Delta T}{G} \tag{6.8b}$$

where $\alpha$ is the absorption factor at 0.75 – 0.80, and $\beta = 5$ to 6 $\text{Wm}^{-2}\ \text{K}^{-1}$ for $\Delta T = T_{\text{o}} - T_{\text{a}} = 30$ K, according to Schlaich (Schlaich, 1995). This is useful as a quick estimate but for accuracy $\Delta T$ should be estimated from Eq. (6.2) accounting for both the heat gained by the air entering the collector area and the simultaneous convective heat loss by the collector to the surrounding.

The turbine efficiency $f_t$ is computed from

$$f_t = \frac{\Delta p_t}{\Delta p_{tot}} \tag{6.9}$$

where $\Delta p_t$ is the turbine pressure drop and $\Delta p_{tot}$ is the total available pressure drop obtained from Eq. 6.1.

To gauge if you have achieved the optimum turbine pressure drop, the range ratio of the turbine pressure drop and the total available pressure drop is given as $\Delta p_t / \Delta p_{tot} = 0.83 - 0.91$, on the basis of data similar to Manzanares prototype solar chimney (Nizetic & Klarin, 2010).

The power coefficient $c_f$, can range from 0.43 to 0.50 (Schubel & Corssley, 2012), where the Betz theoretical maximum is 0.59. For blade, transmission and generator, efficiency is nominally set at 0.8 (Guo et al., 2016).

An approximate optimum velocity at chimney inlet for maximum overall efficiency was derived by Nizetic and Klarin (2010) after making a simplification by assuming that $\Delta T$ is independent of mass flow rate, is given by Eq. (6.10):

$$w_{Copt} \sqrt{h_C \Delta T} \tag{6.10}$$

Equation 6.10 agrees with the maximum theoretical efficiency of Eq. 6.5 that the optimum velocity for optimum overall efficiency is dependent on the chimney height and in addition, it is dependent on the temperature difference between the hot air and the ambient air.

In summary, the efficiencies are tabulated in Table 1.

Thus, given a chimney height, collector area, and solar irradiation rate, the electrical power output can be readily estimated for an initial design, without knowing what the flow rate or velocity should be. The flow rate can be estimated from the mechanical turbine power $W$ prior to transmission, i.e., the efficiency up to turbine only, calculated from the turbine pressure drop and the volumetric flow rate of air:

$$W = \dot{V} \Delta p_t \tag{6.11}$$

The characteristics of the efficiencies do not allow for much room to maneuver and an optimum velocity is unlikely without consideration of economics and maintenance.

**Table 1** Typical efficiencies in a solar chimney power plant (Eq. 6.7)

| Efficiency | Typical range % |
|---|---|
| Solar chimney (Carnot) $\eta_{SC}$ | $33 \times 10^{-6}\ h_c \times 100$ |
| Collector $\eta_{Coll}$ | 75–80 |
| Turbine power ratio $f_t$ | 83–90 |
| Transmission $\eta_{WT}$ | 80 |

## Geometrical Parameters of Solar Chimney for Effective Operation

To estimate a preliminary design power output, the value of the other variable to ascertain is the volumetric flow rate, which is related to the tower diameter.

The conventional solar chimney is tall and slim, from the prototype of 200 m in Spain up to 1 km in height for the proposed tower in Australia. There will be a situation when it may be better to build lower towers for construction and maintenance costs, since the taller a tower is the more difficult it is to build it to run safely.

The tradeoff is the reduced available head and the reduced efficiency in the available head (see Eqs. 6.1, 6.4, and 6.5) to convert to mechanical power at the turbine and electricity, since power output of a turbine is related to the cube of velocity, and the available pressure drop is related to the square of velocity. This is a double-whammy: the drop in performance is exponential. As the diameter gets wider and the height reduces, cold inflow may also become an issue that needs to be resolved. There is an optimum height to diameter for power generation application, but for natural ventilation this problem is less critical.

## The Problem of Cold Inflow/Flow Reversal/Backflow

Equations (6.1)–(6.9) are based on the assumption that the chimney tower does not suffer from cold air inflow, or flow reversal of hot air at the chimney tower exit. Should this occur, the efficiency of the tower drops further due to the loss of chimney height $h_c$. To increase the designed power output at the same height, the diameter of the chimney will need to be increased. This leads to another possible problem: cold air inflow at the top.

To estimate quickly if there is cold inflow, Jörg and Scorer (1967) produced a formula of penetration depth ratio $h_p/D$ into the chimney, based on laboratory experiments of glass tubes with air and water as the test fluids. To check if a buoyancy-driven flow is afflicted by cold inflow at the exit, the critical velocity is evaluated by setting the penetration depth to 0.

A rigorous method to check if there is cold air inflow is to use CFD to simulate the solar chimney. The setup can follow the work presented in Chu et al. (2016). An example of cold inflow is shown in Fig. 6.1.

Cold inflow must not be allowed to take place as its existence precludes the stack effect of the near-field plume or lazy plume and impairs the chimney stack effect (2012). On the other hand, the prevention of cold inflow can mean an enhancement of throughput by anywhere from 5 to 90% (Chu et al., 2009; Chu, 2012).

Once the cold inflow is confirmed to not be taking place by Eq. (6.9) and CFD simulation as exemplified by Fig. 6.2, the effective plume-chimney height $h_o$ can be added to the chimney height in Eq. (6.2). This has more implications than just adding

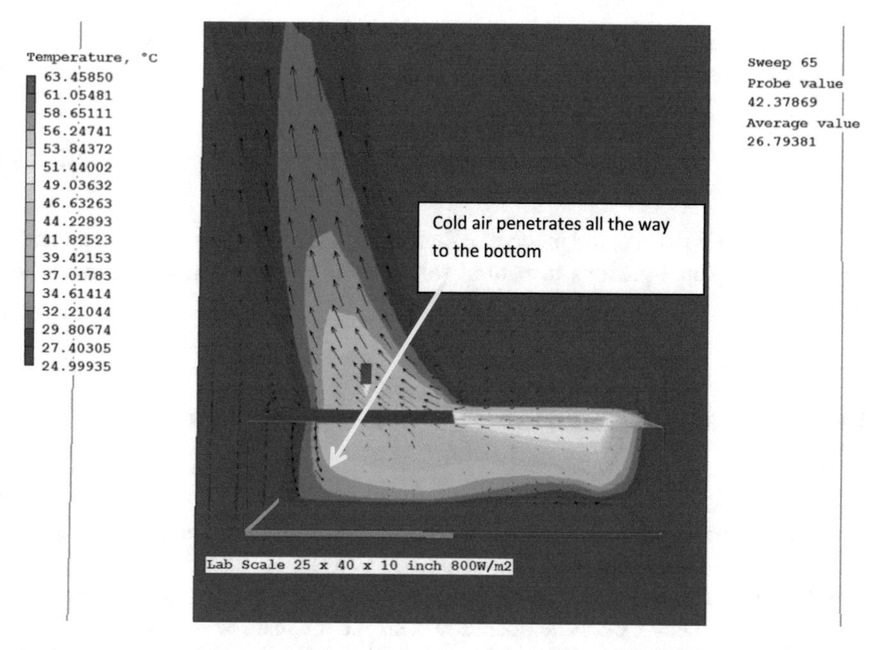


**Fig. 6.1** Flow reversal at a natural draft chimney outlet (right is vertical up) as indicated by temperature profile and the vectors

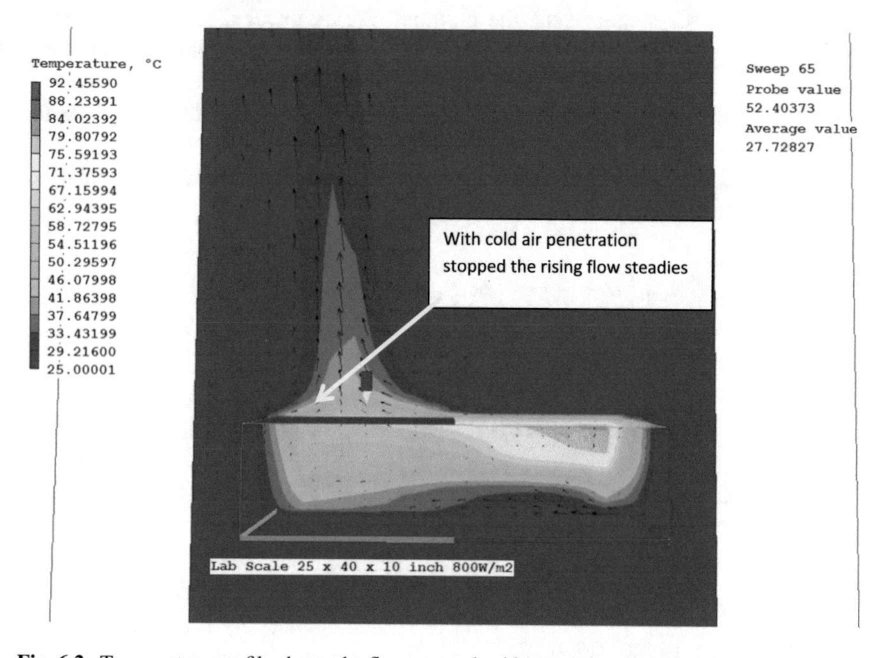


**Fig. 6.2** Temperature profile shows the flow upward achieves non-reversal

a height to a chimney; it also means the chimney exit loss is now turned into a gain! Eq. 6.12 is reconstituted from Chu (2002) equation:

$$h_o = 10.54\left(\frac{\eta^2 L^3 T_o}{\rho^2 g \Delta T}\right)^{\frac{1}{6}} \tag{6.12}$$

Inclusion of Eq. (6.12) in a procedure for the solution of simultaneous heat transfer and pressure drop equations of natural forced draft air-cooled heat exchangers is described in (Chu, 2005).

## Derivation of a Formula for a Quick Design of Solar Chimney

A quick design of the solar chimney for power generation will be outlined below through the simultaneous solution of energy conversion equation (energy) and thermal pneumatic equation (momentum) because there two unknowns: tower inlet temperature (collector outlet temperature) and the air mass flow rate through the tower. These will be solved iteratively and can be easily set up on a spreadsheet.

Begin by setting the values of parameters solar irradiation rate $G$, $\mathrm{Wm}^{-2}$, ambient temperature $T_\mathrm{a}$, °C, height of a solar chimney, $h_\mathrm{c}$, m, tower diameter $D$, m, solar collector area $A_\mathrm{coll}$, $\mathrm{m}^2$, according to the overall specification of the solar chimney tower.

The maximum available energy for conversion to the mechanical work at the turbine is.

Solar irradiation rate × collector efficiency × theoretical chimney efficiency = available pressure energy for input to turbine

$$GA_\mathrm{coll}\left(\alpha - \frac{\beta \Delta T}{G}\right) \times 33 \times 10^{-6} h_C = A_C w \Delta p_\mathrm{tot} \tag{6.13}$$

From Eq. 6.2,

$$\dot{Q} = GA_\mathrm{coll} = \dot{M} c_p (T_o - T_a) + \alpha_H A_\mathrm{coll} (T_o - T_a) \tag{6.2}$$

Rearrange to obtain an expression of $\Delta T = (T_\mathrm{o} - T_\mathrm{a})$ to substitute in Eq. 6.13,

$$\Delta T = \frac{GA_\mathrm{coll}}{\dot{M} c_p + \alpha_H A_\mathrm{coll}} \tag{6.14}$$

Inserting Eq. 6.14 into Eq. 6.13,

$$\frac{\alpha G A_{\mathrm{coll}}}{A_C w}\left(1-\frac{\alpha_H A_{\mathrm{coll}}}{\rho_o w A_C c_p+\alpha_H A_{\mathrm{coll}}}\right)\times 33\times 10^{-6} h_C=\Delta p_{\mathrm{tot}} \tag{6.15}$$

From Eq. 6.1, the thermal pneumatic balance is

$$[h_c(\rho_a-\rho_o)+h_o(\rho_a-\rho_o)]g=\Delta p_{\mathrm{tot}} \tag{6.1}$$

Or

$$[h_c+h_o](\rho_a-\rho_o)g=\Delta p_{\mathrm{tot}}$$

From Eq. 6.4,

$$\Delta p_{\mathrm{tot}}=\Delta p_w+\Delta p_{\mathrm{ex}}+\Delta p_{\mathrm{ten}}+\Delta p_{\mathrm{cen}}+\Delta p_t \tag{6.4}$$

Rearranging,

$$\Delta p_t=\Delta p_{\mathrm{tot}}-(\Delta p_w+\Delta p_{\mathrm{ex}}+\Delta p_{\mathrm{ten}}+\Delta p_{\mathrm{cen}}) \tag{6.16}$$

$\Delta p_t=c_f\frac{\rho_o w^2}{2}$, and $c_f$ typically ranges between 0.25 and 0.45 (theoretical 0.59), Eq. (6.16) can be used to back-calculate from known data to determine $c_f$ values more accurately for predictive simulation. Eq. (6.4) may be written, by ignoring $\Delta p_{\mathrm{cen}}$,

$$\Delta p_{\mathrm{tot}}=\left(f\frac{h_C}{D}+K_{\mathrm{ex}}+K_{\mathrm{ten}}+c_f\right)\frac{\rho_o w^2}{2} \tag{6.17}$$

Inserting Eq. (6.15) to Eq. (6.1),

$$[h_c+h_o](\rho_a-\rho_o)g=\left(f\frac{h_C}{D}+K_{\mathrm{ex}}+K_{\mathrm{ten}}+c_f\right)\frac{\rho_o w^2}{2} \tag{6.18}$$

Inserting Eq. 6.15 to Eq. 6.14,

$$\frac{\alpha G A_{\mathrm{coll}}}{A_C w}\left(1-\frac{\alpha_H A_{\mathrm{coll}}}{\rho_o w A_C c_p+\alpha_H A_{\mathrm{coll}}}\right)\times 33\times 10^{-6} h_C =\left(f\frac{h_C}{D}+K_{\mathrm{ex}}+K_{\mathrm{ten}}+c_f\right)\frac{\rho_o w^2}{2} \tag{6.19}$$

The absorption coefficient $\alpha$ is taken by Schlaich (Schlaich, 1995) to be 0.75. Now $\beta$ is the natural convection heat transfer coefficient of all the solar collector surface, so $\beta=\alpha_{\mathrm{H}}$, and $\alpha_H=1.52\Delta T^{1/3}$ (Eq. 6.3). If the temperature difference is assumed constant, it can be assigned a constant value. $f$ is the Fanning friction factor using the Haaland (Haaland, 1983) equation, and $K_{\mathrm{ex}}=1.0$ and $K_{\mathrm{ten}}=0.5$ (Haaland, 1983). For the case of Manzanares it was found in our calculations as described in the sections following that the entrance and exit loss coefficients needed to be set to zero to match their data. The power coefficient $c_f$ is assigned a value of 0.45.

As an example, the Manzanares solar chimney tower geometrical specification is given as (Michaud, 1999):

Height of a solar chimney, $h_c = 194$ m.

Height of collector, $h_{coll} = 1.8$ m.

Tower diameter $D = 10$ m.

Solar collector area $A_{coll} = 46{,}000$ m$^2$.

The solar irradiation rate for winter is 1040 Wm$^{-2}$ and the ambient temperature = 29 °C (Cao et al., 2013).

The two Eqs. (6.2) and (6.18), heat transfer and pressure drop balance, respectively, will be solved by initially guessing the mean tower air temperature $T_{o,1}$ and after obtaining the $h_o$ from Eq. (6.12), evaluate the mean tower updraft velocity $w$ from Eq. (6.18). The critical velocity for adverse cold air inflow penetration is calculated by Jörg and Scorer's (1967) formula. The value of $w$ will be substituted into Eq. (6.2) and a new $T_{o,2}$ is obtained. The two $T_o$'s are compared. If the deviation is larger than the tolerance margin, $T_{o,1}$ will be substituted by $T_{o,2}$ and the computation repeated in Eqs. (6.2) and (6.18) until the two $T_o$'s agree to within the criterion set. The double loop iterations can be easily solved by Excel add-on Solver, as the example computation below in Table 2 shows. The LHS of Eq. (6.19), which contains an approximate maximum thermal efficiency, serves to check the solution obtained above.

The power output compares closely to the reported value of 36–41 kWe in literature (Cao et al., 2013). The overall efficiency is 0.09% (Eq. 6.7 is approximately 0.17%), which is close to that reported for the Spanish prototype in Manzanares with a range of 0.055 to 0.09 (Guo et al., 2019). The tower exit loss coefficient has to be revised to zero, since the exit loss is probably much less due to the addition of effective plume-chimney height. The turbine efficiency or power ratio is less than the reported 0.83–0.91. The reported power ratios are derived from the thermal pneumatic equations by assuming constant temperature rise and detaching the relationship between temperature rise $\Delta T$ and tower velocity $w_c$. Moreover, it has not been reported as a measurement value in the literature. The model is subject to modification by the user who may have better insight, e.g., by adding entrance and acceleration losses, and this can easily be done on the spreadsheet.

**Table 2** Computation example based on manzanares data

| ***Design of solar Chimney*** | | | ***Remark*** |
|---|---|---|---|
| Height of chimney = | 195.8 | m | Including collector height |
| Chimney diameter = | 10 | m | |
| Flow area AC = | 78.54 | $m^2$ | |
| Collector area = | 46,000 | $m^2$ | Michaud (1999) |
| Absorption coeff. alpha = | 0.75 | | Schlaich (1995) |
| Power coefficient cf = | 0.45 | | Wind turbine maximum (Schubel & Corssley, 2012) |
| $K_{ex}$ = | 1 | | Typical (Sinnott et al., 2019) |
| $K_{ten}$ = | 0.5 | | Typical (Sinnott et al., 2019) |
| $K_{wt}$ = | 0.8 | | Blade, transmission and generator efficiency Guo et al. (2016) |
| Solar irradiation rate = | 1040 | $W/m^2$ | Cao et al. (2013) |
| Ambient temperature = | 29 | deg C | |
| Ambient density = | 1.170 | $kg/m^3$ | |
| Dynamic viscosity = | 2.00E−05 | Pa s | |
| Ambient pressure = | 101,325 | Pa | |
| **Given the above data, calculate the power output of the turbine** | | | |
| **Both velocity and temperature can be varied at the same time to solve the buoyancy-driven flow** | | | |
| To1 = | 56.62 | degC | <--Guess using solver |
| Rhoo1 = | 1.0722 | $kg/m^3$ | |
| uCritical | 0.320 | m/s | Jörg and Scorer (1967) |
| ho = | ***0.91*** | m | Equation 6.12 |
| On the RHS Eq. 6.18, w1 | 12.91 | m/s | <-- Guess using solver |
| Re = | 6.92E + 06 | | |
| $f$ = | 0.00857 | | Friction factor from Haaland equation, |
| | 0.00617 | | Friction factor from Blasius (for checking) |
| LHS Eq. 6.18= | 189.263 | Pa | |

(continued)

**Table 2** (continued)

| | | | | |
|---|---|---|---|---|
| | RHS Eq. 6.18= | 189.267 | Pa | |
| | Error = | 0.00 | % | |
| AlphaH = | | 4.59 | W/(m$^2$ K) | Equation 6.3 |
| To2 = | | 56.63 | deg C | Equation 6.2 |
| Rhoo2 = | | 1.0722 | kg/m$^3$ | Hot air density |
| | To1 = | 56.62 | deg C | |
| | To2 = | 56.63 | deg C | |
| | Error = | 0.01 | K | |
| | Combined error (Velocity and temperature) | 0.01 | % + K | <-- Objective in solver |
| **If guessed temperature matches the calculated temperature,** | | | | |
| | Power output = | 41 | kW | Turbine<br>LHS Eq. 6.13 |
| | Power from irradiation | 195.0 | kW | |
| | Turbine power ratio = | 0.21 | | Based on Eqs. 6.9 and 6.17 |
| | | 0.21 | | Based on Eqs. 6.9 and 6.13 |
| | Electrical power = | ***32.6*** | kWe | agrees with Cao et al. (2013) |
| | Absorbed irradiation = | 35,880 | kW | 75% of irradiation |
| Overall efficiency conversion to electrical power = 0.09% | | | | Eq. (6.7) approx. 0.17%; Agrees with Guo et al. (2019) |

## References

Cao, F., Li, H. S., Zhao, L, Bao, T. Y., Guo, L. J. (2013). Design and simulation of the solar chimney power plants with TRNSYS. *Solar Energy, 98*, 23–33. (Elsevier).

Chu, C. M. (2002). A preliminary method for estimating the effective plume chimney height above a forced-draft air-cooled heat exchanger operating under natural convection. *Heat Transfer Engineering, 23*, 3–13. (Taylor and Francis, London).

Chu, C. M. (2005). Improved heat transfer predictions for air-cooled heat exchangers. *Chemical Engineering Progress, 101*(11), 46–48. (*A.I.Ch.E.*, New York, NY).

Chu, C. C. M., Chu, R. K. H., & Rahman, M. M. (2012). Experimental study of cold inflow and its effect on draft of a Chimney. *Advanced Computational Methods and Experiments in Heat Transfer XII*, 73–79. 9 (WIT Press).

Chu, C. M., & Rahman, M. M. (2009). A method to achieve robust aerodynamics and enhancement of updraft in natural draft dry cooling towers. In *ASME Summer Heat Transfer Conference*, HT2009-88289, 19–23 July, San Francisco, Calif., U.S.A.

Chu, C. M., Rahman, M.M., & Kumaresan, S. (2012). Effect of cold inflow on chimney height of natural draft cooling towers. *Nuclear Engineering and Design, 249*, 125–131. (Elsevier).

Chu, C. M., Rahman, M. M., & Kumaresan, S. (2016). Improved thermal energy discharge rate from a temperature-controlled heating source in a natural draft chimney. *Applied Thermal Engineering, 98*, 991–1002 (Elsevier).

Geankoplis, C. J. (2003). *Transport processes and separation process principles* (4th ed.). Prentice Hall.

Guo, P. H., Li, J. Y., Wang, Y. F., & Wang, Y. (2016). Evaluation of the optimal turbine pressure drop ratio for a solar chimney power plant. *Energy Conversion and Management, 108*, 14–22. (Elsevier).

Guo, P. H., Lia, T. T., Xu, B., Xu, X. H., & Lia, J. Y. (2019). Questions and current understanding about solar chimney power plant: A Review. *Energy Conversion and Management, 182*, 21–33. (Elsevier).

Haaland, S. E. (1983). Simple and explicit formulas for the friction factor in turbulent pipe flow. *Journal of Fluids Engineering*, 89–90.

Jörg, O., & Scorer, R. S. (1967). An experimental study of cold inflow into chimneys. *Atmospheric Environment, 1*(6), 645–646. (Elsevier).

Michaud, L. M. (1999). Vortex process for capturing mechanical energy during upward heat-convection in the atmosphere. *Applied Energy, 62*, 241–251. (Elsevier).

Nizetic, S., & Klarin, B. (2010). A simplified analytical approach for evaluation of the optimal ratio of pressure drop across the turbine in solar chimney power plants. *Applied Energy, 87*, 587–591. (Elsevier).

Schlaich, J. (1995). *The solar chimney: Electricity from the sun* (p. 55). Geislingen: Maurer C.

Schubel, P. J., & Corssley, R. J. (2012). Review wind turbine blade design. *Energies, 5*, 3425–3449. https://doi.org/10.3390/en509342. www.mdpi.com

Sinnott, R. K., & Towler, G. P. (2019). Coulson and Richardson series: chemical engineering design (6th ed.). Elsevier.

Windpower Engineering and Development. https://www.windpowerengineering.com/calculate-wind-power-output/. Accessed September 24, 2020.

# Chapter 7
# Study the Effect of Divergent Section on Solar Chimney Performance

**Ahmed Jawad, Mohd. Suffian bin Misran, Mohammad Mashud, and Md. Mizanur Rahman**

This chapter mainly discusses the divergent and convergent section effect on solar chimney performances. Several experiments were conducted for different heights of solar chimney models, and the results are discussed in this chapter. CAD software was used to design the solar chimney, and dimensional equations were used to achieve the proper dimensions of the solar chimney model. The fabrication process of the model chimneys also explained, along with the limitation that involved during the process. Also, data acquisition instrumentation and experimental procedure were described in detail along with the equations used for the analysis of the experimental. At the end of this chapter, the experimental results were concluded, and a set of recommendations was proposed.

## Introduction

A chimney is a structure that used for many purposed such as power generation, ventilation as well as the exile of hot flue gases. Solar chimneys come in a cylindrical or

A. Jawad (✉) · Mohd. Suffian bin Misran
Faculty of Engineering, Department of Mechanical Engineering, Universiti Malaysia Sabah, Kota Kinabalu, Sabah, Malaysia
e-mail: jawadansari80@gmail.com

Mohd. Suffian bin Misran
e-mail: suffian@ums.edu.my

M. Mashud
Department of Mechanical Engineering, Khulna University of Engineering and Technology, Khulna, Bangladesh

Md. M. Rahman
Department of Mechatronics Engineering, World University of Bangladesh, 151/8, Green Road, Dhaka 1205, Bangladesh



Md. M. Rahman and C.-M. Chu (eds.), *Cold Inflow-Free Solar Chimney*,
https://doi.org/10.1007/978-981-33-6831-6_7

rectangular shape, and the process of removing hot gases through it is called the Stack Effect. The process is explained by natural convection theory in which air moves due to a buoyancy effect or density difference. Natural and forced convection types chimneys are available. The main difference between natural and forced convection chimney is the used of mechanical device to enhance the flow. The application of the chimney is extensive that includes locomotives, cooling towers in industries for removing waste heat and in residential for ventilation. The natural convection process in a solar chimney can also be used for power generation and in residential as an effective alternative cooling system. A Solar chimney generally comprised of three main components; solar collector, chimney and a turbine. It works on an elementary working principle; the air is heated due to the presence of solar radiation in the solar collector, the density difference caused the buoyancy effect and resulted in the flow of air current through the chimney that runs the turbine and generates electricity. The performance of the solar chimney depends on solar collector characteristic, which are the surface area, collector design, the material used and weather conditions (Kasaeian et al., 2011; Hamdam, 2011; Hosien & Selim, 2017; Toghraie et al., 2018; Arzpeyma et al., 2020).

## Literature Review

A solar collector functioned as an air heater in a solar chimney power plant. As a result, the air temperature increased significantly. The hot air in the solar collector has less density air than outside air. The density difference caused the buoyancy effects and resulted in the warm air to flow through the system. Koonsrisuk et al. (2010) published an article and discussed the solar chimney configuration. The study discussed four different cylindrical-shaped solar chimney configurations. The altered parameters are the diameter, height of the cylindrical-shaped chimney, radius and opening height of the solar collector. The airflow was caused in the vertical chimney due to the effect of buoyancy force; thus, the shape of the chimney was not considered (Koonsrisuk et al., 2010). Hu et al. (2017) presented a performance analysis on a divergent solar chimney. The analysis was done numerically with chimney inlet and outlet area ratio ranging from 1.25 to 32 and divergent angles ranging from 1.5 to 6.8° as shape control elements. The study concentrates on the performance of three different solar chimney heights; 100, 200 and 300 m solar chimney. The divergent solar chimney provides higher output power compared to the cylindrical-shaped traditional chimney. The maximum power output measured at the critical inlet and outlet areas ratio of 10 and inclination angle around 2.7°. After these values, the power output drops significantly (Hu et al., 2017). Koonsrisuk and Chitsomboon (2013) investigate the effects of variations flow area ranging from 0.25 to 16 on static pressure, mass flow rate and power generation by the solar chimney. The CFD simulation results show that the divergent-shaped chimney has better performance than a cylindrical-shaped and convergent-shaped solar chimney. The divergent chimney inlet and outlet areas ratio 16 be able to generate 400 times more power than a cylindrical-shaped solar chimney. Ming et al. (2013), published a numerical and mathematical 10 MW solar chimney models outcomes and discuss the airflow rate, heat transfer rate and power generation. Three different chimney configurations (cylindrical, divergent and convergent) were studied in the report. The study uses cylindrical-shaped chimney on the base while the chimney exit diameter varies from 60 to 160 m showed better power output and efficiency amongst the three configurations. Patel et al. (2014), published numerical studies data on solar chimney performance for different divergent chimney configurations. The divergent chimney angle varies from 0 to 3°. The analysis was also done for the based diameter of the chimney varied from 0.25 to 0.3 m. The collector configurations and opening of the collectors are kept constant. Solar chimney with divergent angle 2° and the chimney inlet diameter 0.25 m shows the best performance compared to other configuration in the study.

## Design of Chimney Model

In 2018, a study was performed at Faculty of Engineering, Universiti Malaysia Sabah (UMS), Malaysia. The study aimed to understand the working principle and performance of divergent solar chimney models. The study uses an existing working prototype of a solar chimney with 195 m-high with a diameter of 10 m as the based design to develop models of solar chimney using dimensional analysis. The models enable the research work to be carried out without the expensive construction cost and time constraint. Since the acceleration due to gravity plays a vital role on the airflow in the solar chimney, therefore, Froude number is used for scaling down the prototype to obtain the diameter for the model (Chakrabarti, 2005). According to the definition of the Froude number, it can be expressed as

$$fr = \frac{v}{(gD)^{1/2}} \tag{7.1}$$

This equation can be modified and equate the Froude number of the model and prototype to obtain the dimensions of the model chimney. So Eq. (7.1) can be rewritten as Eq. (7.2)

$$\frac{(V_p^2)}{D_p} = \frac{(V_m^2)}{D_m} \tag{7.2}$$

In Eq. (7.2), the subscription $P$ indicates the prototype and m indicates the model. The air velocity inside the chimney is obtained from the buoyancy Eq. (7.3) as suggested by Schlaich (1995)

$$v = \left(2gh\frac{\Delta T}{T_\infty}\right)^{1/2} \tag{7.3}$$

where $D$ represents the exit diameter of the chimney, $v$ represents the air velocity in the chimney, $g$ represents the gravity, $h$ represents the height of the chimney, $T_\infty$ represents ambient temperature, and $\Delta T$ represents the temperature difference between hot air inside the collector and ambient temperature.

A bell mouth is used at the entrance of the chimney. The primary function of it is to enhance air intake into the chimney. The following Eqs. (7.4) and (7.5) are used to calculate the bell mouth dimension (Idelchik, 1987).

$$BM_r = 0.2 \sim 0.4 \times D \tag{7.4}$$

$$BM_h = 0.2 \sim 0.8 \times D \tag{7.5}$$

In these equations, the variables $BM_r$, $BM_h$ are referred to as bell mouth radius and bell mouth height, respectively. The dimensional analysis leads to the following dimensions for the model chimneys;

- Chimney height: 1 to 2 m
- Chimney exit diameter: 0.15 to 0.3 m
- Bell mouth height: 0.11 to 0.21 m
- Bell mouth diameter: 0.06 to 0.24 m.

The dimensions of the solar chimney are shown in Fig. 7.1, whose throat angle was 2° as it is the recommended angle for maximum output.

**Model fabrication** Fabrication starts with the material selection; zinc sheets were used as it is easy to weld, suitable rigid property and readily available in the market. Another alternative material is an aluminium sheet, it is a cheap option, but it needs carbon welding, which requires an additional process to fabricate. Furthermore, zinc sheet was able to withstand high temperatures with minimum distortion. Fabrication of the chimney was done in the mechanical workshop at Universiti Malaysia Sabah.

**Measuring instruments** Several measurements and recording instruments were used during the experiment. The experiments schematics are as shown in Fig. 7.2. K-type thermocouples were used for measuring temperature in different points of the chimney. A Cole-Parmer USB Data Acquisition Module is connected with thermocouple to record temperature data automatically in the computer; the thermocouples are calibrated with a standard temperature sensor in the ambient conditions to determine the error. The error in thermocouple lies within the range of +3 to −2 K. An electric heating system is used as a heat source, with power rating 2000 W and 240 V, The diameter of the heating system was approximately 0.2 m. A manual voltage regulator is used to control the air temperature in the system. A manual anemometer (centre 330 anemometer) model was used to record the air velocity at the exit of the chimney. The error lies within the range of 2.5% of air velocity with a measurement intensity of 0.025 times/s. A digital clamp meter (UNI-T UT202A) is used to measure the current as well as input power. The errors of the clamp meter reading lie within the range of ±2%. A Flir C2 thermal image camera is used to

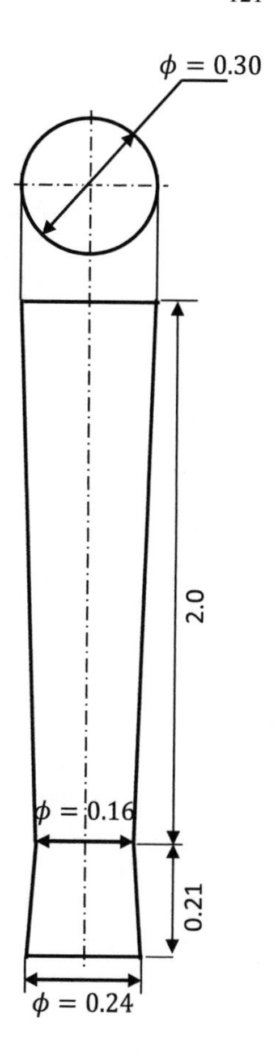


**Fig. 7.1** Solar chimney model with bell mouth

take thermal images during experiments. These images are used to investigate the cold inflow phenomena during the experiment. The temperature measurement range of the thermal image camera is −10 to 150 °C with an accuracy of ±2%. A smoke generator EF-501 is also used to generate smoke during experiments to visualise cold inflow phenomena in the chimney. All the electrical connections were coated with fibreglass to avoid any kind of hazards due to excessive heat

**Experimental procedure** The experiments were conducted in an enclosed laboratory with a ventilator to ensure the ambient temperature remains constant throughout the process. The heating coil is turned on and adjusted to a designated electric load. Then, the air temperature inside the model chimney and flow rate were monitored. The measurement was taken when a steady state was achieved, means there is no significant change in temperature and airflow rate. A total of 108 experiments was carried out covering different heights, heat loads and solar chimney configurations. Each experiment was repeated three times to reduce the error and achieve more accurate values.

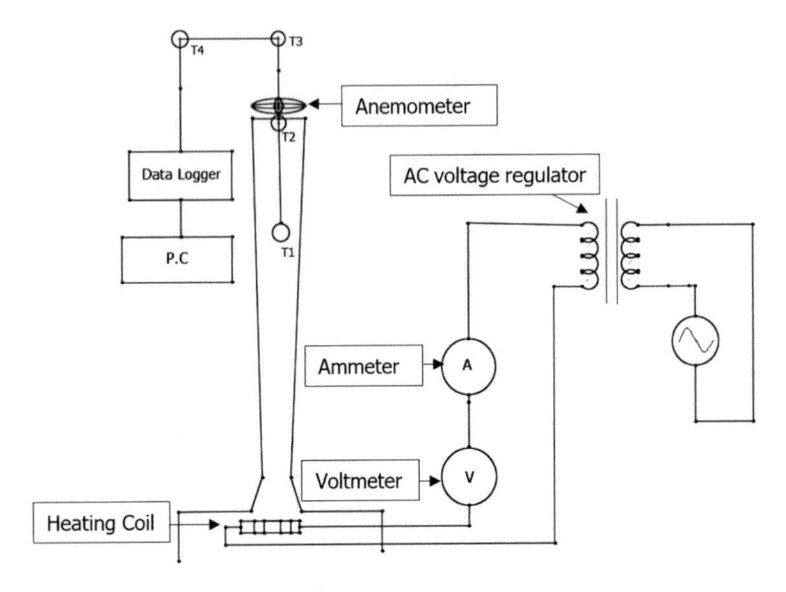


**Fig. 7.2** Schematic diagram of solar chimney experiments set-up

**Experimental results and data analysis** The experimental results were analysed to determine the relationships between four different parameters: temperature, air velocity, theoretical density and theoretical power generation. The mass flow rate and throat velocity were obtained using the mass flow rate continuity equation, as shown in Eq. (7.6). Furthermore, the density equation was obtained from the Excel by using the related function of temperature as shown in Eq. (7.7), and density was derived from the air properties between 0 and 120 °C from air properties at 1 atm pressure (Cengel, 2014).

$$\dot{m} = \rho A v \tag{7.6}$$

$$\rho = 1\text{E} - 05\text{T}^2 - 0.0045\text{T} + 1.29 \tag{7.7}$$

$$V_1 A_2 = V_2 A_2 \tag{7.8}$$

$$P_{\text{wind}} = \frac{1}{2}\rho \times A_{\text{ch}} \times V^3 \tag{7.9}$$

$$P_{\text{electric}} = P_{\text{wind}} \times \eta_t \times \eta_g \tag{7.10}$$

In continuity Eq. (7.8) given above, A represents the area of the chimney, $\rho$ represents density, and $v$ represents the velocity. $T_\infty$ is the ambient temperature in the laboratory used in Eq. (7.7). The velocity at the throat was calculated using the energy balance Eq. (7.8). Equation (7.9) represents the theoretical wind power that spin the turbine, $P_{\text{wind}}$ defines power wind power potential, $\rho$ represents air density, $A_{\text{ch}}$ represents the cross-sectional area of chimney, and $V$ represents the velocity. Subsequently, Eq. (7.10) represents the theoretical electric power generation, whereas $P_{\text{electric}}$ represent electric power generation, $\eta_t$ represents the efficiency of the turbine, and $\eta_g$ is generator efficiency. The results can be used to analyse the following criteria.

**Comparison of the divergent and cylindrical chimney** This study investigates the performance of divergent chimney with a bell mouth design and compares it with equivalent cylindrical chimney. Temperature variances were recorded using a data logger and an anemometer was used to measure the air velocity at the exit of the chimney at four different electric heat load. The obtained velocity and temperature readings from the divergent and cylindrical chimney were then used to calculate theoretical electric power potential for comparison purposes.

**Effect of wire mesh in the divergent-shaped chimney** Cold inflow occurs typically in a tall chimney where the cold air in the surrounding chimney outlet sinks into the centre exit area of the chimney. This sinking of the cold air instantly drops the performance efficiency of the plant. In this experiment, a smoke generator was used to observe the cold air inflow at the exit of the chimney. A thermal camera was also used to observe the chimney outlet in an attempt to capture flow behaviour. A study has reported the use of wire mesh at a chimney outlet able to mitigate the effect of cold inflow. Thus, this study aims to investigate the effect of wire mesh at the outlet of a divergent chimney and its effect on performance.

## Result Discussion

This section discusses the experiments result and categorised them into two main criteria: divergent chimney performance compared to a cylindrical chimney and divergent chimney performance with a wire mesh variance.

**Comparison of divergent and cylindrical chimney** The performance of solar chimney depends on the chimney design, height and solar radiation intensity. Figure 7.3 presents the experiment conducted on the 1 m divergent and cylindrical solar chimney models under different electric heat load. It can be observed that the air velocity rate is much higher in the divergent chimney compared to the cylindrical chimney. The highest average velocities in the divergent and cylindrical chimneys were measured at electric heat load 2 kW at 0.994 m/s and 0.820 m/s, respectively. In contrast, the lowest average velocity of the cylindrical chimney and divergent chimney were observed at 0.580 and 0.716 m/s at an electric heat load of 0.9 kW. A similar trend was observed in the 1.5 and 2 m divergent and cylindrical chimney models. It can be said that divergent chimney improves talk effect and increases air velocity, as shown in Figs. 7.3, 7.4 and 7.5.

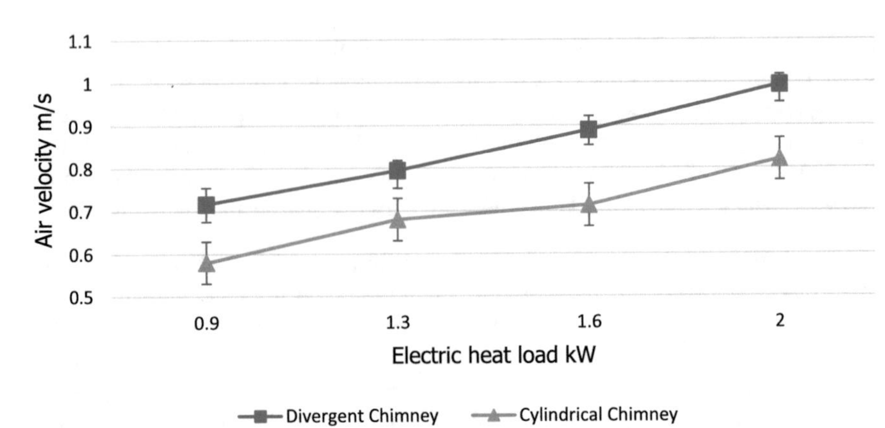


**Fig. 7.3** Air velocity of 1 m chimney models at different electric heat loads

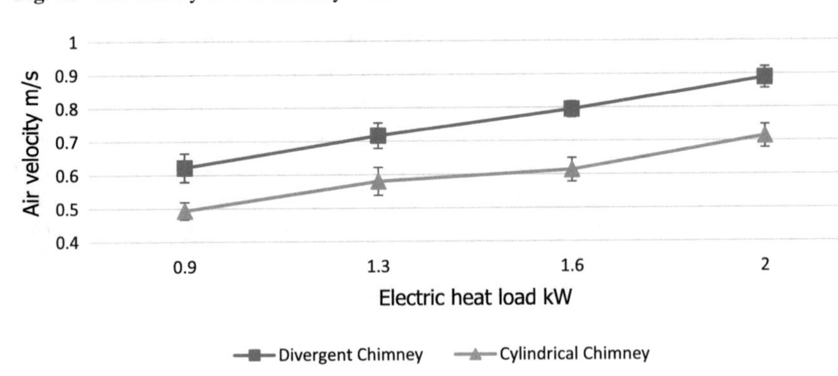


**Fig. 7.4** Air velocity of 1.5 m chimney models at different electric heat loads

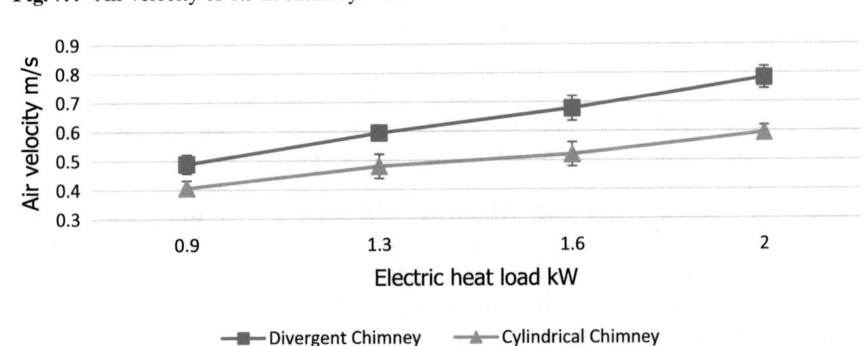


**Fig. 7.5** Air velocity of 2 m chimney models at different electric heat loads

**Table 7.1** Temperature results of the 2 m cylindrical and divergent-shaped chimneys under four different electric heat loads

| Electric heat load | $T_1$ | | $T_2$ | | $T_3$ | |
|---|---|---|---|---|---|---|
| | Cylindrical | Divergent | Cylindrical | Divergent | Cylindrical | Divergent |
| kW | K | K | K | K | K | K |
| 0.9 | 324.31 | 333.24 | 320.20 | 326.54 | 316.06 | 319.80 |
| 1.3 | 329.78 | 340.18 | 324.04 | 335.89 | 319.92 | 328.85 |
| 1.6 | 334.27 | 343.64 | 326.02 | 337.55 | 321.47 | 330.46 |
| 2 | 338.60 | 347.44 | 330.44 | 341.30 | 324.83 | 331.68 |

All experiments conducted on the divergent and cylindrical chimney show that the divergent solar-shaped chimney has higher air velocity compared to the cylindrical chimney at all electric heat load range. The reason for this is that the divergent shape of chimney improves the stack effect, ultimately increases air velocity due to divergence.

The obtained temperature readings from the experiment conducted on the 2 m divergent and cylindrical chimney models are shown in Table 7.1. Temperature readings demonstrate that divergent chimneys have higher average temperatures compared to cylindrical chimneys. The maximum average temperature observed was 347.44 K in the divergent chimney, whereas in the cylindrical chimney maximum temperature recorded was 338.6 K, both on electric heat load of 2 kW. This data is significant because higher inlet temperature leads to a better stack effect and higher velocity in the divergent chimney.

The experiments also show that air velocity is higher in the smaller-scale chimney model due to the fixed collector size at 0.8 $m^2$, for all chimney models. Since a 1 m chimney is smaller in footprint size compared to the 1.5 m chimney (scaled), more space was left behind to heat the air inside the collector resulting in a higher temperature at the inlet. This higher temperature increases the buoyancy effect, which has caused increases air velocity in the short chimney model compares to tall chimney models. Furthermore, the shorter chimney has fewer air frictions and thermal losses due to smaller surface area compared to the large-scaled chimney. Thus, a higher temperature difference compared to ambient temperature in short-scaled models causes faster air velocity due to the buoyancy effect (Schlaich, 1995). Table 7.2 represents the recorded temperature on the experiment performed on 1, 1.5 and 2 m divergent chimney. The temperature in the 1 m divergent chimney has the highest temperature readings compared to all configuration, followed by 1.5 and 2 m variance. Similar trends are exhibited in the cylindrical chimney models, as shown in Table 7.3; the smaller-scale chimney has a higher temperature difference compared to the taller chimney. As discussed, it was due to the fixed heat collector size for the chimney model regardless of the height, and smaller scale of chimney models has less thermal and frictional losses.

Wind power potential was calculated using Eq. (7.9) and subsequently was used to calculate the electric power generation by using Eq. (7.10). The theoretical electric

**Table 7.2** Temperature results of the 1, 1.5 and 2 m divergent-shaped chimneys at various electric heat loads

| Electric heat load | $T_1$ | | | $T_2$ | | | $T_3$ | | |
|---|---|---|---|---|---|---|---|---|---|
| | 1 m | 1.5 m | 2 m | 1 m | 1.5 m | 2 m | 1 m | 1.5 m | 2 m |
| kW | K | K | K | K | K | K | K | K | K |
| 0.9 | 348.44 | 338.07 | 333.24 | 341.88 | 330.06 | 326.54 | 336.53 | 324.96 | 319.80 |
| 1.3 | 354.71 | 345.61 | 340.18 | 350.17 | 339.38 | 335.89 | 345.64 | 334.92 | 328.85 |
| 1.6 | 358.02 | 346.61 | 343.64 | 354.50 | 341.12 | 337.55 | 349.75 | 336.89 | 330.46 |
| 2 | 362.96 | 349.49 | 347.44 | 358.38 | 342.46 | 341.30 | 354.22 | 334.58 | 331.68 |

**Table 7.3** Temperature results of the 1, 1.5 and 2 m cylindrical-shaped chimney at various electric heat loads

| Electric heat load | $T_1$ | | | $T_2$ | | | $T_3$ | | |
|---|---|---|---|---|---|---|---|---|---|
| | 1 m | 1.5 m | 2 m | 1 m | 1.5 m | 2 m | 1 m | 1.5 m | 2 m |
| kW | K | K | K | K | K | K | K | K | K |
| 0.9 | 339.12 | 328.56 | 324.31 | 334.90 | 324.14 | 320.20 | 327.74 | 318.29 | 316.06 |
| 1.3 | 340.44 | 333.79 | 329.78 | 335.75 | 327.80 | 324.04 | 330.03 | 320.09 | 319.92 |
| 1.6 | 345.87 | 335.64 | 334.27 | 339.21 | 330.70 | 326.02 | 333.15 | 323.25 | 321.47 |
| 2 | 349.43 | 339.12 | 338.60 | 341.88 | 334.90 | 330.44 | 336.71 | 327.74 | 324.83 |

power generation of the divergent chimney models is shown in Fig. 7.6, whereas cylindrical chimney theoretical power generation is shown in Fig. 7.7.

Theoretical electric power generation is higher in the divergent solar chimney compared to the cylindrical chimney in all variance. The reason for this behaviour is that the velocity at the throat in a divergent chimney increases due to reduction of the bell mouth area which causes higher torque potential compared to the

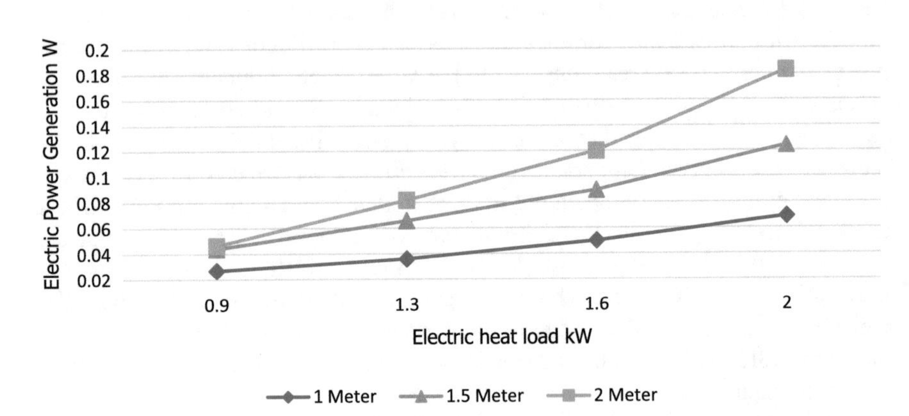


**Fig. 7.6** Theoretical electric power generation of divergent chimney models under different electric heat loads

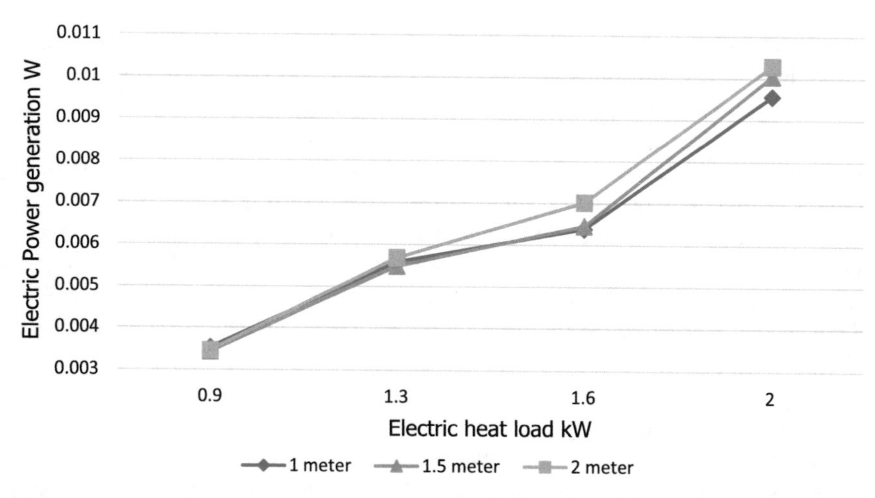


**Fig. 7.7** Theoretical electric power generation of cylindrical chimney models under different electric heat load

cylindrical chimney (Bouabidi et al., 2018; Ahirwar & Sharma, 2019; Das & Chandramohan, 2020). Calculated data shows theoretical electric power generation in the divergent chimney considerably increases with the electric heat load, whereas it barely increases in the cylindrical chimney. The experiments result has generally confirmed the divergent chimney, in general, and has better performance compared to a cylindrical chimney; this is attributed to the pressure recovery mechanism in the divergent chimney due to the particular shape of its diffuser (Koonsrisuk & Chitsomboon, 2013; Hu et al., 2017).

Several studies have reported in which the divergent chimney is 2–26 times more efficient compared to the cylindrical chimney, depending on physical design parameters such as the throat angle, chimney height as well as solar radiation intensity (Hu & Leung, 2017; Hu et al., 2017). This study finds using model experiments approach shows that divergent chimney is 6–18 times more capable of generating theoretical electric power compared to cylindrical chimney variance. Thus, a small-sized divergent chimney has a higher electric power generation tendency compared to a large-sized cylindrical-shaped chimney. Consequently, the cost for constructing a divergent chimney can be lowered while possibly improving the power output of the solar chimney power plant.

## Cold Inflow

Jörg and Scorer (1967) were the first who experimentally observed cold inflow phenomena. Cold inflow, also known as flow reversal, is a phenomenon that happens when cold air attempts to penetrate the chimney, cooling towers, and open-topped vessels due to its higher density. Cold inflow is the phenomenon mostly observed in the cooling towers and chimneys (Moore & Garde, 1983). Cooling towers with low air velocity has a higher possibility of undergoing cold inflow, which can lower the performance (Kloppers & Kröger, 2004). Petrus (2007) discovered that solar chimney power plant has higher possibilities of undergoing cold inflow phenomenon because it has a big diameter outlet. It is unable to avoid fractional flow that causes low air velocity in the chimney due to the large cross-sectional area and thus experiences more cold inflow occurrences. Xu and Zhou (2018) ran CFD simulation on divergent chimneys reported the cold inflow could drop the temperature of hot air at the exit of chimney, thereby decreasing the buoyancy effect which can reduce the performance of the divergent solar chimney power.

Cold inflow can be reduced by increasing the area of the solar collector as it will collect more solar radiation (Petrus, 2007; Xu & Zhou, 2018) and using forced convection like a fan or pump. However, a large-size solar collector has higher heat convective losses (Al-Azawiey et al., 2017) and using a fan will consume electric power and will also increase maintenance cost. A reliable and effective way to minimise the cold inflow is to install a wire mesh at the exit of the chimney. Chu et al. (2012) performed experiments on the laboratory-scaled model on a cooling tower. Experiments result shows the cold inflow can be minimised with the installation of a wire mesh at the exit of the chimney, and a similar solution can be used for the divergent solar chimney.

A smoke test was conducted on different divergent chimney models to visualise cold inflow phenomena. Observations made show the cold air from surroundings enters into the chimney, as shown in Fig. 7.8 in a 1 m chimney. As the diameter increases with higher chimney, the cold inflow increases and occurrences increase as experience by 1.5 m divergent chimney shown in Fig. 7.9.

**Fig. 7.8** Smoke test on 1 m chimney without a wire mesh

**Fig. 7.9** Smoke test on 1.5 m chimney without a wire mesh

A 0.64 mm × 0.64 mm stainless steel wire mesh was installed at the exit of the 1 and 1.5 m divergent chimney, and the smoke test was repeated. It was observed that the cold inflow was decreased, as shown in Figs. 7.10 and 7.11.

The performance of divergent chimney with wire mesh was also investigated. The measured air velocity in a divergent chimney with and without wire mesh is shown in Fig. 7.12. The experimental data unveils the highest average velocity value of air velocity with a wire mesh installed is 0.883 m/s and without wire mesh is

**Fig. 7.10** Smoke test on a 1 m chimney with wire mesh

**Fig. 7.11** Smoke test on a 1.5 m chimney with a wire mesh

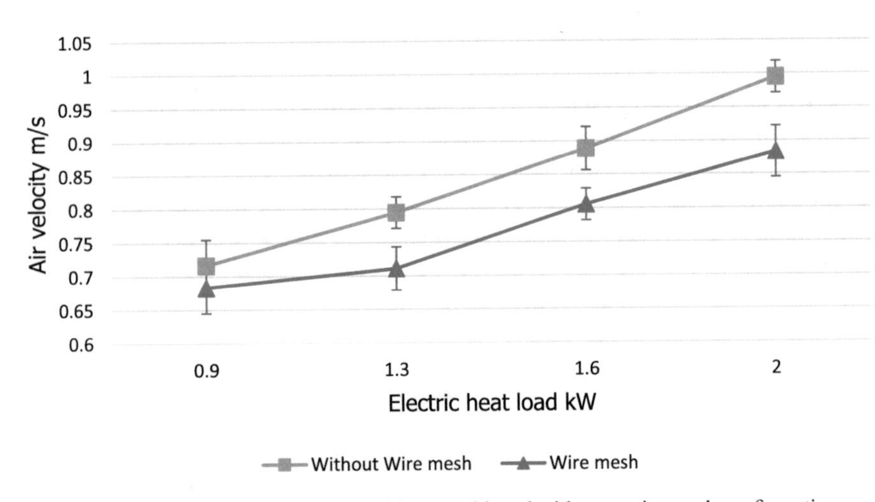


**Fig. 7.12** Air velocity of 1 m divergent chimney with and without a wire mesh configuration

0.994 m/s. Similar trends can be seen with higher chimneys, as shown in Figs. 7.13 and 7.14. A high air velocity was observed in divergent chimneys without a wire mesh configuration. In comparison, lower average air velocity was observed in a wire mesh configuration, in all experimented electric heat loads. Slow air velocities are expected as airflow slows down when they come in contact with the stainless steel wire mesh at the exit of the chimney.

It is observed that shorter divergent chimney models have higher air velocity as compared to large scale such that the 1 m chimney with a wire mesh configuration has higher air velocity than 1.5 and 2 m chimneys. The occurrences have been described

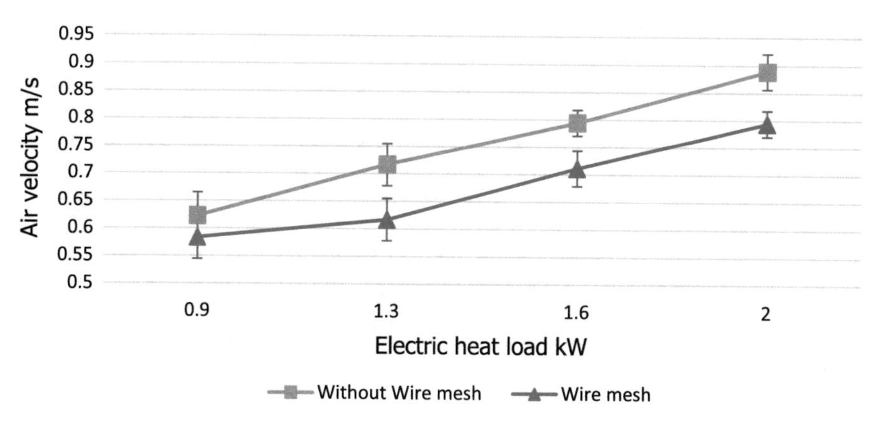


**Fig. 7.13** Air velocity in 1.5 m divergent chimney with and without a wire mesh configuration

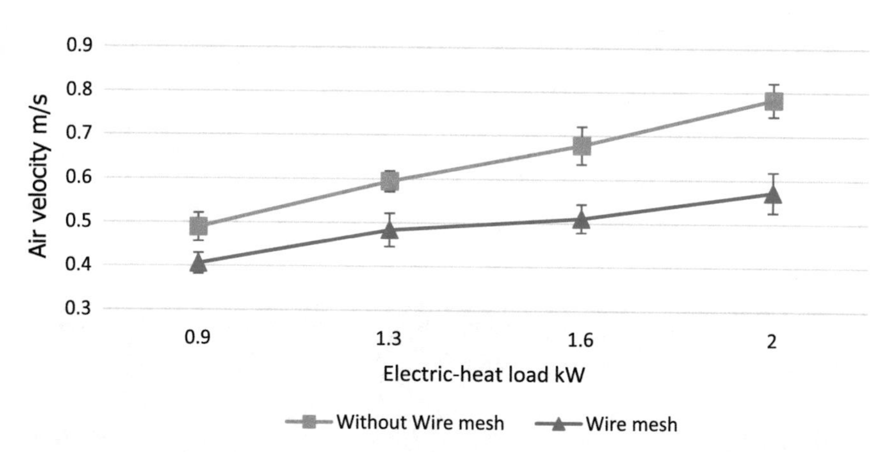


**Fig. 7.14** Air velocity of a 2 m divergent chimney with and without a wire mesh configuration

in detail in Sect. 5.4.1. This is also reflected in the temperature data recorded in the divergent chimney with wire mesh variance, as shown in Table 7.4.

The chimney outfitted with wire mesh minimises the cold inflow phenomenon as evident from the temperature data had minimum deviations compared to without a wire mesh configuration. For a 2 m divergent chimney, temperature data is shown in Table 7.5. The temperature observed at inside ($T_1$), outlet ($T_2$) and 10 cm above outlet ($T_3$) is 335.1, 334.25 and 332 K that has small variances. Compared to a without wire mesh chimney, the temperature at ($T_1$), ($T_2$) and ($T_3$) are 333.24, 326.54 and 319.80 K where variation is high. Temperature is also higher in wire mesh chimney than without wire mesh which aids in the prevention of cold inflow. Similar trends are observed at all electric heat loads.

To further visualise the effect of the wire mesh installation at the exit of the chimney, a thermal camera was used. The images captured using the thermal camera on the outlet of the chimney without a wire mesh are as shown in Fig. 7.15. It

**Table 7.4** Temperature readings of 1, 1.5 and 2 m divergent chimneys with a wire mesh configuration

| Electric heat load | $T_1$ | | | $T_2$ | | | $T_3$ | | |
|---|---|---|---|---|---|---|---|---|---|
| | 1 m | 1.5 m | 2 m | 1 m | 1.5 m | 2 m | 1 m | 1.5 m | 2 m |
| kW | K | K | K | K | K | K | K | K | K |
| 0.9 | 352.27 | 341.03 | 335.10 | 351.29 | 339.01 | 334.25 | 351.12 | 338.12 | 332.00 |
| 1.3 | 359.03 | 348.41 | 342.06 | 357.70 | 347.25 | 341.07 | 356.09 | 345.89 | 338.26 |
| 1.6 | 364.39 | 351.27 | 346.35 | 363.50 | 349.03 | 345.52 | 362.73 | 348.07 | 343.36 |
| 2 | 370.23 | 354.31 | 351.44 | 370.10 | 352.07 | 351.15 | 368.12 | 350.03 | 348.51 |

**Table 7.5** Temperature differences of a 2 m divergent chimney with and without wire mesh configuration

| Electric heat load | Wire mesh | | | Without wire mesh | | |
|---|---|---|---|---|---|---|
| | T1 | T2 | T3 | T1 | T2 | T3 |
| kW | K | K | K | K | K | K |
| 0.9 | 335.10 | 334.25 | 332.00 | 333.24 | 326.54 | 319.80 |
| 1.3 | 342.06 | 341.07 | 338.26 | 340.18 | 335.89 | 328.85 |
| 1.6 | 346.35 | 345.52 | 343.36 | 343.64 | 337.55 | 330.46 |
| 2 | 351.44 | 351.15 | 348.51 | 347.44 | 341.30 | 331.68 |

**Fig. 7.15** Thermal camera images of a chimney without a wire mesh configuration

can be observed that temperature distributions are not uniform. An installed wire mesh screen on the exit of the chimney shows a much better-distributed temperature gradient as shown in Fig. 7.16 to observe the impact of the wire mesh configuration.

Once air particles come in contact with the cylindrical wire mesh, mounted at the exit of the chimney, the air velocity decreases and forms the vortices and low-pressure region on the adjacent side. This phenomenon is identified as flow separation (Houghton, 2012) and is illustrated in Fig. 7.17 (Houghton & Carruthers, 1982).

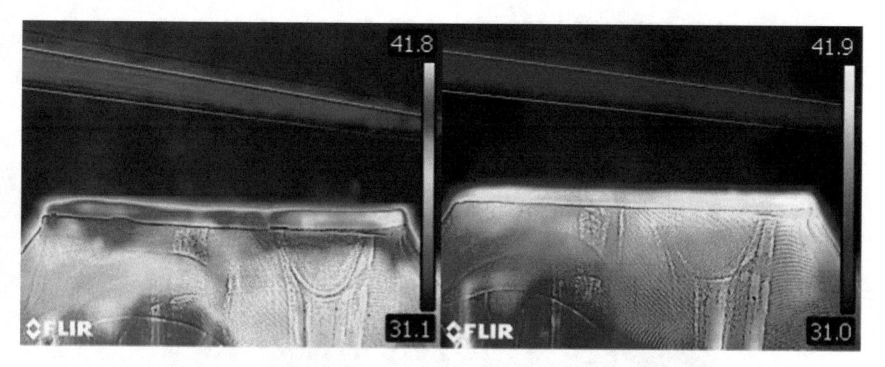


**Fig. 7.16** Thermal camera images of a chimney with a wire mesh configuration

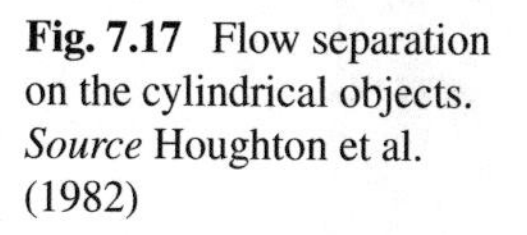
**Fig. 7.17** Flow separation on the cylindrical objects. *Source* Houghton et al. (1982)

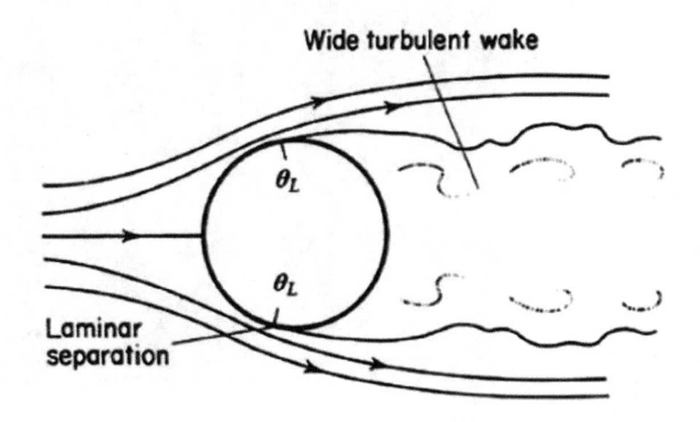


According to the Chu et al. (2012), wire mesh screen prevents cold air inflow to the chimney due to formation of the flow separation and development of eddies on the opposite side increases the turbulence flow, thus improving the convectional heat transfer, the buoyancy and air intake velocity. In another experiment, as shown in Fig. 7.18, it was reported that installation of the wire mesh at the exit of the chimney improves stalk effect and that leads to an increment in air intake by 30%.

The experiments performed on a 2 m divergent chimney with a wire mesh configuration shows that air velocity at the exit of chimney decreases. At the same time, it increases air intake velocity, as shown in Fig. 7.19.

Theoretical electric power potential was calculated based on the consideration of the turbine efficiency of 85% and the generator efficiency of 90%. Theoretical electric power generation of a 1 m chimney with and without a wire mesh is shown in Fig. 7.20. Experimental results show that the theoretical electric power generation can be enhanced by installing a wire mesh at the exit of the chimney. Although the wire mesh slows down air velocity at the outlet of the chimney, the formation of eddies on the adjacent side increases the turbulence flow improves stalk effect increases air intake. Subsequently, the theoretical wind power potential also increased due to an increment in the air intake. Theoretical electric power generation of the 1 m divergent chimney with and without wire mesh is shown in Fig. 7.20. The data indicates that the

Fig. 7.18 Chimney used to investigate the impact of with and without wire mesh configuration

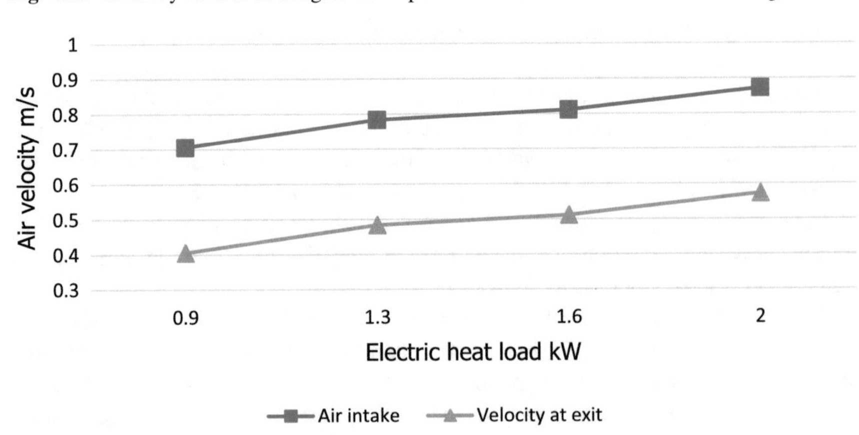


Fig. 7.19 Effects on air intake velocity and air velocity at the exit of 2 m divergent chimney with a wire mesh screen

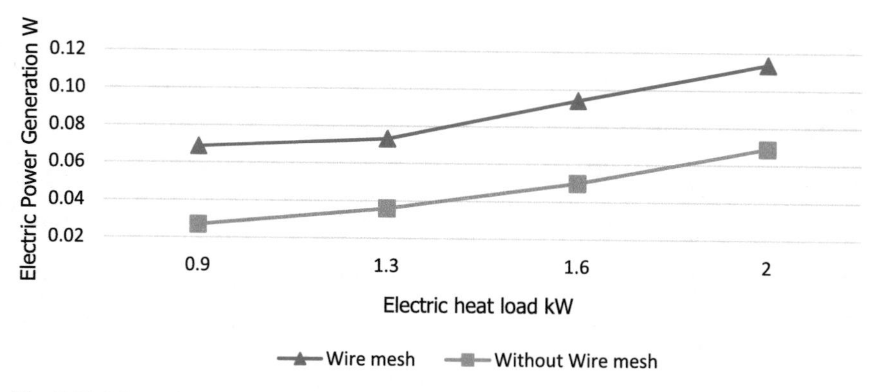


**Fig. 7.20** Theoretical electric power generation in a 1 m divergent chimney with and without a wire mesh

maximum electric power of 0.11 W was achieved with the wire mesh installation. In comparison, without wire mesh maximum electric power was calculated at 0.07 W under the electric heat load of 2 kW.

Theoretical electric power generation for the 1.5 m solar chimney models is shown in Fig. 7.21. The results demonstrated how a wire mesh affect potential power generation capability as it generates electric power almost twice compared to the chimney without a wire mesh configuration. Similar trends can be seen in the 2 m divergent chimney, as shown in Fig. 7.22. The data reveals that divergent solar chimney equipped with a wire mesh have more tendency to generate power. As discussed, the wire mesh prevents cold inflow, increases turbulence flow and increases buoyancy effect (Chi-Ming Chu et al., 2012; Chu et al., 2012). Thus, electric power generation is enhances compared to divergent chimney without a wire mesh and is even more potent compared to a cylindrical chimney.

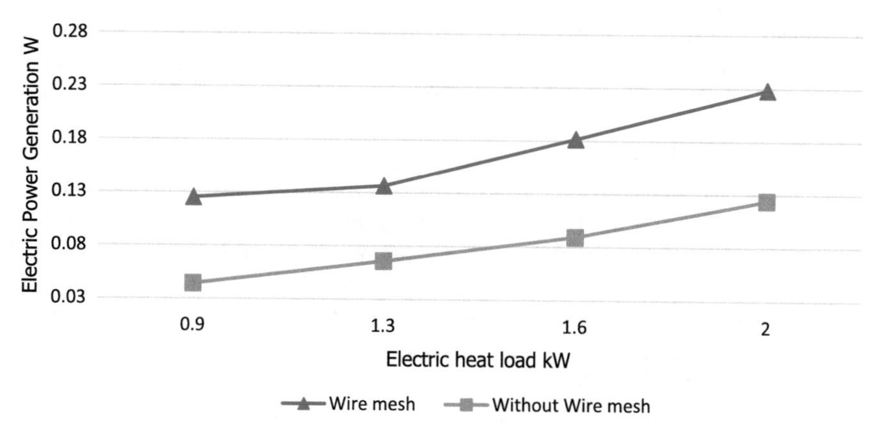


**Fig. 7.21** Theoretical electric power generation in a 1.5 m divergent chimney with and without a wire mesh

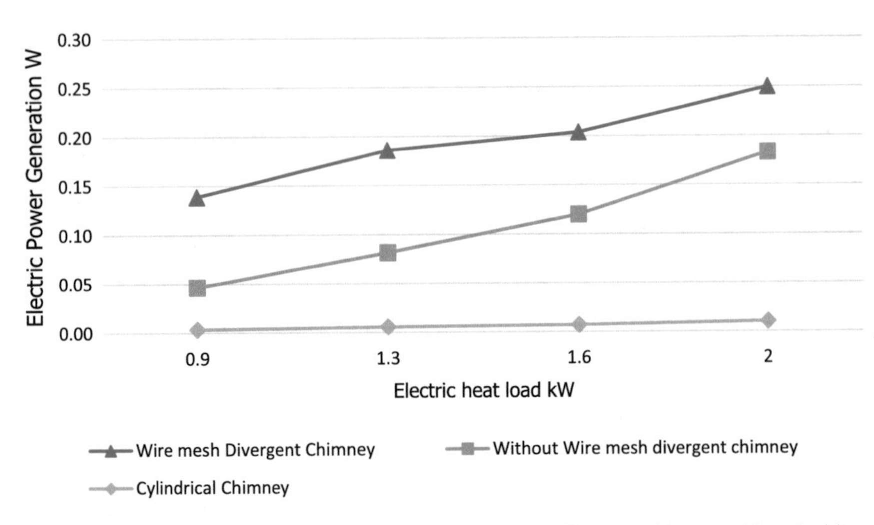


**Fig. 7.22** Theoretical electric power generation in the a 2 m divergent chimney with and without a wire mesh and the cylindrical chimney

## Conclusion

This study has successfully investigated the performance of divergent solar chimney using experiment method. Experiments result shows that divergent solar chimney achieves a better performance of up to 18 times theoretical electric power potential, which is in concurrence with other published research work. An increase of the chimney height and electric heat load translates to power generation potential increased by 2.6 times, while in the cylindrical solar chimney, the power generation potential increased a bit with an increment in the chimney height. Significance power potential difference was observed between the two designs because the divergent shape improves the stack effect due to diffuser shape and velocity increases at the throat due to reduction of the area in the bell mouth section of the divergent solar chimney (Ahirwar & Sharma, 2019; Bouabidi et al., 2018; Das & Chandramohan, 2020).

The effect of cold air inflow was also investigated using the smoke visualisation test method. Observation made shows that there exists cold air inflow in the divergent chimney without a wire mesh. The effect is much more evident in the taller chimney compared to the shorter model. However, a smooth upward flow of the smoke was observed in all divergent chimney models installed with wire mesh which indicates the installation of the wire mesh can improve the divergent solar chimney performance. The installed wire mesh at the exit of divergent chimney models prevents the cold inflow occurrences as hot air contact with the wire mesh causes the formation of eddies due to flow separation thus increases the turbulence flow and buoyancy. Subsequently, it increases air intake by 30% and enhances the power generation. Thus, a divergent solar chimney with a wire mesh variant is a more potent solution for electric power generation compared to a traditional cylindrical solar chimney power plant.

## References

Ahirwar, J., & Sharma, P. P. (2019). Analyzing the effect of solar chimney power plant by varying chimney height. *Collector Slope and Chimney Diverging Angle, 6*(7), 213–219.

Al-Azawiey, S. S., Al-Kayiem, H. H., & Hassan, S. B. (2017). On the influence of collector size on the solar chimneys performance. *MATEC Web of Conferences, 131*. https://doi.org/10.1051/matecconf/201713102011.

Arzpeyma, M., Mekhilef, S., Newaz, K. M. S., Horan, B., Seyedmahmoudian, M., Akram, N., et al. (2020). Solar chimney power plant and its correlation with ambient wind effect. *Journal of Thermal Analysis and Calorimetry, 141*(2), 649–668.

Bouabidi, A., Ayadi, A., Nasraoui, H., Driss, Z., & Abid, M. S. (2018). Study of solar chimney in Tunisia: Effect of the chimney configurations on the local flow characteristics. *Energy and Buildings, 169,* 27–38. https://doi.org/10.1016/j.enbuild.2018.01.049.

Cengel, Y. (2014). *Heat and mass transfer: Fundamentals and applications*. McGraw-Hill Higher Education.

Chakrabarti, S. K. (2005). Physical modelling of offshore structures. In *Handbook of offshore engineering* (pp. 1001–1054). Elsevier. https://doi.org/10.1016/b978-0-08-044381-2.50020-5.

Chi-Ming Chu, C., Kwok-How Chu, R., & Rahman, M. M. (2012). Experimental study of cold inflow and its effect on draft of a chimney. *WIT Transactions on Engineering Sciences, 75,* 73–82. https://doi.org/10.2495/HT120071.

Chu, C., Rahman, M., & Kumaresan, S. (2012). Effect of cold inflow on chimney height of natural draft cooling towers. *Nuclear Engineering and Design, 249,* 125–131. https://doi.org/10.1016/j.nucengdes.2011.08.046.

Das, P., & Chandramohan, V. P. (2020). 3D numerical study on estimating flow and performance parameters of solar updraft tower (SUT) plant: Impact of divergent angle of chimney, ambient temperature, solar flux and turbine efficiency. *Journal of Cleaner Production, 256*. https://doi.org/10.1016/j.jclepro.2020.120353.

Hamdan, M. O. (2011). Analysis of a solar chimney power plant in the Arabian Gulf region. *Renewable Energy, 36*(10), 2593–2598.

Hosien, M. A., & Selim, S. M. (2017). Effects of the geometrical and operational parameters and alternative outer cover materials on the performance of solar chimney used for natural ventilation. *Energy and Buildings, 138,* 355–367.

Houghton, E. L. (2012). Aerodynamics for engineering students. [electronic resource]. *Scribbr* http://search.ebscohost.com/login.aspx?direct=true&db=cat00164a&AN=cran.579071&site=eds-live%5Cnhttp://www.cranfield.eblib.com/patron/FullRecord.aspx?p=879862.

Houghton, E. L., & Carruthers, N. B. (1982). *Aerodynamics for engineering students* (3rd ed.).

Hu, S., & Leung, D. Y. C. (2017). Mathematical modelling of the performance of a solar chimney power plant with divergent chimneys. *Energy Procedia, 110*, 440–445. https://doi.org/10.1016/j.egypro.2017.03.166.

Hu, S., Leung, D. Y. C., & Chan, J. C. Y. (2017a). Numerical modelling and comparison of the performance of diffuser-type solar chimneys for power generation. *Applied Energy, 204,* 948–957. https://doi.org/10.1016/j.apenergy.2017.03.040.

Hu, S., Leung, D. Y. C., & Chen, M. Z. Q. (2017b). Effect of divergent chimneys on the performance of a solar chimney power plant. *Energy Procedia, 105,* 7–13. https://doi.org/10.1016/j.egypro.2017.03.273.

Idelchik, I. E. (1987). *Handbook of hydraulic resistance* (2nd ed.). *Journal of Pressure Vessel Technology*, *109*(2), 260–261. https://doi.org/10.1115/1.3264907.

Jörg, O., & Scorer, R. S. (1967). An experimental study of cold inflow into chimneys. *Atmospheric Environment*, *1*(6), 645–654.
Kasaeian, A. B., Heidari, E., & Vatan, S. N. (2011). Experimental investigation of climatic effects on the efficiency of a solar chimney pilot power plant. *Renewable and Sustainable Energy Reviews, 15*(9), 5202–5206.
Kloppers, J. C., & Kröger, D. G. (2004). Cost optimization of cooling tower geometry. *Engineering Optimization, 36*(5), 575–584. https://doi.org/10.1080/03052150410001696179.
Koonsrisuk, A., & Chitsomboon, T. (2013). Effects of flow area changes on the potential of solar chimney power plants. *Energy, 51,* 400–406. https://doi.org/10.1016/j.energy.2012.12.051.
Koonsrisuk, A., Lorente, S., & Bejan, A. (2010). Constructal solar chimney configuration. *International Journal of Heat and Mass Transfer, 53*(1–3), 327–333.
Ming, T., de Richter, R. K., Meng, F., Pan, Y., & Liu, W. (2013). Chimney shape numerical study for solar chimney power generating systems. *International Journal of Energy Research, 37*(4), 310–322.
Moore, F. K., & Garde, M. A. (1983). Aerodynamic losses of highly flared natural draft cooling towers. In *Third Waste Heat Management and Utilization Conference* (pp. 221–223).
Patel, S. K., Prasad, D., & Ahmed, M. R. (2014). Computational studies on the effect of geometric parameters on the performance of a solar chimney power plant. *Energy Conversion and Management, 77,* 424–431.
Petrus, J. (2007). Optimization and control of a large-scale solar chimney power plant by. *Mechanical Engineering.*
Schlaich, J. (1995). The solar chimney: electricity from the sun. Edition Axel Menges. *Scribbr* https://books.google.com/books?hl=en&lr=&id=CVy6Nh57MdMC&oi=fnd&pg=PA6&dq=Schlaich+J.+The+solar+chimney:+electricity+from+the+sun.+Geislingen:+Maurer+C%3B1995.+pp.+55.&ots=mrrDMe8RKF&sig=-kXdZEPerHLwaFaSmz5_stkOIRE.
Toghraie, D., Karami, A., Afrand, M., & Karimipour, A. (2018). Effects of geometric parameters on the performance of solar chimney power plants. *Energy, 162,* 1052–1061.
Xu, Y., & Zhou, X. (2018). Performance of divergent-chimney solar power plants. *Solar Energy, 170*(June), 379–387. https://doi.org/10.1016/j.solener.2018.05.068.

# Chapter 8
# Solar Chimney and Turbine-Assisted Ventilation System

**Ling Leh Sung, Mohd. Suffian bin Misran, Md. Tarek Ur Rahman Erin, and Md. Mizanur Rahman**

All over the world, the demand for electricity in the household sector is increasing significantly. At the same time, the traditional sources from where the electricity generated are decreasing perceptively. Therefore, alternative sources of energy may be a valid suitable option for this sector (Chik et al., 2012; Fudholi et al., 2018). In this sector, electricity is mainly used for heating and cooling purposes. There are many ways cooling can be done for the buildings. Most of the cooling appliances are needed electricity. The common and old technique for the building cooling is ventilation which is introduced in the house a long time back. A good ventilation system not only responsible for building cooling, but it also improves the quality of air as a result improves the health conditions. Natural and forced ventilation are well-recognized ventilation system that is commonly used for cooling. In forced ventilation or a forced draft ventilation system, a mechanical device such as fan, blower, and exhausts is used to create sufficient draft to create air movement. On the other hand, there are no mechanical devices in a natural ventilation system. Air ventilation process takes place due to the effect of natural convection or buoyancy force (Liu et al., 2017; Micallef et al., 2016; Sundell, 2004; Tamm & Jaluria, 2017). Besides, the building thermal comfort is highly influenced by solar radiation intensities and relative humidity. In a hot and humid climates condition, it has posed difficulties to achieve thermal comfort for building indoor environment (Ariffin et al., 2002). Therefore, most of the commercial and residential buildings are depended on the electrical appliances such as electric fan, air conditioning unit to achieve thermal comfort. Furthermore, the standard of living changes and population increases are also influencing factors that lead the increases in the uses of electrical appliances

L. L. Sung · Mohd. Suffian bin Misran (✉)
Faculty of Engineering, Universiti Malaysia Sabah, Kota Kinabalu, Sabah, Malaysia
e-mail: suffian@ums.edu.my

Md. Tarek Ur Rahman Erin · Md. M. Rahman
Department of Mechatronics Engineering, World University of Bangladesh, 151/8, Green Road, Dhaka 1205, Bangladesh



Md. M. Rahman and C.-M. Chu (eds.), *Cold Inflow-Free Solar Chimney*,
https://doi.org/10.1007/978-981-33-6831-6_8

in the houses. Hence, the energy consumption is reasonably high in the household sector compared to others and is projected to add more gradually (Kamal, 2012). For building, to reduce the energy consumption and to ensure thermal comfort one of the valid suitable options is passive cooling (Santamouris & Asimakopoulos, 1996; Taleb, 2014). Passive cooling is a technique to remove heat from the building interior by using a natural process. Many commonly known techniques are using passive cooling such as shading with overhangs, louvers and awnings, roof cover or roof shading, decent landscaping, thermal resistance building materials, insulation, natural ventilation technique, radiative cooling, evaporative cooling, earth coupling, and desiccant cooling (Breesch et al., 2005; Danny, 2005; Gieseler et al., 2002; Kamal, 2012; Kazanci et al., 2013; Sundell, 2004; Taleb, 2014).

Among all the techniques, natural ventilation is a significant technique for sustainable building development as it relies on wind velocity and stack effects or buoyancy. Furthermore, these processes potentially able to save a significant amount of electric energy by reducing the number of electrical appliances usage. Moreover, it is an old common technique that is used for ventilation purposes (Aynsley, 2014; Hughes et al., 2012; Kleiven 2003; Lomas, 2007).

The advantages of natural ventilation include reduction of cooling cost, improvement of indoor air quality as well as provide a satisfactory level of thermal comfort under certain environmental conditions (Liping & Hien, 2007). Although natural ventilation has significant advantages, it is not performing effectively in the modern building since it fully depends on the buoyancy force. The performance of natural ventilation can be increased by enhancing the pressure difference between inlet and outlet of the chimney through a solar chimney or wind turbine ventilator (Lal et al., 2013; Lien & Ahmed, 2011). Solar chimney performance depends on solar radiation, daily sunlight hours, chimney geometry as well as environmental conditions. The solar chimney performance can be improved by changing glazing, increasing height, changing air gap, integrating Trombe wall with roof solar collector, and chimney wall inclination angle, etc. (Kaneko et al., 2006; Khedari et al., 2000; Lal et al., 2013).

Wind turbine ventilator or roof ventilator is also used to ventilate and induce air movement in a building. It is comprehensively influenced by external airstream and functioned as an alternative solution to powered fans. The ventilator is very effective to remove smoke, damp, and foul air from a building as well as hot air from a building to the outside. Research works related to wind turbine ventilator performance specifically in Sabah Malaysia were limited and not widely available (Lien & Ahmed, 2011; Revel & Huynh, 2004). Furthermore, the efficiency of wind turbine ventilator combined with a rooftop solar chimney is also not well studied. Therefore, this project aims to study the performance of a hybrid wind turbine ventilator and a solar chimney system at various working parameters. The combined solar chimney and wind turbine-assisted ventilation system helps to solve many problems as opposed to solar chimney used for ventilation independently. Some of the problems are restricted day time ventilation, crosswind effect, hot air recirculation, and cold inflow or flow reversal (AboulNaga & Abdrabboh, 2000; Bouchair, 1994; Rahman et al., 2014; Xiong et al., 2019; Zhou et al., 2009). Solar radiation instability affects the airflow rate, causing the system to be unstable and inefficient (Sharma et al., 2007). Also, a solar chimney is not sufficient for a large building since it depends on solar radiation, chimney geometry, and the air gap between collector and glass (Kaneko et al., 2006; Khedari et al., 2000; Lal et al., 2013; Tan & Wong, 2013). Thus, an integrated system such as Trombe wall and rooftop solar chimney performed much better compared to individual and other systems. These integrated systems performance is however purely depending on solar radiation and daylight hours. Therefore, if a solar chimney system is integrated with other systems such as wind turbine ventilator, it can offer a viable solution in terms of operational and performance point of view. Normally turbine ventilator is driven by wind and rotates in its vertical axis to create updraft forces to push the air out from the building. The main advantage of turbine ventilator is that it can facilitate ventilation in the absence of wind flow as well as the presence of buoyancy-driven force (Lien & Ahmed, 2011). As a stand-alone system, the solar chimney and the turbine ventilation have their advantages, and an integrated hybrid system performance is yet to be discovered. So far, research work on the performance of integrated solar chimney and wind turbine ventilator has been published.

## Energy Sustainability for Building

The simplest way of saving energy is to switch off all power when not in use (Oh et al., 2014). Another way of saving energy usage is to recycle all recyclable materials. Paper and glass made from recycled materials take 30% less energy compared to paper and glass made from raw material. While aluminum can be made from recycled materials use 90% less energy, than it does to make one from raw materials (Robert, 2005). Additionally, a significant amount of energy can also be saved by using the concept of energy conservation as shown in Fig. 8.1. An energy conservation strategy is a passive architecture practice that uses basic building elements to address the local climate (Zaki et al., 2010). Under this practice, the building's design considers the

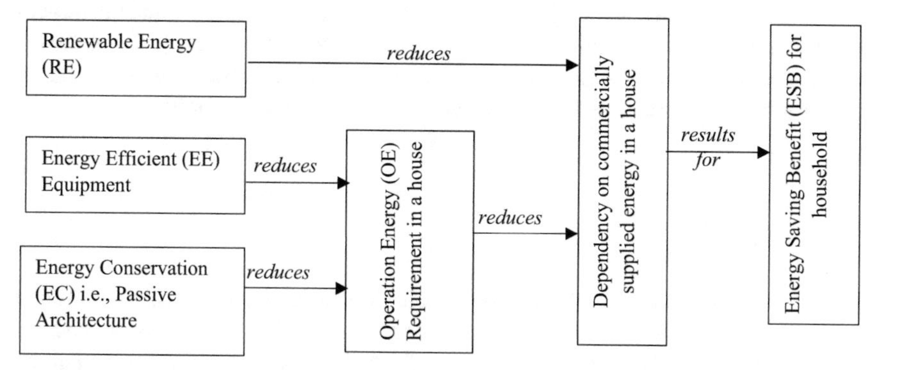


**Fig. 8.1** Concept of reducing building energy. *Source* Zaki et al. (2010)

orientation, exposure, and shape, roof overhangs, window size and placement, natural ventilation, insulation and shutters (Amer, 2006).

The low-energy consumption building is known as low-energy building (LEB), green energy building (GEB), zero energy building (ZEB), and diamond building which have smart cool and energy management system. The main aim of these types of buildings is to reduce a significant amount of energy consumption in sustainable ways. The socioeconomical, technical, and environmental issues are considered when designing these building. In one word, these buildings are known as environmentally friendly and sustainable building (Amara et al., 2015; Collinge et al., 2018; Eichholtz et al., 2013; Ismail & Rashid, 2014; Oh et al., 2014).

## Natural Ventilation

Building in the topical country consumes a large amount of electricity for cooling purpose; therefore, the natural ventilation seems to have good potential among the other strategies for saving energy. Study shows that natural ventilation is important energy conservation and sustainable strategies for a building and can provide indoor thermal comfort and reduce the cooling load effectively and efficiently (Khedari et al., 2000). Furthermore, natural ventilation has other potential advantages in terms of operational cost, energy requirement, economic and environmental benefits compared with mechanical ventilation systems (Khanal & Lei, 2011). Ventilation is also very important for the health of the occupants by improving the quality of indoor air quality. According to the definition of natural ventilation, it is a process of circulating air freely in a building by using natural forces from wind and thermal gradient (Khedari et al., 2000). Natural ventilation process depends on wind intensity and its direction. In the wind-driven natural ventilation process, the wind strikes the building and creates a pressure difference that acts as a driving force. This process is started with the airflowing into the building through the windward opening and leaving the building through the leeward opening. Alternatively, the thermal gradient or the buoyancy-driven ventilation depends on temperature difference and position of the openings. This ventilation is also referred to as stack ventilation (Khanal & Lei, 2011). The natural ventilation systems are named as Trombe wall, turbine ventilator, ridge vents, and soffit vents (Al-Obaidi et al., 2014; Arce et al., 2009).

## Solar Chimney

Solar chimney is not well known in many tropical countries, but this technique already exists since the sixteen century. It is originated from wind towers and stack-assisted chimneys (Tan & Wong, 2014). In 1903, an article about a solar chimney named "*Proyecto de motor solar*" was published in Spain (Arce et al., 2009). The study showed that solar chimney already started developing since the nineteenth century.

Along with the development of Trombe–Michel wall in the nineteenth and twentieth centuries, solar chimneys were getting popular and stack ventilation was further developed as well (Tan & Wong, 2014). However, little information has been found in the literature between 1925 and 1970. This is possibly due to the development of conventional air conditioning systems that led to little interest in the concept (Arce et al., 2009).

Then, the energy crisis becomes apparent and fuel prices are increased. In the early twenty-first century, the interest in passive cooling has been renewed to reduce energy consumption and carbon footprint (Tan & Wong, 2012). Passive cooling systems were a viable option in topical countries since solar radiation is abundant in this region. Among all the ventilation technology, the solar chimney offers a promising solution. It assures a credible energy conservation strategy in many countries. The solar chimney mainly relies on energy from the sun (Khanal & Lei, 2014). Concurrently, the solar chimney is also capable of reducing heat gain or heat transfer through walls and roofs. Thus, the cooling load of air conditioning systems is reduced significantly. Besides, solar chimney is also capable of enhancing natural ventilation and improves the indoor condition to ensure better thermal comfort (Zhai et al., 2011a). It is a type of passive cooling methods under the category of heat dissipation technique. It is also capable to act as passive heating and thermal insulation in the cold countries where thermal control is achieved by the distribution and opening of air vents. However, it has more potential for cooling than for heating (Pacheco et al., 2012; Zhai et al., 2011a).

There are several types of solar chimney configurations that have been demonstrated for building ventilation. Those configurations can be categorized into three main categories as shown in Fig. 8.2.

## Vertical Wall Type

Vertical wall-type solar chimney is the type that is located on a wall of a structure. It is further divided into the Trombe wall and vertical solar wall.

## Trombe Wall Solar Chimney

In the Trombe wall solar chimney, the wall has exterior glazing with an air gap in between the wall. Generally, solar energy passes through the glazing and is absorbed by the massive wall. Some part of the heat energy is transferred into the building through the wall by conduction process, while most of the heat energy heats the air in the air gap. The buoyancy force creates hot air movement in an upward direction and it enters the building through the upper vent of the wall. The cold air from the building is moved toward the air gap at the bottom inlet of the Trombe wall solar chimney. This process continues as long as the temperature difference exists in the Trombe wall

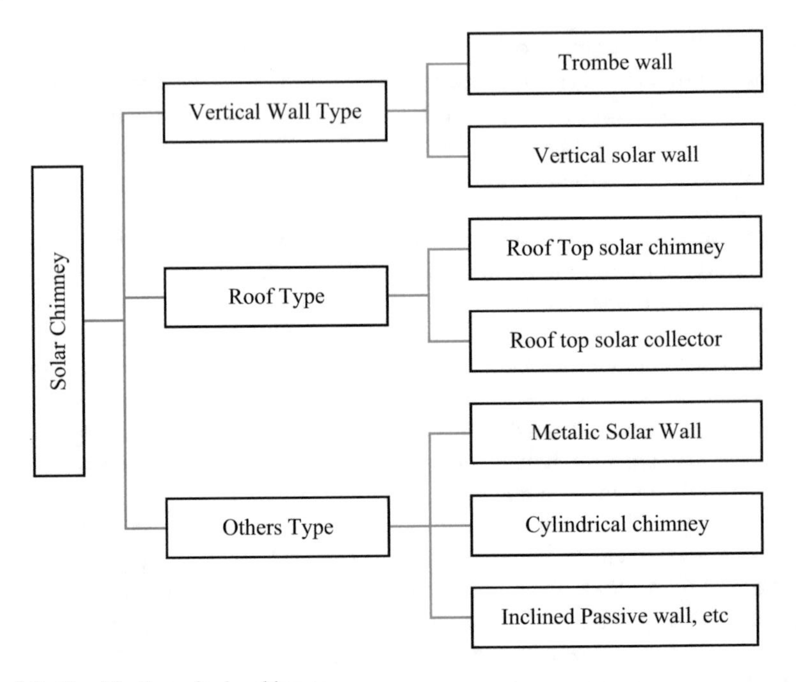


**Fig. 8.2** Classification of solar chimney

and inside of the building (Pacheco et al., 2012). Normally, conventional Trombe wall is used for passive building or space heating, but the conventional Trombe wall disadvantages include obstruction of windows placement and possesses significant thermal mass (Harris & Helwig, 2007); low thermal resistance or excessive heat loss when a small amount of solar energy absorbed by the wall, reversing air circulation during winter, at night or cloudy day, highly dependant on solar intensity and at times unpleasant architectural design (Chan et al., 2010). This is demonstrated by an application of a Trombe wall using the visible black-matt surface of the blackened massive wall for better heat absorption (Haghighi & Maerefat, 2014). Numbers of studies have been carried out to improve the conventional Trombe wall such as zigzag Trombe wall design, solar water wall, solar transwall, solar hybrid wall, Trombe wall with phase-change material, composite Trombe wall, fluidized Trombe wall, and photovoltaic Trombe wall (Gan 1998; Hu et al., 2017; Ong, 2003; Rabani et al., 2015; Saadatian et al., 2012; Zhu et al., 2019).

## Vertical Solar Wall

A vertical solar wall is a common type of solar chimney and widely known by the general public because of its simplicity. The main components of a vertical solar wall are absorber surface, air gap, glazing, and insulation. Both radiation and wind effects are considered in designing this solar wall. Figure 8.3 shows some of the components to discuss the operating principle of a vertical solar wall, and it is a combination of a solar-assisted stack and wind-driven ventilation system. Here, the air temperature increases due to solar radiation and moves upward as a result of air buoyancy. Also, the air is pulled by the outdoor ambient wind which creates low pressure. This cycle repeats as long as the air from the inlet is in contacts with the hot absorber and pressure difference effects are presented in the system (Belfuguais & Larbi, 2011; Chan et al., 2010; Harris & Helwig, 2007; Tan & Wong, 2014).

The vertical solar wall does not require massive wall, and therefore, it is considered to be one of the simplest solar chimneys with no reverse flow circulation at a large air gap up to of 0.3 m. With the system, the building can ventilate hot air which improves indoor air conditions and thermal comfort as oppose to a Trombe wall setup. No electrical or mechanical device is needed with this type of solar chimney

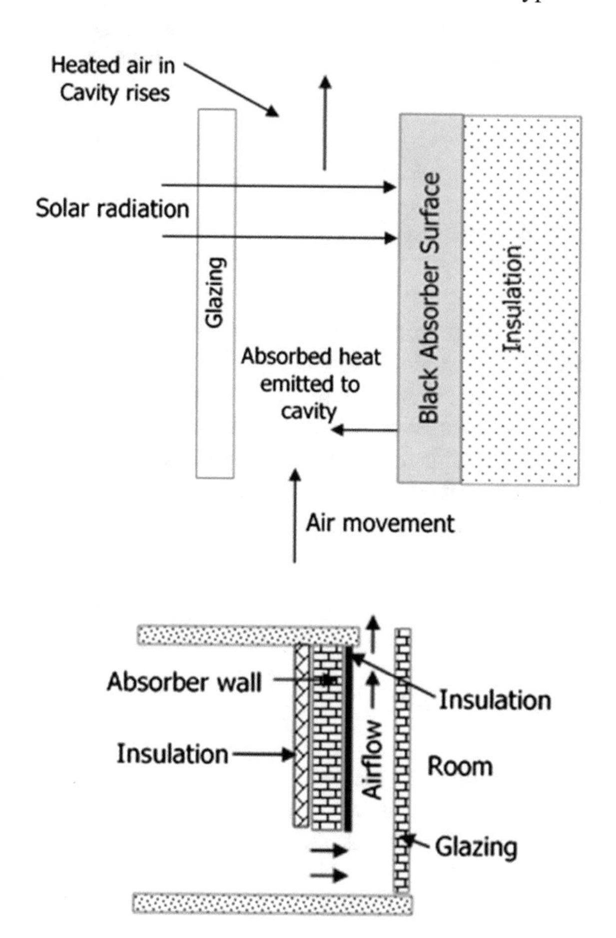


**Fig. 8.3** Schematic operation of vertical solar wall *Source* Harris and Helwig (2007) and Belfuguais and Larbi (2011)

**Table 8.1** Differences between Trombe wall and vertical solar wall

| Characteristics | Wall-type solar chimney | |
|---|---|---|
| | Trombe wall | Vertical solar wall |
| Massive solar wall | Yes | Not necessary |
| Favorable passive system | Passive heating method | Passive cooling method or natural ventilation |
| Room air | Recirculation of room air | Drawing outdoor air into the room |
| Inlet and outlet vents | Both are on the massive wall | Inlet vent on the absorber wall and outlet vent on the top of chimney or glazing |
| Operating principle | Stack effect | Stack effect and wind-driven ventilation |
| Induced airflow | Outside Inside Solar Energy Wall Glazing | Inside Outside |

and is recommended for night passive cooling or multistory solar chimney with little alteration. Even though it got a lot of advantages, this type of chimney challenges the architecture and overall appearances of the building and thus failed to get attention by adopters. The cost of installation of the vertical wall solar chimney is high because of the major structure needs adjustment. In summer, it is not able to provide sufficient ventilation due to high temperature and low temperature difference (Amer, 2006; Chan et al., 2010; Haghighi & Maerefat, 2014; Yusoff et al., 2010; Zhai et al., 2011a; Ziskind et al., 2002). The difference between a vertical solar wall and Trombe wall is shown in Table 8.1.

## Roof Type Solar Chimney

The roof type of solar chimney is located along the roof slope getting its maximum radiation when the sun is high. The roof type solar chimney is further grouped into roof solar chimney and roof solar collector.

## Roof Solar Chimney

The roof solar chimneys are principally similar to a vertical solar wall. The only differences between them are the inclination angles and their location. The vertical solar wall is vertical, while roof solar chimney is inclined and on a rooftop. The main advantages of a roof solar chimney over a vertical solar wall are less construction cost and limited visual obstruction. The rooftop solar chimney is more efficient than vertical wall solar chimney, provides a larger surface area, and can be used effectively to remove unwanted heat gain as well as integrated with photovoltaic systems. The rooftop solar chimney provides free cooling for building throughout the day and night (Chan et al., 2010; DeBlois et al., 2013; Harris & Helwig, 2007; Jianliu & Weihua, 2013). Like other passive cooling systems, the rooftop solar chimney does not require any electrical or mechanical power during operation therefore the technology is considered as green and environmentally friendly technology (Zhai et al., 2011a). One of the recognized disadvantages of this type of solar chimney is that it may not able to produce sufficient ventilation for thermal comfort. Also, the chimney height is the same with the roof height and is a visible part of a building, hence the appearance of the building is going to be affected (Chan et al., 2010; Harris & Helwig, 2007; Mathur et al., 2006; Yusoff et al., 2010). The concept of a rooftop solar chimney is as shown in Fig. 8.4.

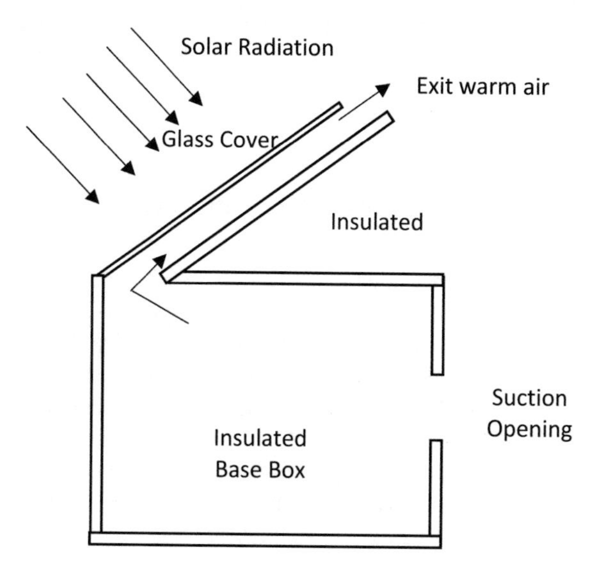


**Fig. 8.4** Structural feature of roof solar chimney. *Source* Jianliu and Weihua (2013)

## Roof Solar Collector

A rooftop solar collector is different from a rooftop solar chimney, but the operating principle is still the same. In the rooftop solar collector, the concrete tiles act as a heat absorber and heat the air present at the airgap, driving the air movement through the stacking effect. The rooftop solar collector has the same benefits and limitations as a rooftop solar chimney except not requiring any glazing which affects the performance (Hall et al., 2011; Hirunlabh et al., 2001a, 2001b; Khedari et al., 1997).

## Other Types of Solar Chimneys

Other types of solar chimney that are not appropriately categorized in the main two categories such as a metallic wall, glazed, cylindrical, double-sided rooftop, and inclined passive wall solar chimney are grouped here. Considerable discussion has been made by the researchers in their publications regarding the alternative type of solar chimneys, albeit all of them are using the same working principle (Al-Kayiem et al., 2014; Chantawong et al., 2006; Hirunlabh et al., 2001a, 2001b; Khanal & Lei, 2011). Among all, the cylindrical solar chimney is easier to be integrated with the existing building because of its shape. This type of solar chimney is mainly used for passive cooling, passive heating, or natural ventilation. It is also widely used for the natural circulation of air in solar dryers (Ekechukwu & Norton, 1995; Khanal & Lei, 2011).

## Performance of Solar Chimney for Ventilation

Solar chimney technology has been actively researched and innovated throughout the years. With recent awareness of global warming and sustainability issues on the rise, people continue to look and explore means of increasing the performance and viability of the solar chimney technology. One of the main challenges of the technology is dependent on weather conditions (Arce et al., 2009; Jianliu & Weihua, 2013; Khanal & Lei, 2011; Khedari et al., 1997; Tan & Wong, 2013). Solar chimney with heat storage system may increase the performance, but it will also increase operation and maintenance cost significantly. Furthermore, stand-alone solar chimney systems are not reliable enough to create natural ventilation for good indoor thermal comfort (Chan et al., 2010).

## Optimize Solar Chimney Parameters

The performance of the solar chimney can be improved by optimizing its geometrical configurations. Geometry parameters that affect the solar chimney performance are: the opening areas, the inlet and outlet positions, the chimney aspect ratio, absorber, insulator, glazing, inclination angle, stack height, air gap depth, etc. (Bansal et al., 2005; Harris & Helwig, 2007; Khanal & Lei, 2011; Tan & Wong, 2013). It is also found that the optimum combinations of geometrical configurations of a solar chimney are location specific. Since the solar chimney performance heavily depends on the local weather. Therefore, meteorological parameters such as wind, solar radiation intensity, and ambient air temperature are considered as important factors that influence the optimum geometrical configurations (Dai et al., 2003; Harris & Helwig, 2007; Jianliu & Weihua, 2013; Maerefat & Haghighi, 2010; Waewsak et al., 2003).

## Position of Inlet and Outlet Apertures

The positions of inlet and outlet apertures have direct influence or impact on the performance of a solar chimney. Depending on the position of the inlet and outlet sections, to either windward or leeward affects the performance. The leeward position of outlet and windward position of an inlet helps to increase the performance of solar chimney (Yusoff et al., 2010). It was also reported by Tan and Wong (2013) that solar chimney's inlet position has limited influence on solar chimney outlet air temperature, heat transfer coefficient, and solar chimney outlet airspeed (Tan & Wong, 2013). The inlet aperture approximate height of 1.2 m has better thermal performance and interior airspeed. The outlet aperture near the ceiling helps to obtain uniform temperature inside the building. Additionally, the vertical distance between the solar chimney apertures also plays an important role in determining the quantity of induced airflow. The airflow rate also increased with the heightened of the height between inlet and outlet apertures (Amori & Mohammed, 2012; Haghighi & Maerefat, 2014; Khedari et al., 2003; Tan & Wong, 2012; Zhai et al., 2011a; Ziskind et al., 2002).

## Area of Inlet and Outlet Apertures

The solar chimney performance also depends on the area of inlet and outlet apertures. A larger area of apertures can reduce the frictional force, reduce entry flow resistance, and increase ventilation rate as well as increase air change per hour (ACH). Studies showed that it is necessary to have a larger area of apertures for better ventilation performance or better specific rate of heat removal per unit volume of airflow rate (Haghighi & Maerefat, 2014; Khedari et al., 2003; Li & Liu, 2014; Maerefat & Haghighi, 2010). The outlet aperture area affects more on indoor conditions compared to the inlet aperture area. The smaller outlet aperture area can avoid undesirable vertical temperature gradient and discomfort sensation. Therefore, it is important to have the correct area of apertures. Most studies suggest that the inlet and outlet apertures' area be equal size and as large as possible to induce the highest airflow rate. Also, a solar chimney with variable apertures dimensions is a viable option to control ACH and inside room temperature (Haghighi & Maerefat, 2014; Khanal & Lei, 2011; Khedari et al., 2000; Maerefat & Haghighi, 2010; Zhai et al., 2011a).

## Insulation

Insulation helps to avoid additional heat gains to the room as well as prevent undesirable overheating. Studies have shown that the solar-assisted ventilation efficiency is reduced by more than 60% if no insulation is provided. The proper insulation maximizes the ventilation rate or the flow rate during day and night. It also improves indoor environmental conditions as well as decrease energy consumption. However, the insulation can cause an increase in thermal resistance as a result of decreasing heat load. Thus, an insulation thickness of 5 cm is considered sufficient to reduce the effect of heat load imbalances (Afonso & Oliveira, 2000; Amer, 2006; Harris & Helwig, 2007; Khanal & Lei, 2011; Zhai et al., 2011a).

## Glazing

Glazing is one of the main components of a solar chimney. It enables collection and use of solar irradiation to drive the airflow rate better and to reduce the convection heat losses. One of the widely used as glazing materials is glass. Different types of glass are used as glazing materials which are identified as heat-absorbing glass, heat-reflecting glass, low radiation glass, clear glass, tinted glass, coated glass, laminated glass, patterned glass, and obscured glass (Chantawong et al., 2006; Khanal & Lei, 2011; Pacheco et al., 2012).

Glazing is also used for amplifying the greenhouse effects that increase the thermal stack within the solar chimney allowing it to increase ventilation rate. In the case of double glazing, studies showed that it can enhance passive cooling, ventilation rate, and reduction of heat losses, but this improvement is not significant enough to make the system cost-effective. Therefore, single glazing with low emissivity is suggested for better performance (Harris & Helwig, 2007; Khanal & Lei, 2011; Tan & Wong, 2013; Zhai et al., 2011a).

## Stack Height

The performance of the ventilation depends on the stack height or the vertical distance travel by the air. This is because stack height corresponds to the air temperature within the solar chimney as well as the temperature difference between inside and ambient air temperature. These temperatures lead to the airflow and stack pressure induced in the solar chimney. However, as thermal energy is being transferred in great quantity, the air temperature in the solar chimney is also increased. As a result, the temperature difference between inside and ambient air temperature is increased. These lead to enhancement of the stack effect and better performance of solar chimney in terms of lesser flow resistance, higher airspeed, higher airflow rate (mass flow rate and volume flow rate), higher pressure head, and higher ACH (Al-Kayiem et al., 2014; Bassiouny & Korah, 2009; Hirunlabh et al., 2001a, 2001b; Khanal & Lei, 2011; Khedari et al., 2000; Maerefat & Haghighi, 2010; Tan & Wong, 2013; Zhai et al., 2011a).

Studies also showed that airflow rate per unit area decreases with the increase of stack height. Therefore, smaller multiple numbers of solar chimneys (approximately 1 m) showed better performance than a single higher solar chimney. It is also reported that the induce airflow rate is reduced when the stack height is less than 3 m. The airflow rate is increased by 80% when the stack height is varied from 1.95 to 3.45 m. The stack height has also a significant effect on solar chimney temperature. The temperature is decreased with the increment of stack height due to a greater heat transfer from a higher airflow rate. The stack height should not exceed a certain height limit where the solar chimney air temperature is less than the ambient temperature (AboulNaga & Abdrabboh, 2000; Afonso & Oliveira, 2000; Al-Kayiem et al., 2014; Amori & Mohammed, 2012; Chungloo & Limmeechokchai, 2007; Ekechukwu & Norton, 1997; Hirunlabh et al., 2001a, 2001b; Jianliu & Weihua, 2013; Khanal & Lei, 2011; Khedari et al., 1997; Mathur et al., 2006; Tan & Wong, 2013).

## Air Gap Depth

In a narrow air gap depth, flow is restricted and friction losses may occur. The conduction and the convection of heat that occur in the air gap and the glazing are high at narrow air gap depth. It is also found that a wider air gap depth has better performance, enhanced ACH, and improved heat removal rate. Although wider air gap depth shows better performance, flow reversal phenomena are observed when the air gap depth increases and exceeds its critical value. Flow reversal significantly reduces the exit air temperature as well as the efficiency of the chimney. According to the Khanal and Lei (2011), the optimum air gap depth 0.3 m can be used for a solar chimney but flow reversal may occur at the depth of greater than 0.2 m. Therefore, the air gap depth 0.2 m is a suitable option to obtain maximum performance of the solar chimney. Khanal and Lei (2014) reported that the air gap depth 0.1 m is of the optimum value. Thus, the air gap depth can be maintained 0.1–0.2 m for optimum performance while avoiding flow reversal phenomena (Arce et al., 2008; Chungloo & Limmeechokchai, 2007; Harris & Helwig, 2007; Hirunlabh et al., 2001a, 2001b; Li et al., 2014; Khanal & Lei, 2011; Khedari et al., 2000; Mathur et al., 2006; Tan & Wong, 2012).

## Aspect Ratio

Aspect ratio is the ratio of stack height and the air gap depth. High aspect ratio can increase the rate of ventilation, increase the air temperature and surface temperature. Study shows that the heat transfer rate declines more dramatically at higher aspect ratio than lower aspect ratio (Haghighi & Maerefat, 2014; Jianliu & Weihua, 2013; Khanal & Lei, 2011; Nouanégué & Bilgen, 2009; Zhai et al., 2011a).

## Width

Solar chimney's width has a direct effect on the interior airflow and airspeed. The performance of the solar chimney increases with width increase. Study shows that it is better to have a larger width rather than larger height when trying to increase solar collection area to improve performance (Afonso & Oliveira, 2000; Tan & Wong, 2013).

## Inclination Angle

The inclined chimney has shown better performance than vertical chimney since the inclination angle plays an important role in the performance of solar chimney. The inclination angle has a significant impact on the amount of heat transmitted through the glazing and affects the solar radiation that reaches the surface of the collector. It also changed the value of heat convection, heat loses, temperature distribution, space flow pattern, and airflow induced in the chimney. Solar chimney with a high inclination angle from the horizontal axis translates to higher stack height. Thus, it increases the stack height effect, reduces the flow resistance, and increases the effective pressure head. However, a smaller inclination angle able to exploit greater exposure of the solar collector to solar radiance and enhanced its energy collection. As a result, higher heat utilization and more intense buoyant airflow will be achieved. Thus, studies recommended the inclination angle at the optimum inclination angle about 70°. Further deviation of the inclination angle from the optimum value will cause the performance of the solar chimney to drop significantly. However, it is also reported that the optimum inclination angle varies according to the latitude of the location and the days of the year (Al-Kayiem et al., 2014; Amer, 2006;

Amori & Mohammed, 2012; Bassiouny & Korah, 2009; Bouchair & Fitzgerald, 1988; Gunerhan & Hepbasli, 2007; Haghighi & Maerefat, 2014; Harris & Helwig, 2007; Hirunlabh et al., 2001a, 2001b; Jianliu & Weihua, 2013).

## Integrated Solar Chimney

The integration of different types solar chimney is an alternative option to improve solar chimney performance. One of the integrated configurations is the multistory solar chimney. The solar chimney has an inlet opening on each floor and one outlet opening only on the third floor. The multistory solar chimney has also found an inlet and outlet openings on each floor (Zhai et al., 2011a). Rooftop solar chimneys can be integrated with different numbers? to improve performance. These multiple steps configurations are proposed to decrease the airflow rate per unit area but increase the length of the roof solar collector (Hirunlabh et al., 2001a, 2001b). In the year 2000, AboulNaga and Abdrabboh (2000) recommended a combined wall–roof solar chimney. This integration improved night ventilation and increased the induced airflow rate by three times greater than a single solar roof chimney (AboulNaga & Abdrabboh, 2000). Tan and Wong (2012) suggested a combined wall–roof solar chimney. This integration is operating well in the hot and humid tropics as well as in the cooler days. It can improve thermal acceptability and makes the overall condition acceptable and comfortable for users (Tan & Wong, 2012). In the year 2003, Khedari et al. conducted a study on integration roof solar collector with modified trombe wall on an air-conditioned building. It was found that the integration system is very efficient for decreasing the air-conditioned load and able to reduce the average daily electrical consumption by 10–20%. This system is effective for hot climate and can be used for both air-conditioned and ventilated building (Khedari et al., 2003). Khedari et al. (2000) proposed an integrated system which consists of roof solar collector (RSC), Trombe wall (TW), modified Trombe wall (MTW), and metallic solar wall (MSW). This integration is reduced heat gain value significantly and ensured thermal comfort by reducing room temperature close to comfort level (Khedari et al., 2000). To improve the building temperature, one of the valid suitable options is to minimize the absorption capacity of solar heat flux as well as improve natural ventilation.

## Integration of Solar Chimney with Other Technologies

Solar chimney integrated with other ventilation technologies is one of the suitable alternative options for further improvement of the indoor air quality as well as to improve thermal comfort. To ensure the thermal comfort, a solar chimney can be used for ventilation in the sanitary areas of a low-cost residential building to precool the air (Chan et al., 2010; Macias et al., 2009). Earth to air heat exchanger (EAHE) is another precooling air technology that can be integrated with a solar chimney to achieve thermal comfort. This type of arrangement helps to reduce high-level energy consumption in the building for cooling (Guramun et al., 2019; Li et al., 2014; Maerefat & Haghighi, 2010). There are some other integrated systems such as solar chimney with an evaporative cooling cavity, solar adsorption cooling cavity, cool metal ceiling, pond roof, or Skytherm system that can be used in building to improve the ventilation and achieve better thermal comfort (Chungloo & Limmeechokchai, 2007; Dai et al., 2003; Maerefat & Haghighi, 2010; Zhai et al., 2011a). A solar chimney (Trombe wall) can also be integrated with sackcloth cooling concepts for heating, cooling, and ventilation in various climates condition. This integrated system has different openings that open and close at different climates situation. This system provides good performance in winter as well as summer. One of the disadvantages of this system is this system increased the cost of the building by 20% (Zhai et al., 2011b). A study also showed that integration of roof solar collector and a vertical stack can enhance the stack ventilation performance in both semi-clear and overcast conditions. This integration is also suitable in the hot and humid climate, where stand-alone stack ventilation is inefficient due to the small temperature difference between the inside and outside (Yusoff et al., 2010).

An arrangement can be made with a photovoltaic (PV) system and a roof solar collector (RSC). This system is not just enhanced the performance of RSC, but also reduced heat gain as well as improving the ventilation rate. This integrated system has 2–4 time's higher ventilation than a single RSC system (Khedari et al., 2002; Zhai et al., 2011a). Another approach is to integrate a rooftop turbine ventilator with the solar chimney enhance overall ventilation performance. Turbine ventilator can be used to assist the delivery of fresh air in a roof space as well as in the building throughout the entire year.

## Estimated Volumetric Airflow Rate from Chimney

Karapantsio et al. (2007) conducted a numerical study on solar chimney system and estimated natural airflow inside the solar chimney. The temperature and monthly average daily total irradiation on the horizontal flat plate were estimated 28.9 °C and 23.1 MJ/m$^2$. The velocity profile of the 1 and 2 m height solar chimney is typical of non-interacting boundary layers. The velocities were varied approximately from 0.4 to 0.8 m s$^{-1}$ near to the wall, close to zero at the center since the absorber wall is hotter than other parts. The numerical study also showed that in the 1 m height chimney, the velocity and temperature profile were not affected by the tilt angle. These results cannot be accepted since the maximum energy is absorbed by the absorber does not match with the maximum airflow rate (Karapantsio et al., 2007).

Kaneko et al. (2006) published a paper on the ventilation performance of a solar chimney. The study was done both numerically and experimentally in Osaka, Japan. The length of the solar chimney was 1.3 m, oriented with south facing and tilted 45° with the horizontal axis. During the experiment, the maximum solar radiation and temperature were recorded 906 W m$^{-2}$ and 34.3 °C. The experiment showed that the prototype chimney can supply an airflow rate between 100 m$^3$/h and 400 (Kaneko et al., 2006). Utama et al. (2014) recommend that airspeed varied from 0.15 to 1.5 m s$^{-1}$ is recommended for ventilation in the tropical country like Malaysia. The study also stated that for the residential house, Eqs. 8.1–8.3 can be used to calculate the solar-induced ventilation potential.

$$T_{\mathrm{abs}} = T_s\left[(1-\eta)\frac{\eta_{\mathrm{optical}}\mathrm{CR}_{\mathrm{theoretical}}}{\varepsilon_{\mathrm{abs}}\mathrm{CR}_{\mathrm{ideal}}}\right]^{1/4} \tag{8.1}$$

$$\text{Flow rate } (Q) = C_d A\left[\frac{2\Delta P}{\rho}\right]^{1/2} \tag{8.2}$$

$$\text{Air change per hour (ACH)} = 60\left[\frac{Q}{\mathrm{vol}}\right] \tag{8.3}$$

where the absorber temperature is $T_{\mathrm{abs}}$; the sun temperature is $T_s$; the optical efficiency from the reflector is $\eta_{\mathrm{optical}}$; Emissivity factor is $\varepsilon_{\mathrm{abs}}$; coefficient of discharge is $C_d$; the opening area is A. The optical efficiency can be used 1 for very good and efficient absorbent. The emissivity factor for the lowest reflectivity can be used 1. The study (Utama et al., 2014) suggests the value for optical efficiency and emissivity factor as 0.1 and 0.02, respectively.

To achieve the maximum airflow rate in the solar chimney, the mass flow rate is calculated using Eq. 8.4.

$$\dot{m} = C_d \rho A_o \sqrt{\frac{2g\Delta h(T_f - T_r)}{(1+Ar^2)T_r}} \tag{8.4}$$

where $A$ is the area and the notation, "o" indicates outlet, and "r" indicates ration between outlet and inlet; $\Delta h$ is the vertical distance between inlet and outlet of the solar chimney; $T$ is the temperature and the notation "f" indicates ambient temperature and "r" indicates room temperature (Alexandra & Muhammed, 2015; Mathur et al., 2006; Neves et al., 2011). Neves et al. (2011) reported that the average coefficient discharge value is 0.12 for maximum global solar irradiation of about 966 W m$^{-1}$ and maximum dry bulb temperature is 30.8 °C (Neves et al., 2011). The air change per hour can be estimated from Eqs. 8.5 and 8.6 (Alexandra & Muhammed, 2015; Mathur et al., 2006).

$$\text{Air velocity at chimney outlet } (u) = \frac{\dot{m}}{\rho A_o} \tag{8.5}$$

$$\text{Volume flow rate } (\dot{V}) = \frac{\dot{m}}{\rho} \quad \text{and} \quad \mathrm{ACH} = \frac{\dot{V}\times 3600}{\text{Ventilated volume } (v)} \tag{8.6}$$

The airflow rate can be estimated from the effect of buoyancy, the air passes through the chimney can be estimated from Eq. 8.7, as suggested by Saleem et al. (2016).

$$\dot{m} = C_d \frac{\rho_f A_o}{\sqrt{1 + A_o/A_i}} \sqrt{\frac{2gl(T_f - T_a)}{T_a}} \tag{8.7}$$

Saleem et al. (2016) recommend the value of the coefficient of discharge for the solar chimney as 0.64. It is also found that the air velocity inside the solar chimney varies between 0.25 and 0.39 m/s when the solar radiation was 650 W/m$^2$ and the height of solar chimney was 2 m. The study also showed that a 2 m height chimney able to create 0.011 kg/s mass flow rate at solar radiation of 450 W/m$^2$.

Adam et al. (2002) mentioned that the flow rate in the solar chimney depends on the flow rate of a different layer. The average density can be calculated from the following equations. The equation for the average density

$$\boldsymbol{\rho}_a = 353.25/\boldsymbol{T}_{fa}$$

The equation for the density of air at the out of the chimney

$$\boldsymbol{\rho}_{fa(n)} = 353.25/\boldsymbol{T}_{fa(n)}$$

The airflow rate at the outlet of the chimney can be determined by using the following equation

$$Q_{oa} = C_{do} A_a \sqrt{\frac{2}{\rho_{fa(n)}} \left[\rho_e - \overline{\rho_{fa}}\right] g L \sin\varphi + P_m} \tag{8.8}$$

The airflow rate can be determined by using the following equation

$$Q_{ia} = C_{di} A_a \sqrt{\frac{2}{\rho_r}(-P_m)} \tag{8.9}$$

Considering the arrangement resembles of vertical solar wall and roof solar chimney, a zero-dimensional calculation can be used as it offers a flexible and less expensive approach. Therefore, validated zero-dimensional models developed for vertical solar wall and roof solar chimney by Mathur et al. (2006) and Jianliu and Weihua (2013), respectively, are used in this section. This is to find out whether those models are suitable to predict the performance of the proposed solar chimney or not.

For vertical solar wall, the theoretical mass flow rate ($\dot{m}$), volumetric flow rate $(\dot{V})$, and air velocity ($v$) inside the solar chimney are estimated through Eqs. 8.10–8.12 (Mathur et al., 2006):

$$\dot{m} = \left(\frac{C_d \rho_{f1} A_o}{\sqrt{(1 + A_r^2)}}\right)\left(\sqrt{\frac{2gL_s(T_f - T_r)}{T_r}}\right) \tag{8.10}$$

$$\dot{V} = \frac{\dot{m}}{\rho_{f1}} = \left(\frac{C_d A_o}{\sqrt{(1 + A_r^2}}\right)\left(\sqrt{\frac{2gL_s(T_f - T_r)}{T_r}}\right) \tag{8.11}$$

$$v = \frac{\dot{m}}{\rho_{f1} A_0} = \left(\frac{C_d}{\sqrt{(1 + A_r^2}}\right)\left(\sqrt{\frac{2gL_s(T_f - T_r)}{T_r}}\right) \tag{8.12}$$

Jianliu and Weihua (2013) develop mathematical models to predict the theoretical mass flow rate ($\dot{m}$), volumetric flow rate $(\dot{V})$, and air velocity ($v$) inside a rooftop solar chimney. Mathematically, it is expressed as shown in Eqs. 8.13–8.15:

$$\dot{m} = C_d \rho_{f1} A_o \left(\sqrt{\frac{2gL_s \sin\theta(T_f - T_r)}{(1 + A_r^2)T_r}}\right) \tag{8.13}$$

$$\dot{V} = \frac{\dot{m}}{\rho_{f1}} = C_d A_o \left(\sqrt{\frac{2gL_s \sin\theta(T_f - T_r)}{(1 + A_r^2)T_r}}\right) \tag{8.14}$$

$$v = \frac{\dot{m}}{\rho_{f1} A_0} = C_d \left(\sqrt{\frac{2gL_s \sin\theta(T_f - T_r)}{(1 + A_r^2)T_r}}\right) \tag{8.15}$$

Therefore, the estimated generated by the solar chimney is shown in Eq. 8.16.

$$P_{\text{solar}} = \rho Q g h_{\text{solar chimney}} \tag{8.16}$$

## Description of Prototype Model Configuration

The scaled prototype was designed and developed in the Faculty of Engineering, Universiti Malaysia Sabah (UMS), Malaysia. The experiment was carried out inside a laboratory under a controlled condition. The experiments were done with various air gap depth, inclination angle, and opening areas of the scaled prototype. The proposed solar chimney consists of two parts, the outer and inner part. Referring to Fig. 8.5, two opposite sides of the outer part are covered by transparent glasses. The glasses act as glaze to allow solar radiation to pass through it. So, the solar energy is absorbed by the absorber in the inner part.

Four sides of the inner part are made up by absorbers on two opposite sides and perforated sheets on the other two opposite sides as shown in Fig. 8.6. The purpose

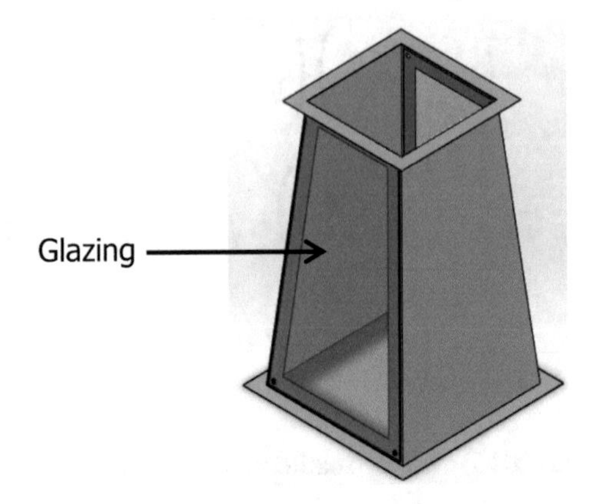


**Fig. 8.5** Outer part of solar chimney

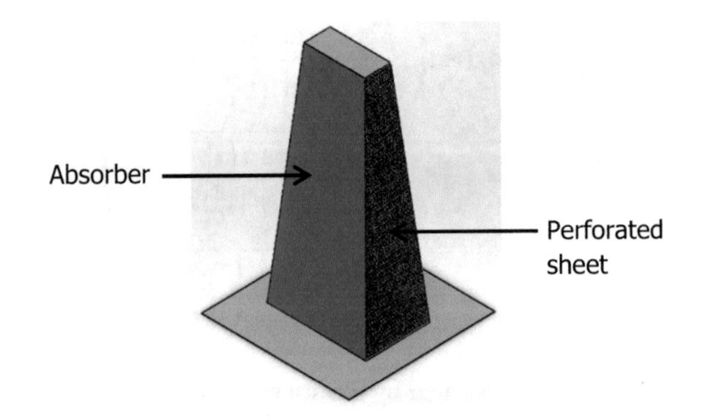


**Fig. 8.6** Inner part of solar chimney

of the perforated sheet is to allow air to flow through the chimney. Figure 8.7 is showing the proposed solar chimney and its section view by assembling the outer part and inner part. To prevent air leakage, a rubber gasket is placed between the inner and outer part. Bitumen flashing tapes are used to seal the gaps between the glazing and the outer part, while the remaining small holes or gaps are sealed with modeling clay.

A total of 16 different configurations was fabricated to investigate the effect of inclination angle and air gap depth on the performance of a solar chimney. The inclination angle and air gap depth vary from 75° to 90° and 10 mm to 16 mm, respectively. This in turn affects the opening areas as well. Therefore, the effect of opening areas on the performance of the chimney is examined at the same time. The inlet and outlet area of the chimney are varied from 0.0224 $m^2$ to 0.6 $m^2$ and 0.1 $m^2$ to 0.14 $m^2$, respectively. Additionally, all configurations have the same stack height which is 1 m. Both the outer and inner parts of the chimney are made up of galvanized iron (GI) as that material has high solar absorptivity and low emissivity

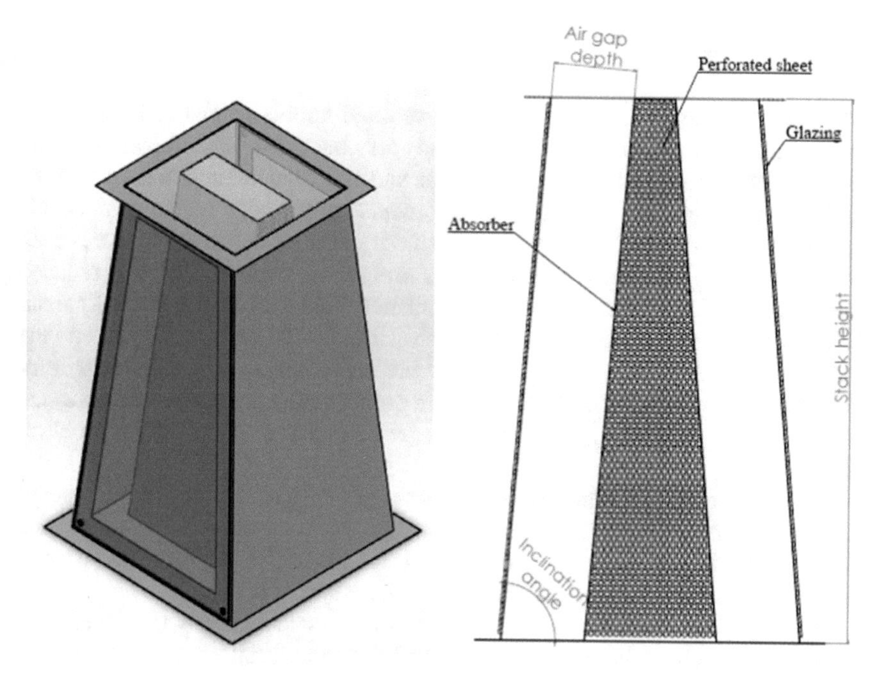


**Fig. 8.7** Solar chimney and its section view

(Cengel, 2006: 866). A preliminary study showed that the highest daily solar average global radiation recorded for Kota Kinabalu, Sabah, Malaysia, is about 495.90 W/m$^2$. Thus, the heating value of 500 W/m$^2$ was used during the experiment

## Experimental Set Up

At the starting of the experiment, the absorber is heated by using an electric heater under constant heat flux (500 W/m$^2$) to simulate uniform solar radiation on the absorber. Numbers of K type thermocouple with the measurement range of 0–250 °C are used for temperatures measurement. There are four thermocouples that are placed at the four corner side of the heat absorber and one thermocouple is placed at the center of the absorber. There are two thermocouples that are placed at the air gap, in the exit and entrance of the chimney. Also, a thermocouple is placed out of the chimney to measure the ambient temperature.

The temperature data are automatically collected and recorded into a computer via two units of USB-based 8-channel thermocouple input module for every 10 s. The inlet air velocity is measured by an air velocity meter and the air velocity data is recorded in every 2 min.

## Experimental Procedure

The solar chimney without wire mesh screen and turbine ventilator is used here as a regular solar chimney. The experiments are conducted for different heat load and different air gap. The air mass flow rate and exit temperature are measured for different heat loads and air gap. The performance of the other chimney models is compared with this combination. The wire mesh screen as shown in Fig. 8.8 is used for the experiments. Studies showed that a wire mesh screen significantly reduced the effect of cold inflow and enhanced ventilation (Chu et al., 2012, 2016; Rahman et al., 2014). The pore size of the wire mesh screen is 0.64 mm × 0.64 mm and the material is stainless steel. It is found that wire mesh significantly helps to prevent cold inflow as well as distribute temperature evenly during operation. The pore size of the wire mesh is selected based on the previous study done by Rahman et al. (2018).

The air mass flow rate and exit air temperature of the solar chimney are measured with the presence of wire mesh. The experimental results are compared with baseline data to determine the effects of wire mesh screen on solar chimney air mass flow rate and exit air temperature. The experimental setup is shown in Fig. 8.9.

In this study, the experiment was also done on the solar chimney integrated with turbine ventilator as shown in Fig. 8.10 and the experimental setup is shown in Fig. 8.11.

The air mass flow rate and exit air temperature of the integrated solar chimney are measured for different heat loads and air gap. The measured experimental results are compared with baseline data to determine the effects of turbine ventilator on solar chimney air mass flow rate and exit air temperature.

All the experiments start from ambient temperature and the absorber is heated under constant heat flux (500 W/m$^2$). The air velocity meter is placed at the center

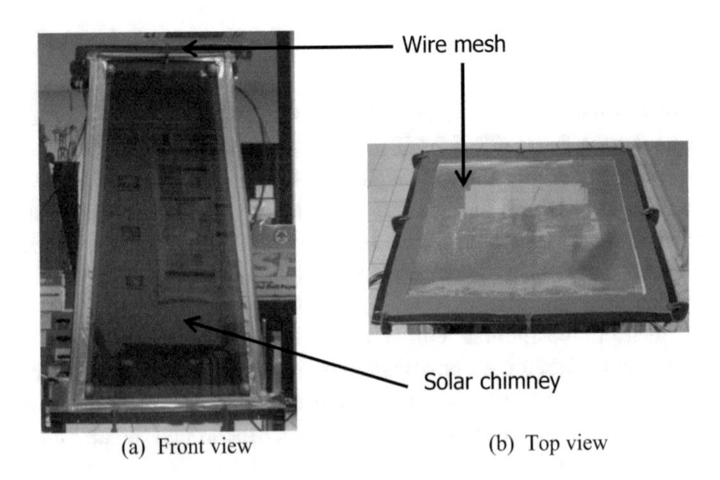


**Fig. 8.8** Front and top view of solar chimney and wire mesh screen

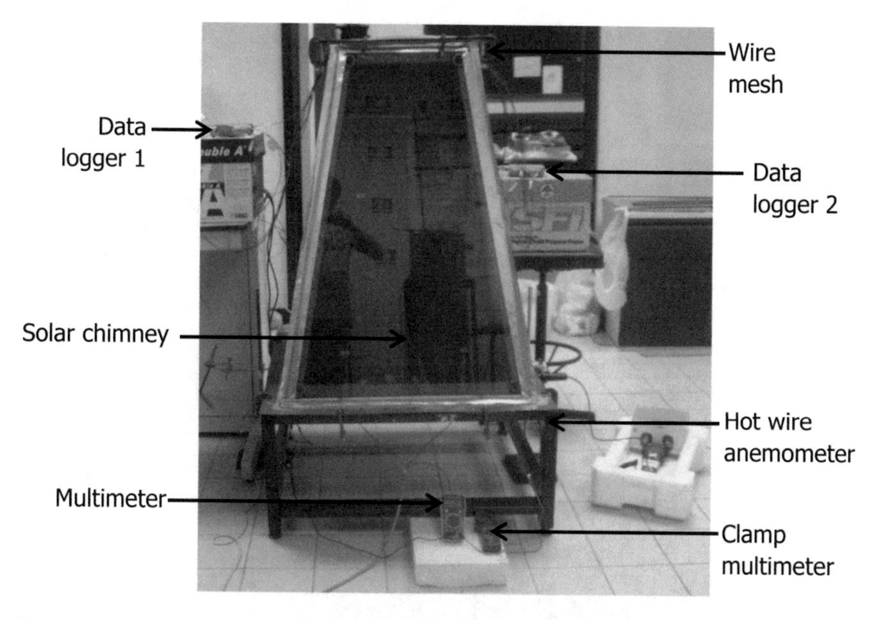


**Fig. 8.9** Experimental setup for solar chimney and wire mesh screen

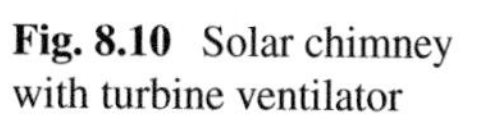

**Fig. 8.10** Solar chimney with turbine ventilator

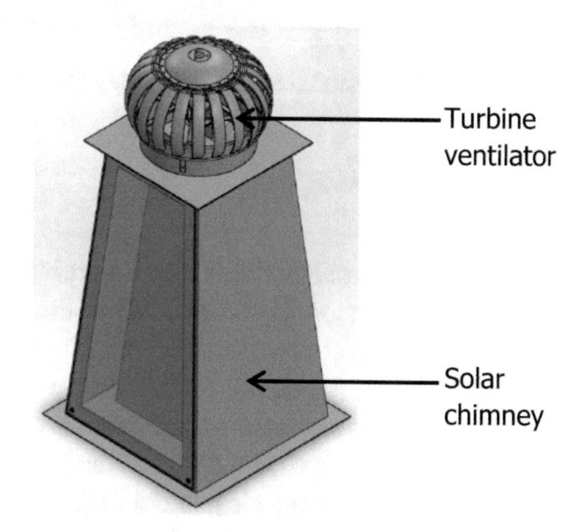


of the solar chimney's entrance to measure inlet air velocity. An axial fan is used to simulate wind to rotate the turbine ventilator. The axial fan is placed in such a way that the effect of wind-induced by the axial fan on the chimney is negligible. The rotational speed of the turbine ventilator is measured with a digital tachometer. A digital multimeter and a digital clamp meter are also used to measure voltage-current respectively.

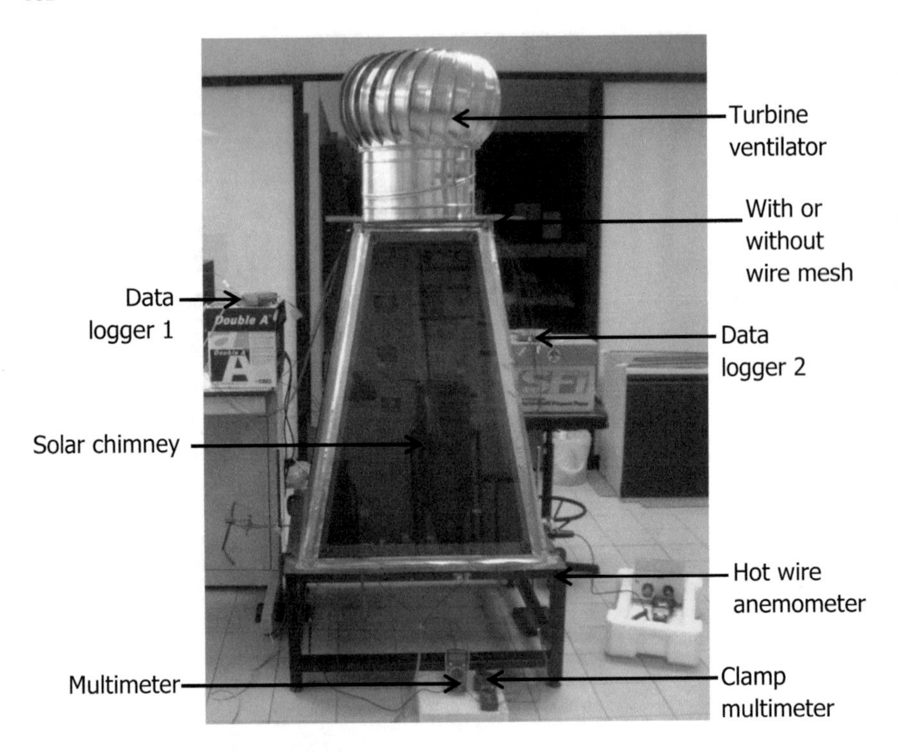


**Fig. 8.11** Experimental setup solar chimney with turbine ventilator

## Effect of Air Gap

The performance of solar chimney is varying with the chimney's air gap. In Table 8.1, it is found that the volume flow rate of the solar chimney with an inclination angle of 75° is increasing consistent with a larger air gap. This finding is consistent with observations made by other researchers where the flow restriction, frictional losses as well as conduction and convection effect in the air inside the air gap are significant at narrow air gap depth. As a result, the performance of 75° solar chimney model is found better with larger air gap depth (Chungloo & Limmeechokchai, 2007; Harris & Helwig, 2007; Hirunlabh et al., 2001a, 2001b; Khanal & Lei, 2011; Khedari et al., 2003; Li et al., 2014; Mathur et al., 2006; Tan & Wong, 2013; Zhai et al., 2011a).

Similarly, the mass flow rate of 80° and 85° solar chimney is also increasing when the air gap is increased. The flow rate is reduced when the air gap is more than 14 cm. This is only possible when flow reversal or cold inflow phenomenon is significant in the chimney. The flow reversal or cold inflow is observed in the solar chimney when the exit pressure lower than ambient pressure. This is only possible when the air gap depth exceeds optimum air gap depth (Arce et al., 2008; Harris & Helwig, 2007; Khanal & Lei, 2012, 2014; Tan & Wong, 2013). Another possible reason for cold inflow in the chimney is greater heat losses at a wider air gap depth (Hirunlabh et al., 2001a, 2001b). Therefore, an optimum air gap for different inclination angle of the proposed solar chimney is proven to exist.

The mass flow rate in the 90° solar chimney model is decreasing when the air gap is getting larger. This can be due to the optimum air gap at 90° solar chimney to be lower than 10 cm. As a result, the flow reversal effect leads to a significant drop in the performance of solar chimney after the air gap is further increased, passing the optimum air gap depth. Moreover, the greater heat losses at a wider air gap can also be another reason for causing this event (Hirunlabh et al., 2001a, 2001b).

## Effect of Inclination Angle

The stack height in all solar chimney models is kept constant. Therefore, the size of the solar chimney is decreased when the absorber inclination angle is increased. The mass flow rate is decreased when the inclination angle is increased from 80° onward for all air gap as shown in Table 8.1. This is possible due to the reduction of solar chimney size and the reduction of exposure area of the solar collector to solar radiance. Thus, lower energy collection, lower heat utilization, and lower volumetric flow rate are yielded when the inclination angle is increasing (Bassiouny & Korah, 2009; Tan & Wong, 2013). The mass flow rate is increased with the inclination angle from 75° to 80° for the solar chimney with 10 cm, 12 cm, and 14 cm air gap depth. This might be due to the greater convection heat transfer and higher heat losses to the surrounding for the inclined cavity (Harris & Helwig, 2007). Studies showed that the reduction of inclination angle (or increment in the exposure area of solar collector) is not able to counteract the heat transfer and losses after certain inclination angle (Bassiouny & Korah, 2009; Harris & Helwig, 2007; Khanal & Lei, 2012).

The solar chimney models with 10, 12, and 14 cm air gap have the same optimum inclination angle at 80° as the highest mass flow rate is achieved. There is no optimum inclination angle is found in the chimney models with air gap 16 cm. It is assumed that the optimum inclination angle for the proposed solar chimney for 16 cm air gap to be at a lower value than the minimum value used in this study since the mass flow rate of the proposed solar chimney is decreasing drastically at higher air gap. The inclination angle has a significant effect on solar chimney performance at higher air gap depth. The performance of the proposed solar chimney is improved with the inclination angle and reached a peak performance at an optimum inclination angle. Higher angle than the optimum value lowered the performance in the model chimney (Bassiouny & Korah, 2009; Harris & Helwig, 2007; Khanal & Lei, 2012). Suitable inclination angle should be selected carefully especially for a solar chimney with bigger air gap depth. This is because the effect of inclination angle on the performance of solar chimney is greater at higher air gap depth.

## Effect of Outlet Apertures Area

The variation of mass flow rate to the outlet apertures area is shown in Fig. 8.12. The outlet aperture area has more effect on the flow rate of a solar chimney than inlet aperture area (Haghighi & Maerefat, 2014). The mass flow rates in the model chimney with absorber angles 75°, 80°, 85°, and 90° have significant relation with outlet apertures area. The mass flow rate is increased with angles and decreased after critical value except for the chimney models with absorber angle 75° and 90°. In the chimney model absorber angle 90°, the flow rate decrease with the area. This is because the chimney with absorber angle 90° is considered as vertical wall solar chimney. In the vertical wall solar chimney, the mass flow rate is the function of heat flow rate and air gap and the thermal efficiency is the function of absorber heat gain performance but there is no relation with an air gap (Burek & Habeb, 2007). Alternatively, in the chimney model absorber angle 75°, the mass flow rate increased with the outlet aperture area. This is because the flow rate is increased with inclination angles between its optimum rage (45–75°). In this inclination angle, the absorber received the highest energy from the solar radiation. Beyond this angle, the amount of radiation capture by the absorber is reduced significantly (Haghighi & Maerefat, 2014; Khedari et al., 2003; Li et al., 2014; Maerefat & Haghighi, 2010).

From Fig. 8.12, it is also found that the performance of the proposed solar chimney with 80° and 85° is reduced after a certain outlet aperture area is reached. This is

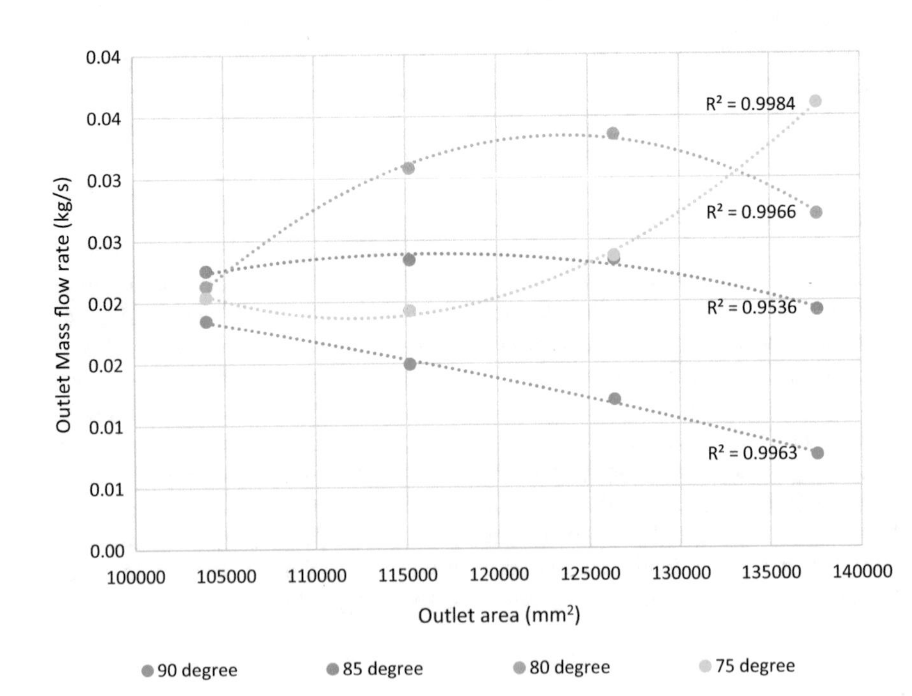


Fig. 8.12 Effect of outlet area on mass flow rate

possibly due to a larger air gap as well as a larger outlet aperture area. Large air gap and big outlet aperture area will have high chances for flow reversal effect as well as heat loss. Similar trends were observed in both the outlet aperture area and the air gap. Hence, the performance of the proposed solar chimney is reduced after an outlet aperture area is reached, suggesting of an existence of an optimum value for the outlet aperture area at a different inclination angle in the proposed solar chimney. The flow reversal effect and heat losses affect the performance of the 90° proposed solar chimney with an increase in the aperture area. Also, the inlet aperture area with equal or larger than the outlet aperture area always performs better compared to a larger outlet aperture area as reported by other researchers (Khanal & Lei, 2011; Khedari et al., 2000; Zhai et al., 2011a).

## References


AboulNaga, M. M., & Abdrabboh, S. N. (2000). Improving night ventilation into low-rise buildings in hot-arid climates exploring a combined wall–roof solar chimney. *Renewable Energy, 19*(1–2), 47–54.

Afonso, C., & Oliveira, A. (2000). Solar chimneys: Simulation and experiment. *Energy and Buildings, 32*(1), 71–79.

Al-Obaidi, K. M., Ismail, M., & Rahman, A. M. A. (2014). A review of the potential of attic ventilation by passive and active turbine ventilators in tropical Malaysia. *Sustainable Cities and Society, 10,* 232–240.

Al-Kayiem, H. H., Sreejaya, K. V., & Gilani, S. I. U. H. (2014). Mathematical analysis of the influence of the chimney height and collector area on the performance of a roof top solar chimney. *Energy and Buildings, 68,* 305–311.

Amer, E. H. (2006). Passive options for solar cooling of buildings in arid areas. *Energy, 31,* 1332–1344.

Amara, F., Agbossou, K., Cardenas, A., Dubé, Y., & Kelouwani, S. (2015). Comparison and simulation of building thermal models for effective energy management. *Smart Grid and Renewable Energy, 6*(04), 95.

Amori, K. E., & Mohammed, S. W. (2012). Experimental and numerical studies of solar chimney for natural ventilation in Iraq. *Energy and Buildings, 47,* 450–457.

Arce, J., Jiménez, M. J., Guzmán, J. D., Heras, M. R., Alvarez, G., & Xamán, J. (2009). Experimental study for natural ventilation on a solar chimney. *Renewable Energy, 34*(12), 2928–2934.

Ariffin, A. R., Rao, A., & Nila, S. P. (2002). Thermal comfort and evaporative cooling of external walls in an equatorial climate. In A. M. A. Rahman (Ed.), *Development of passive solar design and technology in tropical climates* (pp. 23–34). Pulau Pinang: The Universiti Sains Malaysia Co-operative Bookshop Ltd.

Aynsley, R. (2014). Natural ventilation in passive design. *Environment Design Guide, 80,* 1.

Bansal, N. K., Mathur, J., Mathur, S., & Jain, M. (2005). Modeling of window-sized solar chimneys for ventilation. *Building and Environment, 40*(10), 1302–1308.

Bassiouny, R., & Korah, N. S. (2009). Effect of solar chimney inclination angle on space flow pattern and ventilation rate. *Energy and Buildings, 41*(2), 190–196.

Breesch, H., Bossaer, A., & Janssens, A. (2005). Passive cooling in a low-energy office building. *Solar Energy, 79*(6), 682–696.

Bouchair, A., & Fitzgerald, D. (1988). The optimum azimuth for a solar chimney in hot climates. *Energy and Buildings, 12*(2), 135–140.

Bouchair, A. (1994). Solar chimney for promoting cooling ventilation in southern Algeria. *Building Services Engineering Research and Technology, 15*(2), 81–93.

Collinge, W. O., Rickenbacker, H. J., Landis, A. E., Thiel, C. L., & Bilec, M. M. (2018). Dynamic life cycle assessments of a conventional green building and a net zero energy building: Exploration of static, dynamic, attributional, and consequential electricity grid models. *Environmental Science and Technology, 52*(19), 11429–11438.

Chen, K., Wang, J., Dai, Y., & Liu, Y. (2014). Thermodynamic analysis of a low-temperature waste heat recovery system based on the concept of solar chimney. *Energy Conversion and Management, 80,* 78–86.

Chan, H. Y., Riffat, S. B., & Zhu, J. (2010). Review of passive solar heating and cooling technologies. *Renewable and Sustainable Energy Reviews, 14*(2), 781–789.

Chantawong, P., Hirunlabh, J., Zeghmati, B., Khedari, J., Teekasap, S., & Win, M. M. (2006). Investigation on thermal performance of glazed solar chimney walls. *Solar Energy, 80*(3), 288–297.

Chik, N. A., Rahim, K. A., Saari, M. Y., & Alias, E. F. (2012). Changes in consumer energy intensity in Malaysia. *International Journal of Economics and Management, 6*(2), 221–240.

Chungloo, S., & Limmeechokchai, B. (2007). Application of passive cooling systems in the hot and humid climate: The case study of solar chimney and wetted roof in Thailand. *Building and Environment, 42*(9), 3341–3351.

Chu, C. C. M., Chu, R. K. H., & Rahman, M. M. (2012). Experimental study of cold inflow and its effect on draft of a chimney. *Advanced Computational Methods and Experiments in Heat Transfer XII, WIT Transactions on Engineering Sciences, 75,* 73–82.

Danny, P. (2005). Literature review of the impact and need for attic ventilation in Florida homes. Revised Draft Report, May 31 2005, Submitted to Florida Department of Community Affairs, FSEC-CR-1496-05.

DeBlois, J. C., Bilec, M. M., & Schaefer, L. A. (2013). Design and zonal building energy modeling of a roof integrated solar chimney. *Renewable Energy, 52,* 241–250.

Dai, Y. J., Huang, H. B., & Wang, R. Z. (2003). Case study of solar chimney power plants in Northwestern regions of China. *Renewable Energy, 28*(8), 1295–1304.

Eichholtz, P., Kok, N., & Quigley, J. M. (2013). The economics of green building. *Review of Economics and Statistics, 95*(1), 50–63.

Ekechukwu, O. V., & Norton, B. (1997). Experimental studies of integral-type natural-circulation solar-energy tropical crop dryers. *Energy Conversion and Management, 38*(14), 1483–1500.

Fudholi, A., Zohri, M., Jin, G. L., Ibrahim, A., Yen, C. H., Othman, M. Y., et al. (2018). Energy and exergy analyses of photovoltaic thermal collector with ∇-groove. *Solar Energy, 159,* 742–750.

Guramun, S., Misaran, M. S., Ibrahim, M. K. W., & Rahman, M. M. (2019). Trends of hybrid earth-air-pipe (EAP) photovoltaic cooling system for efficiency improvement: A review. *Journal of Mechanical Engineering Research and Development, 42*(4), 191–195.

Gieseler, U. D. J., Bier, W., & Heidt, F. D. (2002). Cost efficiency of ventilation systems for low-energy buildings with earth-to-air heat exchange and heat recovery. In *Proceedings of the International Conference on Passive and Low Energy Architecture (PLEA),* Toulouse (2002).

Gan, G. (1998). A parametric study of Trombe walls for passive cooling of buildings. *Energy and Buildings, 27*(1), 37–44.

Gunerhan, H., & Hepbasli, A. (2007). Determination of the optimum tilt angle of solar collectors for building applications. *Building and Environment, 42*(2), 779–783.

Hall, R., Wang, X., Ogden, R., & Elghali, L. (2011). Transpired solar collectors for ventilation air heating. *Proceedings of the Institution of Civil Engineers-Energy, 164*(3), 101–110.

Hirunlabh, J., Wachirapuwadon, S., Pratinthong, N., & Khedari, J. (2001a). New configurations of a roof solar collector maximizing natural ventilation. *Building and Environment, 36*(3), 383–391.

Harris, D. J., & Helwig, N. (2007). Solar chimney and building ventilation. *Applied Energy, 84*(2), 135–146.

Hughes, B. R., Calautit, J. K., & Ghani, S. A. (2012). The development of commercial wind towers for natural ventilation: A review. *Applied Energy, 92,* 606–627.

Haghighi, A. P., & Maerefat, M. (2014). Solar ventilation and heating of buildings in sunny winter days using solar chimney. *Sustainable Cities and Society, 10,* 72–79.

Hu, Z., He, W., Ji, J., & Zhang, S. (2017). A review on the application of Trombe wall system in buildings. *Renewable and Sustainable Energy Reviews, 70,* 976–987.

Hirunlabh, J., Wachirapuwadon, S., Pratinthong, N., & Khedari, J. (2001b). New configurations of a roof solar collector maximizing natural ventilation. *Building and Environment, 36*(3), 383–391.

Ismail, M. A., & Rashid, F. A. (2014). Malaysia's existing green homes compliance with LEED for homes. *Procedia Environmental Sciences, 20,* 131–140.

Jianliu, X., & Weihua, L. (2013). Study on solar chimney used for room natural ventilation in Nanjing. *Energy and Buildings, 66,* 467–469.

Kamal, M. A. (2012). An overview of passive cooling techniques in buildings: Design concepts and architectural interventions. *Acta Technica Napocensis: Civil Engineering & Architecture, 55*(1), 2012.

Kaneko, Y., Sagara, K., Yamanaka, T., Kotani, H., & Sharma, S. D. (2006, May). Ventilation performance of solar chimney with built-in latent heat storage. In *Proceedings of 10th International Conference of Thermal Energy Conference (ECOSTOCK).*

Kazanci, O. B., Skrupskelis, M., Olesen, B. W., & Pavlov, G. K. (2013). *Solar sustainable heating, cooling and ventilation of a net zero energy house.* Paper presented at Clima 2013, Prague, Czech Republic.

Khanal, R., & Lei, C. (2011). Solar chimney—A passive strategy for natural ventilation. *Energy and Buildings, 43*(8), 1811–1819.

Khanal, R., & Lei, C. (2014). An experimental investigation of an inclined passive wall solar chimney for natural ventilation. *Solar Energy, 107,* 461–474.

Khedari, J., Boonsri, B., & Hirunlabh, J. (2000). Ventilation impact of a solar chimney on indoor temperature fluctuation and air change in a school building. *Energy and Buildings, 32*(1), 89–93.

Khedari, J., Hirunlabh, J., & Bunnag, T. (1997). Experimental study of a roof solar collector towards the natural ventilation of new houses. *Energy and Buildings, 26*(2), 159–164.

Khedari, J., Rachapradit, N., & Hirunlabh, J. (2003). Field study of performance of solar chimney with air-conditioned building. *Energy, 28*(11), 1099–1114.

Kleiven, T. (2003). *Natural ventilation in buildings: Architectural concepts, consequences and possibilities.* Institutt for byggekunst, historie og teknologi.

Lal, S., Kaushik, S. C., & Bhargav, P. K. (2013). Solar chimney: A sustainable approach for ventilation and building space conditioning. *International Journal of Development and Sustainability, 2*(1), 277–297.

Li, Y., & Liu, S. (2014). Experimental study on thermal performance of a solar chimney combined with PCM. *Applied Energy, 114,* 172–178.

Li, H., Yu, Y., Niu, F., Shafik, M., & Chen, B. (2014). Performance of a coupled cooling system with earth-to-air heat exchanger and solar chimney. *Renewable Energy, 62,* 468–477.

Lien, J., & Ahmed, N. (2011). Wind driven ventilation for enhanced indoor air quality. In *Chemistry, emission control, radioactive pollution and indoor air quality*. InTech

Liping, W., & Hien, W. N. (2007). Applying natural ventilation for thermal comfort in residential buildings in Singapore. *Architectural Science Review, 50*(3), 224–233.

Lomas, K. J. (2007). Architectural design of an advanced naturally ventilated building form. *Energy and Buildings, 39*(2), 166–181.

Liu, G., Xiao, M., Zhang, X., Gal, C., Chen, X., Liu, L., et al. (2017). A review of air filtration technologies for sustainable and healthy building ventilation. *Sustainable Cities and Society, 32,* 375–396.

Mathur, J., Bansal, N. K., Mathur, S., & Jain, M. (2006). Experimental investigations on solar chimney for room ventilation. *Solar Energy, 80*(8), 927–935.

Micallef, D., Buhagiar, V., & Borg, S. P. (2016). Cross-ventilation of a room in a courtyard building. *Energy and Buildings, 133,* 658–669.

Macias, M., Gaona, J. A., Luxan, J. M., & Gomez, G. (2009). Low cost passive cooling system for social housing in dry hot climate. *Energy and Buildings, 41*(9), 915–921.

Maerefat, M., & Haghighi, A. P. (2010). Passive cooling of buildings by using integrated earth to air heat exchanger and solar chimney. *Renewable Energy, 35*(10), 2316–2324.

Nouanégué, H. F., & Bilgen, E. (2009). Heat transfer by convection, conduction and radiation in solar chimney systems for ventilation of dwellings. *International Journal of Heat and Fluid Flow, 30*(1), 150–157.

Oh, T. H., Lalchand, G., & Chua, S. C. (2014). Juggling act of electricity demand and supply in Peninsular Malaysia: Energy efficiency, renewable energy or nuclear? *Renewable and Sustainable Energy Reviews, 37,* 809–821.

Ong, K. S. (2003). A mathematical model of a solar chimney. *Renewable Energy, 28*(7), 1047–1060.

Pacheco, R., Ordóñez, J., & Martínez, G. (2012). Energy efficient design of building: A review. *Renewable and Sustainable Energy Reviews, 16*(6), 3559–3573.

Rahman, M. M., Chu, C. M., Kumaresen, S., Yan, F. Y., Kim, P. H., Mashud, M., et al. (2014). Evaluation of the modified chimney performance to replace mechanical ventilation system for livestock housing. *Procedia Engineering, 90,* 245–248.

Rahman, M. M., Misaran, M. S. B., Jamanun, M. J. B., & Jawad, A. (2018). Estimate the ventilation effect from wire mesh screen assisted solar chimney. *Journal of Energy and Power Engineering, 12,* 127–131.

Rabani, M., Kalantar, V., Dehghan, A. A., & Faghih, A. K. (2015). Empirical investigation of the cooling performance of a new designed Trombe wall in combination with solar chimney and water spraying system. *Energy and Buildings, 102,* 45–57.

Revel, A., & Huynh, P. (2004) Characterising roof ventilators. In *Australasian Fluid Mechanics Conference*, The University of Sydney.

Robert, S. (2005). Energy *alternatives.* Heinemann Educational Publishers.

Sharma, S. D., Kotani, H., Kaneko, Y., Yamanaka, T., & Sagara, K. (2007). Design, development of a solar chimney with built-in latent heat storage material for natural ventilation. *International Journal of Green Energy, 4*(3), 313–324.

Saadatian, O., Sopian, K., Lim, C. H., Asim, N., & Sulaiman, M. Y. (2012). Trombe walls: A review of opportunities and challenges in research and development. *Renewable and Sustainable Energy Reviews, 16*(8), 6340–6351.

Sundell, J. (2004). On the history of indoor air quality and health. *Indoor Air, 14*(Suppl. 7), 51–58.

Tamm, G., & Jaluria, Y. (2017). Flow of hot gases in vertical shafts with natural and forced ventilation. *International Journal of Heat and Mass Transfer, 114,* 337–353.

Santamouris, M., & Asimakopoulos, D. (Eds.). (1996). *Passive cooling of buildings.* Earthscan.

Taleb, H. M. (2014). Using passive cooling strategies to improve thermal performance and reduce energy consumption of residential buildings in UAE buildings. *Frontiers of Architectural Research, 3*(2), 154–165.

Tan, A. Y. K., & Wong, N. H. (2013). Parameterization studies of solar chimneys in the tropics. *Energies, 6*(1), 145–163.

Tan, A. Y. K., & Wong, N. H. (2012). Natural ventilation performance of classroom with solar chimney system. *Energy and Buildings, 53,* 19–27.

Waewsak, J., Hirunlabh, J., Khedari, J., & Shin, U. C. (2003). Performance evaluation of the BSRC multi-purpose bio-climatic roof. *Building and Environment, 38*(11), 1297–1302.

Xiong, X., Fulpagare, Y., Sun, C., & Lee, P. S. (2019, May). Numerical study of a new rack layout for better cold air distribution and reduced fan power. In *2019 18th IEEE Intersociety Conference on Thermal and Thermomechanical Phenomena in Electronic Systems (ITherm)* (pp. 399–404). IEEE.

Yusoff, W. F. M., Salleh, E., Adam, N. M., Sapian, A. R., & Sulaiman, M. Y. (2010). Enhancement of stack ventilation in hot and humid climate using a combination of roof solar collector and vertical stack. *Building and Environment, 45*(10), 2296–2308.

Zaki, W. R. M., Nawawi, A. H., & Ahmad, S. S. (2010). Economic assessment of Operational Energy reduction options in a house using Marginal Benefit and Marginal Cost: A case in Bangi, Malaysia. *Energy Conversion and Management, 51*(3), 538–545.

Zhai, Z. J., Johnson, M. H., & Krarti, M. (2011a). Assessment of natural and hybrid ventilation models in whole-building energy simulations. *Energy and Buildings, 43*(9), 2251–2261.
Zhai, X. Q., Song, Z. P., & Wang, R. Z. (2011b). A review for the applications of solar chimneys in buildings. *Renewable and Sustainable Energy Reviews, 15*(8), 3757–3767.
Zhou, X., Yang, J., Ochieng, R. M., Li, X., & Xiao, B. (2009). Numerical investigation of a plume from a power generating solar chimney in an atmospheric cross flow. *Atmospheric Research, 91*(1), 26–35.
Zhu, N., Li, S., Hu, P., Lei, F., & Deng, R. (2019). Numerical investigations on performance of phase change material Trombe wall in building. *Energy, 187,* 116057.
Ziskind, G., Dubovsky, V., & Letan, R. (2002). Ventilation by natural convection of a one-story building. *Energy and Buildings, 34*(1), 91–101.

# Chapter 9
# Effect of Wire Mesh Screen on the Natural Draft Chimney

**Chee Kai Shyan, Mohammad Mashud, Md. Mizanur Rahman, and Fadzlita Mohd. Tamiri**

Natural draft chimney (NDC) or solar chimney has long been implemented into thermal power plants to facilitate the discharge of waste heat into the atmosphere. Although other forms of cooling systems exist, cooling towers are generally being used by the industry as they provide many advantages over the other means of cooling. A natural draft cooling tower requires zero energy to operate as it does not require the use of mechanical parts such as fans that would otherwise produce unwanted mechanical noise in order to induce air flow, thus making it significantly more environmentally friendly. Besides, due to the lack of mechanical parts, the natural draft cooling tower only requires very little maintenance. A typical cooling tower in power plants is capable of having capacity of over hundreds of megawatts (Chu et al., 2012: 125). With such high capacity, the efficiency of the cooling towers is always the main concern.

The NDC is typically being used by power production plants, and the efficiency of a cooling system usually involves the coefficient $\dot{Q}_R/P$ (El-Wakil, 1985: 324). As such, a reduction in the $\dot{Q}_R/P$ coefficient would mean that a greater amount of energy is being transformed into usable energy from the fuel, implying that less heat was lost to the environment. Since the cooling tower is dealing with such a large energetic flow in the power plant, minute improvements made upon reducing the coefficient $\dot{Q}_R/P$ would result in very significant fuel savings and thus reducing the amount of radiation emitted (Smrekar et al., 2006: 1088).

C. K. Shyan · F. Mohd. Tamiri
Faculty of Engineering, Universiti Malaysia Sabah, Kota Kinabalu, Sabah, Malaysia

M. Mashud (✉)
Department of Mechanical Engineering, Khulna University of Engineering and Technology (KUET), Khulna 9203, Bangladesh
e-mail: mdmashud@me.kuet.ac.bd

Md. M. Rahman
Department of Mechatronics Engineering, World University of Bangladesh, Dhaka 1205, Bangladesh



Md. M. Rahman and C.-M. Chu (eds.), *Cold Inflow-Free Solar Chimney*,
https://doi.org/10.1007/978-981-33-6831-6_9

Over recent years, many methods have been experimented on aiming to increase the efficiency of the NDCs. These methods include the optimization of the heat transfer along the cooling tower using water distribution across the plane area (Smrekar et al., 2006: 1088), by optimizing the design of the solar-enhanced natural draft dry cooling tower (Zou et al., 2013: 945). Chu et al. (2012) experimented on the effect of cold inflow on chimney height for a NDC which is very heavily based on the study of Jörg and Scorer (1967) where the inflow of exterior fluid into a chimney from which buoyant emerges was investigated in a water tank. Hosseini et al. (2017: 296) investigated upon the rectangular fin geometry effect on a solar chimney.

The aim of this project is to study the effect of wire mesh screen on cold inflow in the NDC where the enhancement provided by a wire mesh screen implemented into the NDC will be studied. Cold inflow is a phenomenon where the cooling towers experience unstable flow with external airflow, leading to undesirable downdraft (Zhai & Fu, 2006: 1008). In recent decades, metal foams and wire mesh are tested extensively in heat exchangers to increase the efficiency (Boomsma et al., 2003; Fu et al., 2017; Kurian et al., 2016; Ma et al., 2016). A recent study conducted by Fu et al. (2017) is to investigate the enhancement in heat transfer of sprayed wire mesh heat exchanger. In their experiment, aluminium wires with different types of diameter were being used to fabricate the heat exchangers which would increase the surface area for heat transfer, and they concluded that every type of aluminium wire mesh used for heat transfer had increased the efficiency where the maximum enhancement was done using the 20PPI SPW heat exchanger which increased the heat transfer value by 25.9%. Kurian et al. (2016) experimented on using different types of materials to fabricate the wire mesh and metal foam for their copper tube heat exchanger where the materials are aluminium foam of porosity 0.94, a stainless steel wire mesh and a bare copper tube heat exchanger as the Control of their experiment. They concluded that the exchanger embedded in wire mesh showed better performance compared with the other types with a maximum increase of 28% on the air side Nusselt number. On the other hand, Ma et al. (2016) investigated on the flow and heat transfer characteristics in double-laminated woven wire mesh where their focus is more towards the flow of air through Dutch-woven wire mesh of different porosity. The conclusion made in said study states that the increase in average porosity of the wire mesh would increase the overall Nusselt number. Boomsma et al. (2003) performed experiments using metal foams manufactured from aluminium as a compressed open-cell aluminium foam heat exchanger and found that it generated thermal resistances that are two to three times lower than the best commercially available heat exchanger tested.

As stated previously, many researches had been done by researchers aiming to improve the efficiency of the NDC and also on heat exchangers by incorporating wire mesh into the heat exchanger. In this study, the aim was to introduce wire mesh into the heat exchanger of a NDC to improve upon the efficiency of the cooling tower. Although data obtained from previous researches was promising and Boomsma et al. (2003) even stated that their design, through the use of compressed metal foam had decreased the heat exchangers thermal resistance by two to three times compared to those commercially available ones. However, their design was only compared with one previous source while the source is compared to being a research done by a corporation itself, named Asea Brown Boveri where research done by corporation is commonly known to be biased or skewed. Moreover, their research was specifically compared with a company's research chosen specifically by them, begging a lot of questions, reducing the accountability of their research findings. Besides questionable comparison with previous research being chosen, another research upon NDC done by Smrekar et al. (2006) had made conclusions based solely on their own research and also without the use of a CFD software to verify their results, this arouses the suspicion towards their findings, and as a result, further investigation is needed to verify their claims made upon the method which they utilized to improve the efficiency of a NDC.

Research done by Chu et al. (2012) to test the effect of cold inflow on chimney height claimed that the wire mesh used in their experiment behaved abnormally, giving opposite result to what was originally expected; however, a few explanations were given as justification. The lack of use of flow simulation software, specifically computational fluid dynamics (CFD) to justify their explanation, was also mentioned by them. This leads to the objective of this project, where simulation upon similar chimney design will be conducted and compared with their results.

## Cooling System in Power Plants

A thermal power plant converts heat into mechanical energy to generate electricity, and in the process of producing power, a substantial amount of waste heat is produced and is essential that the excess heat is expelled to the surrounding environment. As such, the cooling system of a thermal power plant is very crucial to the performance of the power cycle and overall efficiency of the power plant. Failure in providing adequate cooling to the power generation process would inevitably lead to the decreased electricity produced as well a tremendous loss in the economic sense (Sun et al., 2017). Therefore, it is essential for power plants to employ an efficient cooling system to avoid such detrimental economic punishment.

Cooling systems are employed by power plants into two main types: wet cooling system and dry cooling system. Wet cooling system, as the name suggested, uses liquids, typically flowing or circulating water to remove heat, while the dry cooling variant uses mainly air for the heat exchange.

### *Wet Cooling System*

Most power plants that employ a wet cooling system utilize the wet cooling towers. A typical wet cooling tower design uses method of wet cooling known as evaporative cooling. Evaporative cooling is known to be a cooling technique that is environmental-friendly as it only uses water and air as the working fluids while only requiring low amounts of energy to pump the water and force the air into the cooling pad; therefore, the resulting operating cost to employ this cooling method is relatively low when compared to that of mechanical compression method used for air cooling (Xu et al., 2016). As a result, evaporative cooling technique used in wet cooling towers may prove to be a better and more compared to mechanical cooling (Al-Badri & Al-Waaly, 2017). Although many other technologies are available for the same heat rejection process, wet cooling towers may still be a better option due to the flexibility they possess, capable of handling substantial heat loads that are larger than 352 kW (Naik & Muthukumar, 2017). When utilized in the wet cooling tower, the evaporative method has a significantly higher cooling capacity as well.

The working principle behind evaporative cooling is such that the ambient air is humidified while the warm water used in the cooling tower is cooled due to heat and mass transfer between the two mediums as they interact with one another. The driving potential for the heat transfer between the ambient air and water is the temperature difference while the mass transfer occurs as a result of vapour pressure difference (Naik & Muthukumar, 2017).

The mathematical model for the heat and mass transfer processes involving the wet cooling tower is complex, and the first mathematical model for these processes was proposed by Merkel (1925). Merkel's model predicted the performance of the counterflow cooling tower, and to further simplify his work, Merkel was able to describe the rate of change of the properties of the air and water into an equation using partial differential equations. The equation is one dimensional and can be solved by hand while being able to simultaneously describe the heat and mass transfer from a surface in terms of only a coefficient, the area of the surface and the enthalpy driving potential (Mansour & Hassab, 2014). Zivi and Brand (1956) solved Merkel's equation and developed a model for the cross-flow cooling tower; however, the model is two dimensional and has to be numerically solved by a computer. A more common effective model for the cross-flow cooling towers is the effectiveness—NTU method adopted by Jaber and Webb (1989) where it can be used for both the counterflow and cross-flow variants while simultaneously being one dimensional. Since then the analysis on various types of cooling towers ranging from counterflow to cross-flow had been widely researched and was heavily based upon Merkel's theory with the counterflow variant gaining more attention as compared to the other types of wet cooling tower (Hajidavalloo et al., 2010). The most prominent difference between the counterflow and cross-flow cooling towers is the direction of the water flow, which is illustrated in Figs. 9.1 and 9.2 (Liu et al., 2017).

However, a wet cooling tower does not come without its own drawbacks. The utilization of wet cooling tower produces visible plume expelled through the cooling tower. Under the right conditions, the plume may cause fogging or icing hazards to the surroundings. A more nocuous effect of using the wet cooling tower is the water consumption in large quantities, making the use of wet cooling towers in regions with low water supply problematic. Currently, more than 95% of thermal power plants in the USA employs wet cooling system which is accounted for the freshwater withdrawal of over 40% of the country's fresh water (EPRI, 2013).

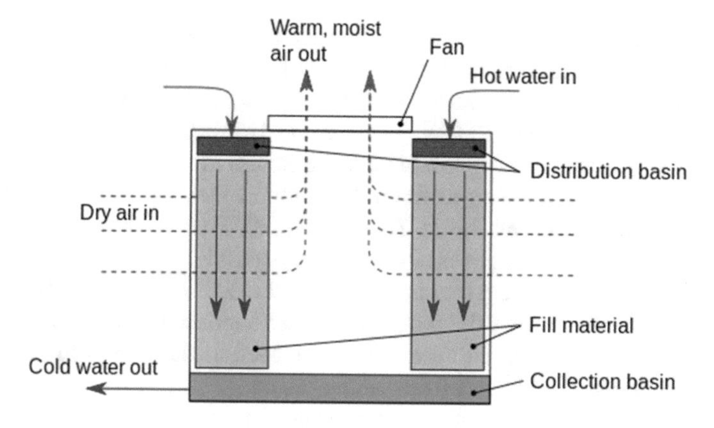


**Fig. 9.1** Schematic of cross-flow cooling tower

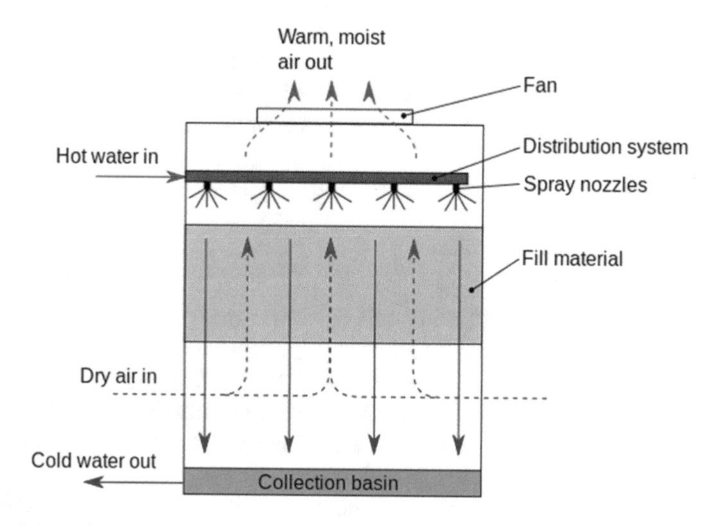


**Fig. 9.2** Schematic of counterflow cooling tower

## *Dry Cooling System*

Dry cooling towers differentiate from the wet cooling tower by separating the working fluid from ambient air such that there is no evaporation of the working fluid. Dry cooling towers are less preferred alternatives to the wet cooling systems due to their ability to handle higher heat loads (Naik & Muthukumar, 2017). However, the major drawbacks discussed in the previous section may outweigh its benefits, making the dry alternative more attractive in terms of its reliability and low resource consumption. With water being an invaluable resource, it may be more beneficial for power plants to switch to dry cooling, where the water consumption is much lower if not entirely unrequired. Nonetheless, air-dry bulb temperature is always higher compared to the wet bulb temperature. Consequently, a dry cooling system is almost always larger in size as compared to the wet cooling towers (Arie et al., 2017).

Most steam power plants operating in dry areas utilize the dry cooling towers as their means to expel waste heat (Seifi et al., 2017). However, the performance of dry cooling tower is sensitively dependent upon the environment conditions, specifically the wind condition of the surroundings, with report showing the total generation capacity by up to 40% (Ding, 1992). Zhai and Fu (2006) stated that the conventional design of dry cooling towers does not adequately consider the impact of wind upon the efficiency. In their previous research, Fu and Zhai (2001) showed that heat transfer patterns and air flow start to exhibit different behaviour once the wind speed is larger than 10 m/s. As a result, recent researches have been focused on introducing wind breakers to mitigate the effects of surrounding wind.

Common dry cooling towers are differentiated through their air flow generation method, mainly by mechanical draught or natural draft. Dry cooling tower running on natural draft is the more common option as it is more environmental friendly due to the fact that they consume little to no water while air flow is generated through natural draft without the need for mechanical parts that would otherwise require maintenance. The aforementioned natural draft is created through convection that occurs naturally due to pressure difference in the cooling tower.

## Cold Inflow

Cold inflow or flow reversal is a commonly acknowledged problem in the NDCs, and it is a phenomenon where the cooling towers experience unstable flow with external airflow, leading to undesirable downdraft (Zhai & Fu, 2006). Cold inflow through the top exit of the NDC will lower the overall chimney efficiency and increase the pressure loss (Andreozzi et al., 2009). Figure 9.3 is a simple yet adequate illustration for the visualization of cold inflow in a chimney.

Jörg and Scorer (1967) successfully demonstrated that cold inflow or flow reversal from the chimney exit is able to occur even in a still ambience. They also found that the exit dimension of the chimney is largely responsible for the cold inflow of air into the cooling tower, causing a decrease in buoyancy force within the chimney. They also discovered that the cold air will almost certainly be reaching the bottom of the cooling tower if the reverse flow of air was to penetrate below the neck of the chimney. This is due the decreasing velocity as it flows downwards or that the dead air region at the bottom would cause the cold air to sink.

Andreozzi et al. (2010) also conducted a study on the effects of cold inflow of the NDC and concluded that the reversal of air flow in the chimney is dependent on several factors the Reynolds number, Froude number, chimney height to diameter ratio, upstream velocity profile, heat transfer through the chimney wall and wall roughness.

In a research by Khanal and Lei (2012), they quantitatively investigated the effect of flow reversal upon the exiting mass flow rate of hot air. They proposed a novel

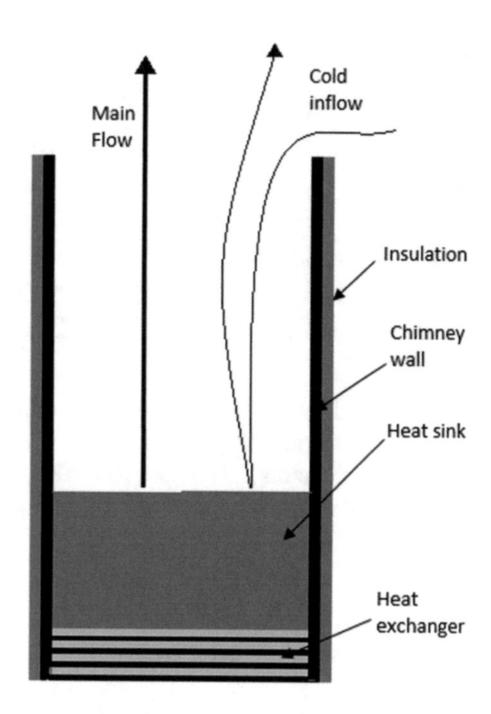


**Fig. 9.3** Illustration of cold inflow. *Source* Fisher et al. (1997)

concept of an inclined passive wall chimney. As supposed to a conventional chimney, their proposed concept is a more superior solution in enhancing the ventilation efficiency by reducing the flow reversal of cold air into the chimney. They concluded that the penetration depth of the reverse flow is dependent on the Rayleigh number, and the inclined passive walled design is found to be applicable to chimney designs with high Rayleigh number.

Li et al. (2017) tested the effects of cold inflow in an actual sized natural draft dry cooling tower for small power plants which is 20 m in height. The cooling tower is experimented in different ambient air conditions. Cold air incursion from the top exit of the tower is repeatedly observed, causing the air temperature inside the tower to decrease significantly. Further analysis found that the cold air incursion decreases the driving force of the tower while forming unwanted flow resistance for the upward airflow passing through the heat exchanger. This decrease in cooling tower performance is correlated with the reciprocal of a densimetric Froude number, $\left(\frac{1}{\mathrm{Fr}_D}\right)$ based on the cooling tower outlet diameter.

$$\frac{1}{\mathrm{Fr}_D} = \frac{(\rho_a - \rho)g}{\rho v^2}$$

Research to lessen the impact of cold inflow on the NDC has also been conducted by Chu et al. (2012). Their proposed solution is to install a layer of wire mesh screen at the top exit of the chimney to lessen the impact of cold inflow. In their model, when

**Table 9.1** Findings from previous research regarding cold inflow

| Author(s) | Year | Findings regarding cold inflow |
|---|---|---|
| Li et al. | 2017 | Flow reversal is correlated with the reciprocal of densimetric Froude number of chimney outlet diameter |
| Chu et al. | 2012 | Effects of cold inflow can be mitigated with wire mesh protection |
| Khanal et al. | 2012 | Flow reversal is dependent on Rayleigh number |
| Andreozzi et al. | 2010 | Flow reversal is dependent on Reynolds number, Froude number, chimney dimensions and flow velocity |
| Jörg et al. | 1967 | Chimney dimension is largely responsible for the cold inflow of air |

not protected by wire mesh, the solid chimney height was impaired by 90%. The impair in performance dropped to only 60% when done with wire mesh installed. However, the authors state that the effects of flow reversal in practice might not have been as determental as result suggested since the order of magnitude of Froude number is three times lesser than normal (Table 1).

## Natural Convection Process

Natural convection occurs when a hot surface comes in contact with cooler ambient air. Heat transfer occurs between the hot surface and the cold ambient air, and the heated air becomes lighter as hotter air is less dense than colder air. The temperature difference between the hotter air around the hot surface and the colder air above causes a density difference and thus creates a buoyancy force between the air above and below. The heated air around the hot surface would rise, and ambient colder air from the sides would continuously replace the heated air (Çengel & Ghajar, 2015a, 2015b).

In a NDC, the density difference of the air around the heat exchanger and the ambient air causes a pressure difference which would cause air to flow upwards. If heat is continuously provided by the heat exchanger, natural convection process would occur indefinitely. This continuous heat transfer would cause cold air to be heated, rising upwards, producing a steady flow of air over the heat exchanger as long as cold ambient air is available. Hence, natural convection process would occur and continue without ceasing as long as the air temperature around the heat exchanger is higher than ambient air and that the upwards flow of air remains steady.

## Natural Draft Chimney

A NDC typically runs without the need of mechanical parts and is self-sufficient due to the use of natural convection to generate air flow. The NDC is considered to be a valuable solution mainly due to the growing environmental concern of the public. Besides, the NDC is able to operate quietly while having an extensive longevity compared with other types of cooling tower because of the lack of mechanical parts that would otherwise generate additional noise and heat while simultaneously requiring constant maintenance (Chu et al., 2012). The NDC's main driving force is the buoyant force formed from the density difference between the outside (ambient air) and the inside (hot air) air, taking full advantage of the force to generate airflow within the tower shell instead of needing dozens of axial fans (Kong et al., 2017).

Over recent years, extensive research has been done on the NDC to improve the efficiency in order to become a better replacement for the less environmental-friendly wet cooling tower. Chu et al. (2015) introduced a wire mesh screen at the exit of the chimney to act as a flow resistor to impede to cold inflow or flow reversal into the chimney from the exit. They concluded that the heat discharge rate of the cooling tower increased by up to 34%. Through geometrical manipulation, researchers are able to optimize the configuration of the NDC, and these researches are mainly applicable in the electronics industry (Auletta & Manca, 2002). Bouchair (1994) worked on the geometrical configuration to optimize the solar chimney effect.

Smrekar et al. (2006) experimented on optimizing the heat transfer along the cooling tower packing utilizing a suitable water distribution across the plane area within the chimney. His result is shown in Fig. 9.4 where it is able to provide a means of evaluating the degree to which the water droplets and the cooling air are mixed in the cooling tower.

Many researchers have acknowledged the adverse effects that crosswind or airflow outside of the cooling tower has on the overall performance of the NDC. Zhai and Fu (2006) performed numerical and experimental study of turbulent flow around the dry cooling tower. Their researched showed that installing wind breakers in a way so that the lateral sections of the cooling towers are vertical to the direction of the airflow would be able to recover the decreased capacity by 50%. Seifi et al. (2017) studied the effects of utilizing external wind breakers to mitigate the external airflow's effects on

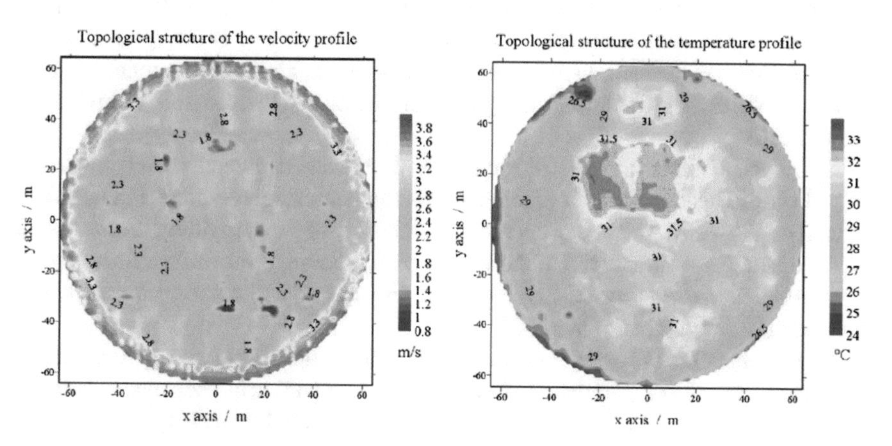


Fig. 9.4 Topological structure of velocity and temperature at exit. *Source* Smrekar et al. (2006)

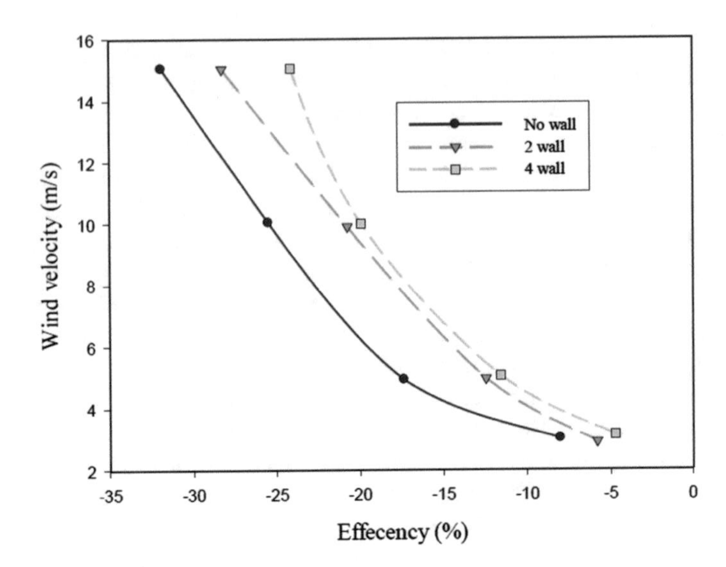


**Fig. 9.5** Variation of efficiency of cooling tower with varying numbers of windbreaker. *Source* Seifi et al. (2017)

the performance of the cooling tower and found that the introduction of windbreakers is able to decrease the wind speed around the chimney while increasing efficiency, as shown in Fig. 9.5.

## Heat Exchanger

Compact heat exchanger is used extensively in various applications, ranging from cooling devices to waste heat recovery systems (Shah et al., 1980). To achieve the goal of increasing surface area density for compact heat exchangers, they are often designed with small diameter passages; however, this would eventually lead to blockage of the channel due to fouling and thus increasing this pressure drop (Kurian et al., 2016).

To induce a higher rate of heat exchange, researchers had manipulated the geometry of various heat exchangers to achieve a greater heat exchange effect. Fins attached to wall of heat exchangers may encourage heat exchange by providing more area for heat exchange and also separate the working and cooling fluid while enhancing the flow turbulence (Fu et al., 2017). However, the cost of maintenance for these fins geometries is high (Kim et al., 2000).

Heat exchangers of varying geometries were continuously studied by researchers. Boomsma et al. (2003) studied and compared a compact heat exchanger equipped with compressed metal foam with a commercially available heat exchanger using

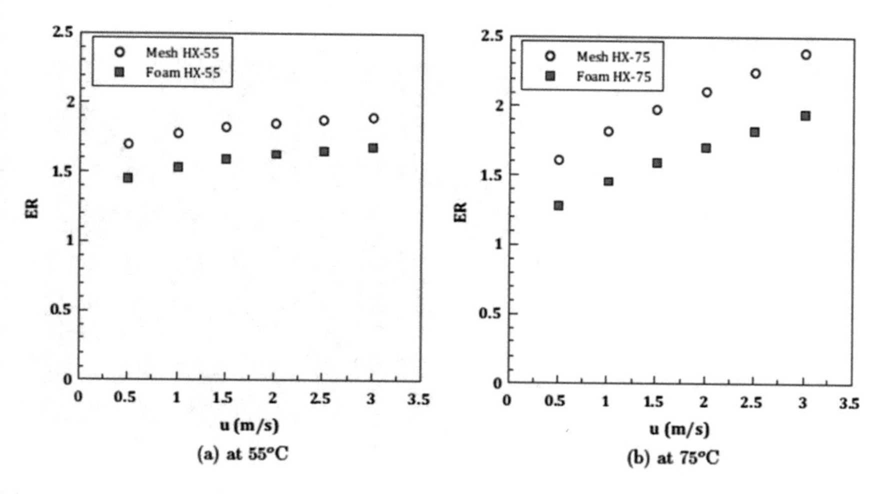


**Fig. 9.6** Variation of enhancement ratio with velocity for metal foam and wire mesh. *Source* Kurian et al. (2016)

identical heat transfer design application and found that the thermal resistance generated by the metal foam compact heat exchanger is two to three times lower than the commercially available compact heat exchanger.

Kurian et al. (2016) research the effectiveness of metal wire mesh screen used as a heat exchanger by comparing its performance with metal foam heat exchangers. They found that metal wire mesh screen heat exchangers showed significant improvement in performance than a metal foam heat exchanger under identical conditions. The enhancement ratio for both metal foam and wire mesh experiments is shown in Fig. 9.6.

## Wire Mesh Screen

Wire mesh screen is the main focus of this research where the pressure drop that it induces when air flows through the wire mesh screen is able to increase the speed of the convection flow within the chimney. This decrease in pressure occurs when the fluid flow passes the wire, forming a boundary layer behind the wire, thus decreasing the flow area and increasing flow velocity and pressure. A research published in 2015 showed that when multiple wire mesh screens of identical parameters are stacked together, the pressure drop that they caused when fluid flows across them can be calculated with high precision (Sun et al., 2015). The research states that the pressure drop when fluid flows through a series of wire mesh screens stacked together is related to parameters of the wire mesh such as the stack distance, $\sigma$; diameter of wire, d; vertical and horizontal separation between wires, h and w, respectively. All

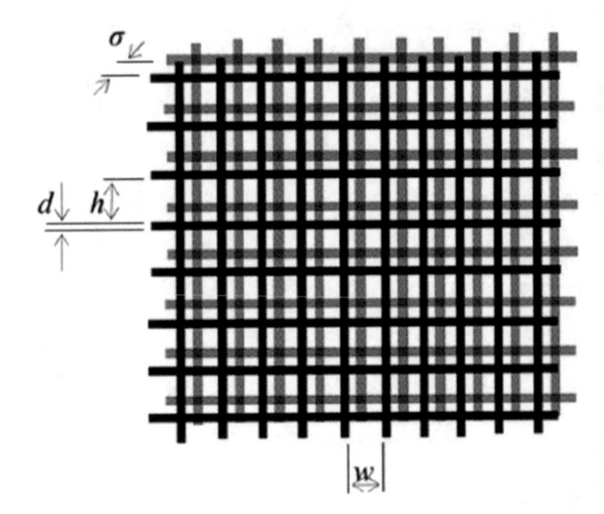


**Fig. 9.7** Wire mesh screen theoretical model

the affecting parameters are represented in Fig. 9.7, while the pressure drop can be seen in Fig. 9.8 as the flow passes through a series of wire mesh screen, the pressure decreases significantly.

A wire mesh screen, when installed at the exit of a chimney, is capable of impeding the backflow of air when used in a NDC. Chu et al. (2015) demonstrated this effect by first simulating the flow reversal phenomenon at the exit of the chimneys that is operating under natural convection. Wire mesh screen is then simulated as flow resistor in CFD and was found to be able to impede flow reversal.

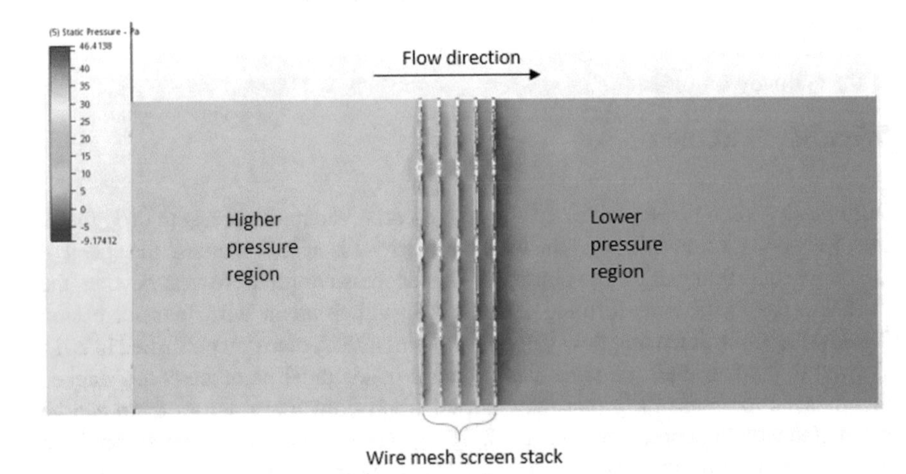


**Fig. 9.8** Flow across five layers wire mesh screen

From the research result of Kurian et al. (2016) and Chu et al. (2015), installing wire mesh screen onto the exit of the NDC may be able to simultaneously encourage heat exchange and act as a flow resistor for flow reversal.

## Summary

Hence, in order to conduct an experimental study for the improvement of the efficiency of the NDC, the following conclusions may be drawn.

The amount of thermal power plants in the world is exceeding high; therefore, a reliable and environmentally friendly cooling system is in dire need, especially in areas drier areas.

Malaysia is a country with an above average amount of thermal power plant running, compared to the world; hence, any efficiency improvement to the NDC would have a big impact to the country's power generation and economy.

Over the years, researchers have studied to improve the efficiency of compact heat exchangers by manipulating its geometry, mainly to increase the contact surface area and to increase flow turbulence. It is concluded that a wire mesh screen is by far the most efficient improvement to heat exchangers in the two respect. Since a heat exchanger is a very pivotal part of the NDC, the efficiency of the chimney heavily relies on the performance of the heat exchanger.

Countless research has been done upon the NDC, and the most common problem faced by the chimney is the effects of airflow outside the cooling tower, mainly cold inflow or flow reversal. Cold inflow is an undesirable effect where cold air would flow into the chimney from its exit, hindering the performance and lowering the efficiency.

In this research, the main focus is on implementing wire mesh screen at the exit of the chimney for two purposes as follows:

- To induce higher heat exchange rate in the NDC in order to increase the overall efficiency.
- To act as a flow resistor to impede the flow reversal of ambient cold air through the chimney exit.

## Overview

This research started with the analysis of problems of the topic and then followed by the construction of the design procedure. The design procedure consists of the CAD design of the model using SOLIDWORKS software. The design of the cooling tower model was based on the model used by Chu et al. (2015) in their research as a reference and also for result and data comparing and verification purposes.

After modelling the cooling tower, a series of simulations and modifications were done in succession to verify the credibility of Autodesk CFD as compared to other

CFD softwares, which was PHOENICS in this case. Once verified, further simulations may be carried out, specifically the mesh independence study to find out the most optimal mesh size for this software for this particular case. Results from the mesh independence study were then analysed by considering the accuracy of the result, amount of accuracy increment and time to complete, specifically in that order of decreasing priority.

With an optimum mesh size for the obtained, decisions were made by choosing the wire mesh screen with different parameters such as the pore size, wire diameter and the number of stacks. These parameters are found theoretically to be able to cause a pressure drop of the fluid flow across the wire mesh screen. Simulations were conducted after the parameters were set, and data collected was compared with the Control.

With the data collected from the simulations, the data was then analysed and discussed for each varying behaviour between each sets of simulations.

The processes of this project follow the task flow chart presented in Fig. 9.9.

## Design Tools

### *SOLIDWORKS*

In the design of the NDC, the CAD software, SOLIDWORKS was used for the designing and visualization of the model prior to running any simulation in another software, namely Autodesk CFD 2018.

Within SOLIDWORKS, the design started by creating each "Part" of the model separately, "Part" in the design would mean different part of the cooling tower, which includes the chimney wall, wire mesh screen and heat exchanger. The dimension of each "Part" is the design to be able to fit in an exact way with each other. Once each "Part" is completed, they will then be loaded to the assembly and be assembled into the completed NDC via the "Assemble" function. The main advantage of having each part designed separately is that their dimension or any other specifications may be altered individually at any point of the project, leaving the other parts unaffected. Moreover, any flaws in the design may be easily pinpointed if done so. SOLIDWORKS is also able to show the individual parts or the assembled cooling tower in two-dimensional view for more clarity in terms of the dimension and specification. The assembled model is then able to be exported to other software capable of running simulation such as SOLIDWORKS Flow simulation.

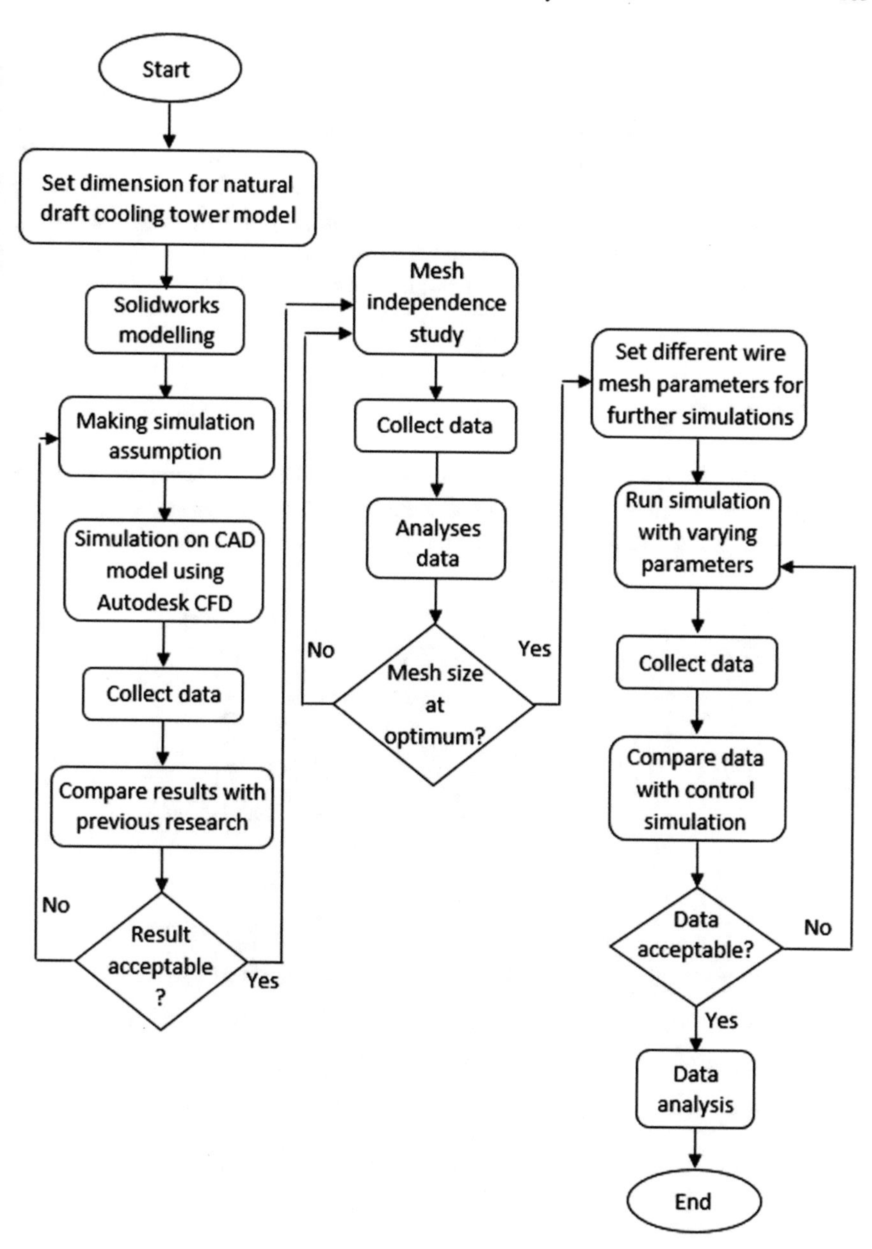


Fig. 9.9 Task flow chart

## Model Design

The model designed was to be used to demonstrate the change in heat discharge rate from the assigned heat source within the NDC using CFD software Autodesk CFD 2018. To ensure that the data obtained was accurate and not skewed due to poorly designed shapes or cross-current flow over the cooling tower, the complexity of the chimney's design was kept to the minimal. The complete chimney model with wire mesh screen and heat exchanger assembled is shown in Fig. 9.10.

### *Cooling Tower*

The cooling tower was a vertically straight cylindrical tube with 102 m outer diameter and 100 m inner diameter with a height of 200 m. The chimney was being designed as a straight vertical tube due to the fact that only the top and bottom exits of the chimney

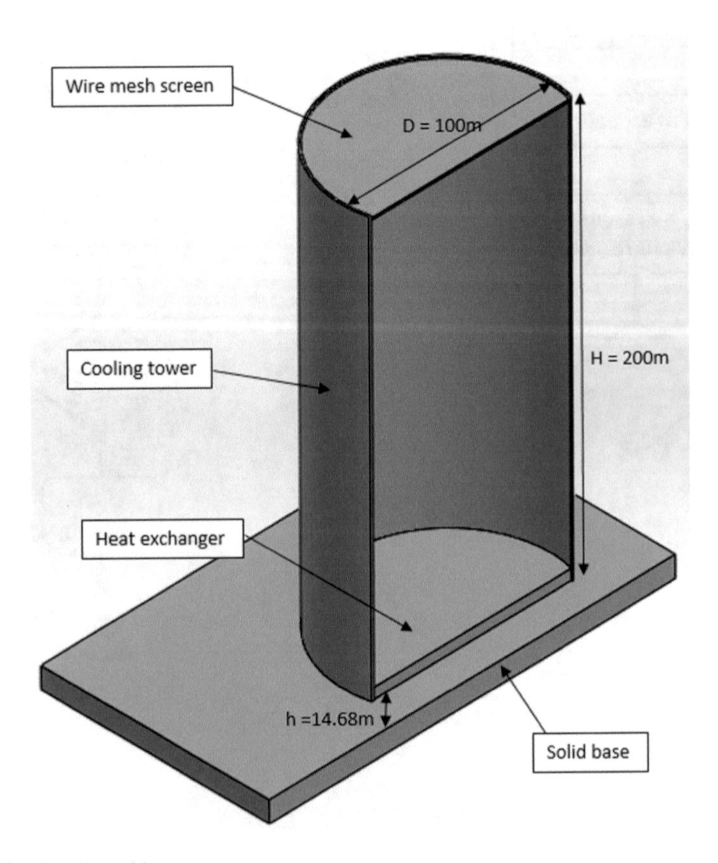


**Fig. 9.10** Complete chimney

are of interest during the simulation; therefore, any heat exchange enhancement effect should not be affected by the awkward shapes of the chimney.

The chimney is assumed to be supported at a height of 14.68 m as it is the highest ground-to-chimney height of any NDC (Busch et al., 2002). The supposed supporting columns are left out in the design to reduce the unwanted complexity of the model for CFD simulation. The ground below the NDC is solid base of 200 m square area. During simulation, the surrounding environment is set to be under atmospheric pressure with the ambient air temperature at 30 °C.

## Heat Exchanger

A cylindrical heating source is set at the 0.5 m above the bottom inlet of the cooling tower. It is a porous and frictionless heating source that is of 5 m in height and has a circular diameter of 100 m, which is an exact fit to the inner diameter of the cooling tower itself.

## Wire Mesh Screen

The wire mesh used for the simulation is modelled as a circular plate not unlike the shape of the previously mentioned heat exchanger. The wire mesh screen has a thickness of 1 m and is porous and causes a pressure drop to the fluid flowing through the wire mesh. The reason behind the simplification of the model is that any wire mesh is able to be represented by a region of pressure drop. With the reducing of model complexity, simulation time can also be reduced.

The research done by Sun et al. (2015) is able to relate the wire mesh parameters with the pressure drop caused. The relationship between the parameters and the pressure drop may be represented by Eq. (9.1)

$$\Delta P = 10n \cdot \mathrm{Re}_d^{0.77} \cdot \mathrm{Re}_\sigma^{-0.09} \cdot \mathrm{Re}_l^{-1.03} \cdot \frac{\rho v^2}{2} \tag{9.1}$$

where $n$ is the number of wire mesh layers, $\rho$ is the fluid density, and $v$ is the fluid velocity. $\mathrm{Re}_d$, $\mathrm{Re}_\sigma$ and $\mathrm{Re}_l$ are the Reynolds numbers computed using the wire diameter $d$, layer spacing $\sigma$ and wire mesh sizing $l$, respectively.

$$\mathrm{Re}_d = \frac{\rho V d}{\mu};\quad \mathrm{Re}_\sigma = \frac{\rho V \sigma}{\mu};\quad \mathrm{Re}_l = \frac{\rho V l}{\mu} \tag{9.2}$$

where $\mu$ is the dynamic viscosity and the wire mesh sizing $l$ is a function of vertical and horizontal separation of wire $h$ and $w$, respectively, and the wire diameter, $d$.

**Table 9.2** Wire mesh screen configuration specifics

| Configuration | Number of layers, $n$ | Wire diameter, $d$ (mm) | Separation between wires, $w$ (mm) | Theoretical pressure drop, $\Delta P$ (Pa) |
|---|---|---|---|---|
| Control | N/A | N/A | N/A | N/A |
| 1 | 5 | 0.5 | 0.01 | 14.55 |
| 2 | 5 | 0.5 | 0.005 | 30.46 |
| 3 | 5 | 5 | 0.01 | 105.11 |
| 4 | 20 | 0.5 | 0.01 | 58.20 |

$$l = \frac{h \cdot w}{h + w + d} \tag{9.3}$$

In this research, the vertical and horizontal separations between the wires were chosen to be identical; therefore, $h$ is equalled to $w$.

Equation (9.3) becomes

$$l = \frac{w^2}{2w + d} \tag{9.4}$$

The simulation of the NDC is done with two modes: Mode 0 where the model is simulated within still air environment and Mode 1 where the model being simulated in crosswind condition. Each mode had one Control simulation and four configurations; the Control is set to have the model simulated with no wire mesh screen installed. The four configurations, Configurations 1–4, will have wire mesh screens of varying pressure drop be installed to the top exit of the chimney. Each configuration will have a different parameter being altered and is summarized in Table 9.2.

A model with five layers of wire mesh stacked was modelled and simulated in the CFD software to verify the homogeneity of the researchers' equation with Autodesk CFD.

## Conservation Equations

The governing equations for the simulation are Reynolds-averaged Navier–Stokes equation. Assuming that it is at a steady state, with two-dimensional axisymmetric flow that is with zero tangential velocity, the two-dimensional single-phased governing equations should be adequate to describe the model.

Continuity:

$$\frac{1}{r}\frac{\partial(\rho r v)}{\partial r} + \frac{\partial(\rho w)}{\partial z} = 0 \tag{9.5}$$

Radial momentum:

$$\begin{aligned}\frac{1}{r}\frac{\partial\left(\rho r v^2\right)}{\partial r}+\frac{\partial(\rho w v)}{\partial z}&=\frac{1}{r}\frac{\partial}{\partial r}\left(r\rho v_e\frac{\partial v}{\partial r}\right)+\frac{\partial}{\partial z}\left(\rho v_e\frac{\partial v}{\partial z}\right)-\frac{\partial p}{\partial r}\\&+\frac{1}{r}\frac{\partial}{\partial r}\left(r\rho v_e\frac{\partial v}{\partial r}\right)+\frac{\partial}{\partial z}\left(\rho v_e\frac{\partial w}{\partial r}\right)\\&-2\rho v_e\frac{v}{r^2}-\frac{2}{3}\frac{1}{r}\frac{\partial}{\partial r}(r\rho v_e[\nabla\cdot U])\end{aligned}\tag{9.6}$$

Axial momentum:

$$\begin{aligned}\frac{1}{r}\frac{\partial(\rho r v w)}{\partial r}+\frac{\partial\left(\rho w^2\right)}{\partial z}&=\frac{1}{r}\frac{\partial}{\partial r}\left(r\rho v_e\frac{\partial w}{\partial r}\right)\\&+\frac{\partial}{\partial z}\left(\rho v_e\frac{\partial w}{\partial z}\right)-\frac{\partial p}{\partial z}+(\rho_r-\rho)g+\frac{1}{r}\frac{\partial}{\partial r}\left(r\rho v_e\frac{\partial v}{\partial z}\right)\\&+\frac{\partial}{\partial z}\left(\rho v_e\frac{\partial w}{\partial z}\right)-\frac{2}{3}\frac{\partial}{\partial z}(\rho v_e[\nabla\cdot U])\end{aligned}\tag{9.7}$$

Energy equation:

$$\frac{1}{r}\frac{\partial}{\partial r}(\rho r c_p T)+\frac{\partial}{\partial z}(\rho w c_p T)=\frac{1}{r}\frac{\partial}{\partial r}\left(r k_e\frac{\partial T}{\partial r}\right)+\frac{\partial}{\partial z}\left(k_e\frac{\partial T}{\partial z}\right)\tag{9.8}$$

## CFD Modelling

### *Simulation Assumption*

Natural convection within a NDC is difficult to under natural conditions; therefore, few assumptions were made to ensure that the flow simulation is affected by only the necessary variables. The assumptions made in the CFD modelling were as follows:

i. The heat transfer between the air and heat exchanger was steady throughout.
ii. The heat transfer occurred simultaneously.
iii. The chimney wall was adiabatic.
iv. The wire mesh screen is modelled as a porous plate with pressure drop formula represented by Eq. (9.1).

## *Solution Domain and Boundary Conditions*

The solution domain for the simulation of the Control and all configurations was set to a cuboid of 200 m square area base with height of 400 m. The chimney is located at the centre of the solution domain. The entire solution domain's material is set to be air (variable) where variable in Autodesk CFD means that the properties of air, such as density and pressure, will change according to the surrounding conditions, namely heat in this case. The boundary condition for the surfaces on the sides of the solution domain is set to 30 °C in temperature and zero Pa in gauge pressure as the surrounding air of the model is assumed to be 30 °C and at atmospheric pressure of 1 atm.

The chimney body and the solid base below the solution domain are set to be made of concrete; however, since the wall and base are assumed to be adiabatic, the material is suppressed. The suppression of the material will disable the meshing of those regions when assigning meshes to the model. This is done to disallow any heat transfer between and through these regions. Since these regions are considered adiabatic, no boundary conditions are assigned to them.

The heat exchanger in all the simulations was assumed and modelled to be a 5-m-thick porous cylindrical plate with no pressure drop across the heat exchanger and the heat exchange occurred instantaneously. As such, the cylindrical volume that represents the heat exchanger has its material set to being air and the boundary condition of the top surface is set to be 50 °C, and this is done to ensure that the air that passes through the heat exchanger is subjected to an instantaneous heat exchange that raises the air temperature to 50 °C.

For the simulations upon the chimney models with wire mesh screen installed (i.e. configurations 1–4), it has its solution domain and boundary condition identical to the Control. The only difference is the addition of the wire mesh screen. Similar to the heat exchanger, the wire mesh screen is modelled as a porous cylindrical volume of 1 m thick. However, the material for this volume is set to resistance where a resistance to a flow can be established across the flow through the material. In this case, a resistance in the opposite direction of the air flow is assigned and the material would cause a pressure drop in the flowing fluid. The change in pressure across the resistance material, $\Delta P$ is affected by the velocity of the fluid, $v$ and loss coefficient, $K$ since the density of the fluid is not changing drastically to cause a significant effect. According to the Autodesk CFD software, the pressure loss due to the resistance of the material is governed by the following equation:

$$\Delta P = K\rho\frac{v^2}{2} \tag{9.9}$$

The calculation for the supposed pressure drop due to the geometry of the wire mesh screen was calculated using Eq. (9.1). Due to the homogeneity of Eqs. (9.1) and (9.9), the loss coefficient due to the wire mesh screen, $K$, is

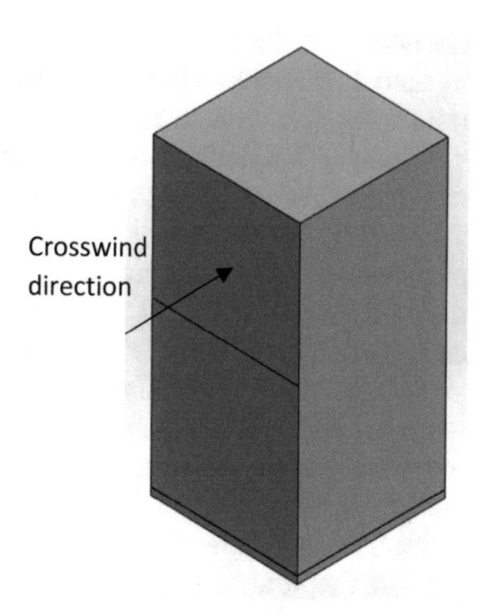


**Fig. 9.11** Crosswind boundary condition

$$K = 10n \cdot \mathrm{Re}_d^{0.77} \cdot \mathrm{Re}_\sigma^{-0.09} \cdot \mathrm{Re}_l^{-1.03} \tag{9.10}$$

During the simulations of Mode 1, where the crosswind was to be introduced, an air flow of 10 m/s in the negative $z$-direction (perpendicular to the chimney exit plane) was applied at the one of the upper and outer solution domain surface. This is done to simulate the crosswind condition of actual natural draft chimney under unfavourable weather conditions. Figure 9.11 vividly illustrates the crosswind boundary condition direction and region within the model.

The solution mode for all the configurations was set to steady state and is considered turbulent for the entire simulation span due to the chimney having high Reynolds number of $3.27 \times 10^7$. The turbulence model chosen for all the configurations was the K-epsilon model.

## Simulation Strategy and Computation

In this section, the steps taken in applying the mesh size of the model were discussed, and the strategy on deciding the final and most efficient mesh size was explained.

### *Computational Mesh Sizing*

For meshing the model, Autodesk CFD is equipped with an automatic sizing function where a comprehensive topological interrogation of the analysis geometry determines the mesh size and distribution on all edges, surfaces and volumes. However, this function only serves to provide the model with a basic mesh and further refinements are needed if the accuracy of the result is to be improved. Numerous refinement method is available within the software, and specifically in this research, surface refinement, gap refinement and regional refinement were used.

Meshing the model using automatic sizing provides the model with the fundamental meshes needed for a basic flow simulation to be performed. The default mesh provided by this function is basically described to be coarse; however, this function may also be tweaked to increase the total number of elements and nodes, thus refining the entire mesh uniformly. This method of refining is functional yet taxing on the computer, increasing computational time.

Surface and gap refinements are refinement methods that could be employed when automatic sizing is applied, where the former refines the mesh by increasing the number of layers of mesh on the wall, while the latter refines and decreases the gap size between the mesh. This method provides more Control to the refinement; however, it also refines the entire model uniformly without emphasizing any particular region.

The most efficient refinement method is the regional meshing method, where the refinement regions are targeted and may be non-uniform. In this method, regions for refinements are chosen using cuboid, spherical or cylindrical region to highlight the targeted regions for refinement and are able to decrease the mesh size by the specified amount. This method is prominently used in this research to refinement only regions connected to the exits of the chimney where these regions are more impactful to the simulation.

Figure 9.12 illustrates the model's meshing after applying all the aforementioned refinements where the model is subjected to automatic sizing with default mesh size of 1 and is then refined with surface and gap refinement. After the non-targeted refinements, cylindrical regional refinement was applied to the top and bottom exits of the chimney. The top exit had its original local mesh size of 2500, decreased to 300 and the bottom exit from 2500 to 400. The model's mesh after refinements.

## *Mesh Sensitivity Analysis*

In the research, mesh sensitivity analysis was carried out upon Mode 0's Control simulation. The analysis was done to efficiently improve the mesh size of the model without performing unnecessary refinements towards the model. The procedure of the mesh sensitivity analysis taken is as follows:

i. A coarse mesh sizing was done by selecting the automatic sizing function while the results were then compared to the result of Chu et al. (2016).
ii. Each mesh size, after automatic sizing was subjected to regional refinements where to local mesh of the top and bottom exit, was refined.
iii. The same model with identical boundary conditions was cloned, the mesh size was reduced to 0.8 of the previous mesh size, and the results were compared with the same set of data.

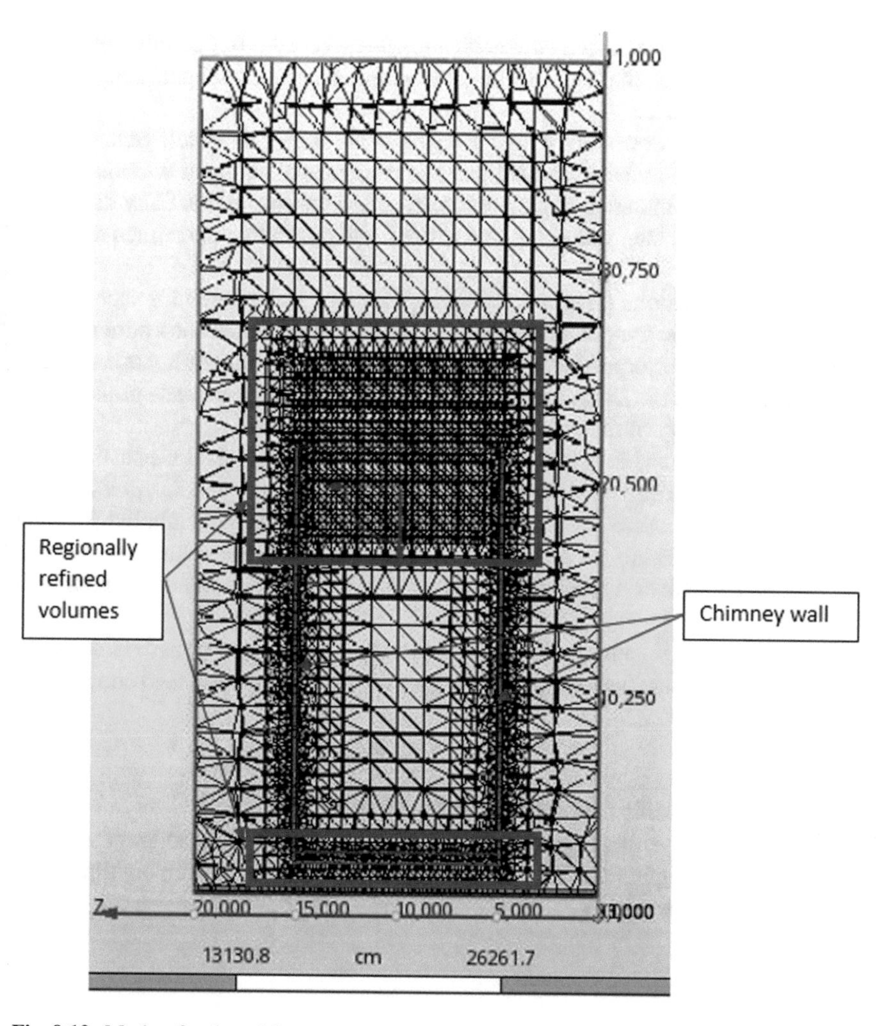


**Fig. 9.12** Mesh refined model

iv. Any improvement over the first set of simulation was recorded.
v. Steps i–iv were repeated until the improvement became lower than 10% before the analysis is completed.

## Effects of Varying Wire Mesh Screen Parameters

After the completion of model validation, further studies upon the presence of wire mesh screen and its varying parameter were simulated and studied. Velocity collected from the Control simulation of Mode 0 was used to generate the loss coefficient, *K*, according to the wire mesh screen specification using Eq. (9.10). The value generated was then assigned to the each of the configurations, from configurations 1 to 4, according to Table 9.2.

All simulations were solved under steady state while the intelligent solution Control was enabled where the automatic convergence assessment was enabled as well to allow the simulation to stop automatically once converged. Only data from the final iteration or step was saved due to the simulation being performed at steady state.

Once the simulations were completed, an YZ plane was created for each configuration, showing the temperature profile of the fluid flow across the entire solution domain. Velocity vector arrows were added to the plane to further illustrate the fluid flow of the simulation. The amount of arrows were set to fine and made proportional to the actual velocity of the corresponding region.

Another plane is added to the model, set to an XZ plane and positioned at the outlet of the chimney. The bulk data for mass flow rate, outlet *y*-axis velocity, pressure and temperature was gathered at the chimney exit. The bulk data calculation function provided by the software calculates the area averaged data of the selected parameters and exports the data as a comma separated value (CSV) file, readable by Microsoft Excel.

Mass flow rates, $\dot{m}$, and change in temperature, $\Delta T$, obtained from the simulation were used to calculate the heat gained by air of the chimney using the formula

$$Q = \dot{m} \cdot C_p \cdot \Delta T \tag{9.11}$$

where $C_p$ is the specific heat capacity of dry air at 30 °C.

Data from each configurations was being analysed both isolated and compared amongst each configuration and Control. Any points of interest were highlighted in Chap. 4 and discussed.

## Overview

Simulation results from the mesh sensitivity analysis, both modes with their respective Control and configurations, were presented and discussed in this chapter.

## Mesh Sensitivity Study

As mentioned in Chap. 3, a mesh sensitivity analysis was studied using the Control simulation from Mode 0 where the data obtained was compared with previous research (Chu et al., 2016). From the Control simulation, the collected data was mass flow rate, outlet velocity, temperature change and heat gained by air. With the

**Table 9.3** Number of elements and nodes

| Iterations | Number of elements | Number of nodes |
|---|---|---|
| 1 | 72,897 | 291,590 |
| 2 | 116,636 | 485,984 |
| 3 | 210,593 | 809,974 |
| 4 | 299,325 | 1,349,958 |
| 5 | 642,579 | 2,294,928 |

**Table 9.4** Mass flow rate from mesh sensitivity study

| Iterations | Mass flow rate (kg/s) | | Percentage error (%) |
|---|---|---|---|
| | Simulation | Verification | |
| 1 | 65,397 | 54,740 | 19.4684 |
| 2 | 59,632 | 54,740 | 8.93679 |
| 3 | 57,671 | 54,740 | 5.3544 |
| 4 | 55,964 | 54,740 | 2.23602 |

mesh size starting from 1, which was identified as coarse meshing size, each iteration of subsequent simulation decreases the mesh size by 0.2 which yields four iterations where the fifth iteration was neglected due to the sharp increase in number of elements and nodes to an absurdly high amount. Table 9.3 shows the parameters collected from the mesh sensitivity study (Tables 9.4, 9.5, 9.6 and 9.7; Figs. 9.13, 9.14, 9.15 and 9.16).

The data from the mesh sensitivity study shows that the mesh sensitivity starts to converge at the fourth iteration where the mesh size is at 0.4 before regional refinement. The reason for not performing the fifth iteration was due to the amount of

**Table 9.5** Outlet velocity from mesh sensitivity study

| Iterations | Outlet velocity (m/s) | | Percentage error (%) |
|---|---|---|---|
| | Simulation | Verification | |
| 1 | 7.12 | 6 | 18.6667 |
| 2 | 6.41 | 6 | 6.83333 |
| 3 | 6.33 | 6 | 5.5 |
| 4 | 6.12 | 6 | 2 |

**Table 9.6** Temperature change from mesh sensitivity study

| Iterations | Temperature change (°C) | | Percentage error (%) |
|---|---|---|---|
| | Simulation | Verification | |
| 1 | 19.65 | 16.81 | 16.8947 |
| 2 | 18.32 | 16.81 | 8.98275 |
| 3 | 17.38 | 16.81 | 3.39084 |
| 4 | 16.98 | 16.81 | 1.0113 |

Table 9.7 Heat gained by air from mesh sensitivity study

| Iterations | Heat gained by air (MW) | | Percentage error (%) |
|---|---|---|---|
| | Simulation | Verification | |
| 1 | 1.295331 | 0.927541 | 39.6522 |
| 2 | 1.101198 | 0.927541 | 18.7223 |
| 3 | 1.010341 | 0.927541 | 8.92685 |
| 4 | 0.957871 | 0.927541 | 3.26995 |

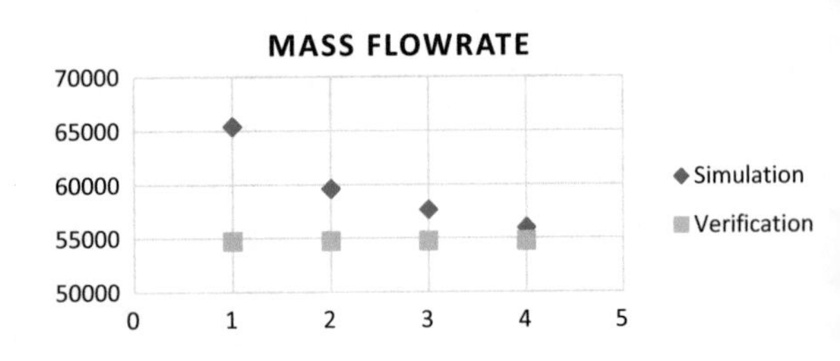


Fig. 9.13 Mass flow rate from mesh sensitivity study

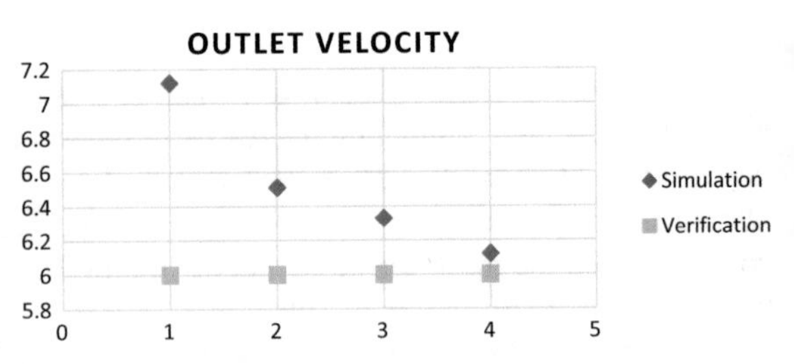


Fig. 9.14 Outlet velocity from mesh sensitivity study

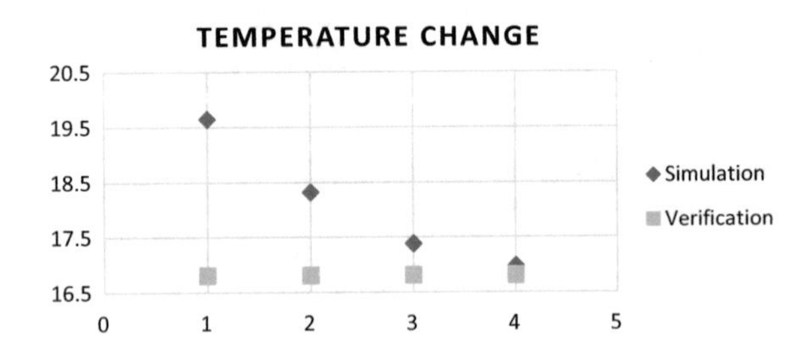


Fig. 9.15 Temperature change from mesh sensitivity study

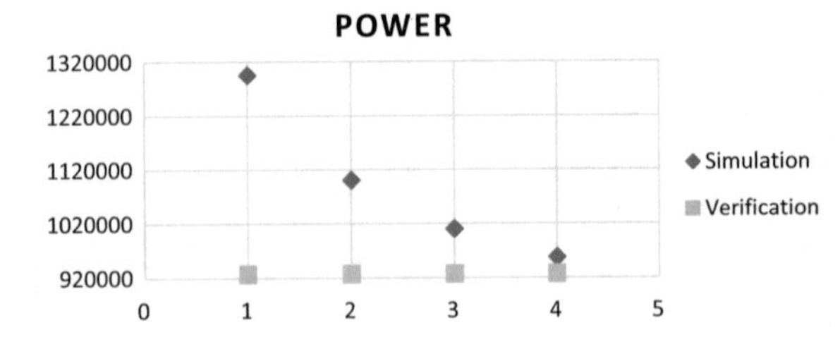


Fig. 9.16 Heat gained by air from mesh sensitivity study

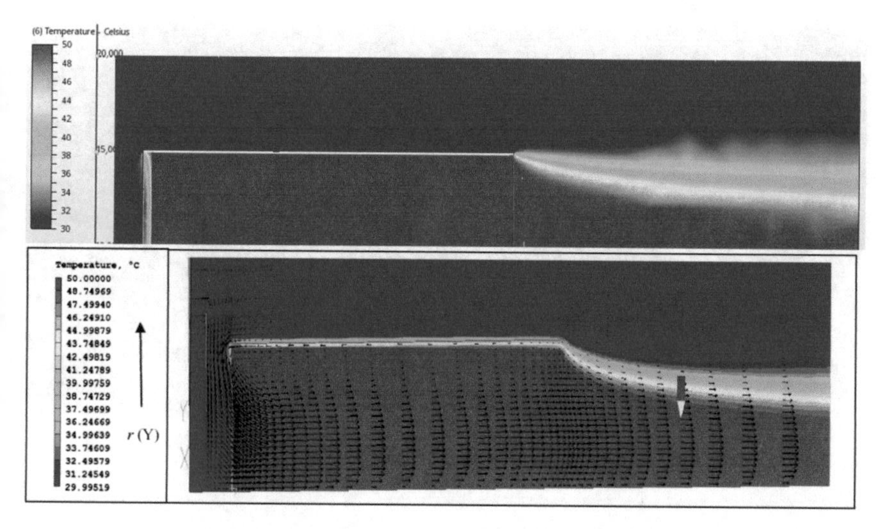


Fig. 9.17 Visual comparison between control simulation (top) and Chu et al. (2016) (bottom)

elements and nodes it possesses, which would require a very long time to simulate and was deemed too resource heavy to perform. Besides, during the fourth iteration, all the compared parameters had fallen below 10% error as compared to the verification data from Chu et al. (2016). Therefore, the fifth iteration was chosen to be neglected to save computational time and resources. Figure 9.17 shows the visual comparison between the simulation and verification of the simulation's temperature profile.

## Simulation Data

In this section, the simulation data obtained from the Control and all configurations was analysed and discussed based on their relevance and impact. Hypotheses were given to explain the abnormal flow behaviours caused by the presence or parameters of the wire mesh screen.

## Small-Scaled Wire Mesh Simulation

Before performing simulations with the chimneys where wire mesh screen was installed at the top exit, a small-scale simulation was performed upon the wire mesh screen to validate and to ensure that the wire mesh screen simulated using Autodesk CFD agrees with the findings of the research, Sun et al. (2015) who proposed the pressure drop formula, as given in Eq. (9.1).

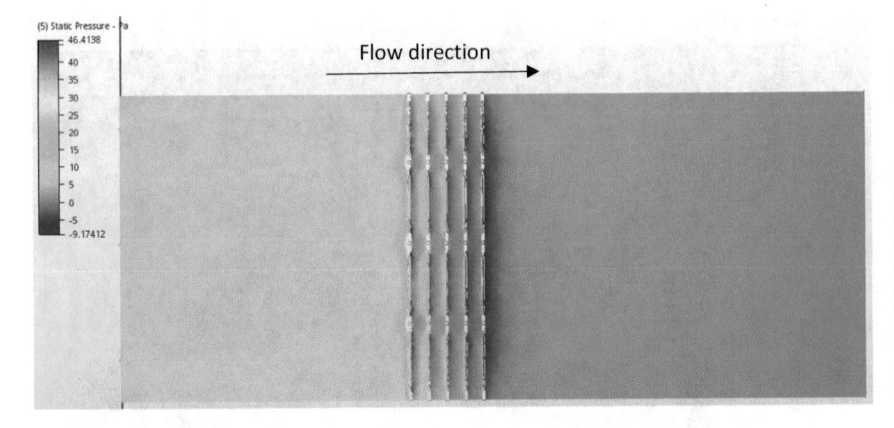


**Fig. 9.18** Pressure drop across wire mesh screen stacks

Figure 9.18 illustrates the simulation done to show the pressure drop across a five-layer wire mesh screen stack. The wire mesh screen has parameters of

i. Wire diameter, $d = 0.5$ mm
ii. Horizontal and vertical wire separation = 10 mm
iii. Number of stacks, $n = 5$
iv. Stack distance, $\sigma = 5$ mm.

Wire mesh screen stacks with such specifications would theoretically yield a pressure drop of 12 Pa with an inlet velocity of 6 m/s. Figure 9.19 is the plot of the pressure recorded across the entire pipe length.

The pressure drop across the wire mesh screen is calculated to be 11.63 Pa, which is only an 8% error compared to the theoretical value, thus validating the wire mesh screen's pressure drop capability.

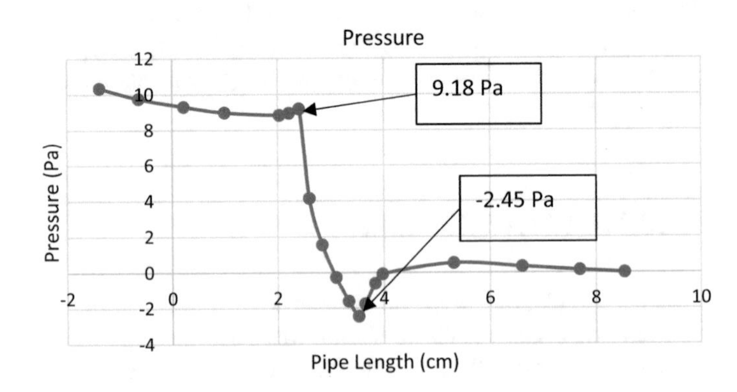


**Fig. 9.19** Pressure across pipe length

## *Mode 0*

Mode 0 was a set of simulations with Control and four configurations which were all simulated without the presence of crosswind at the top of the solution domain. Each configuration will have their heat gained by air compared with the Control of this mode.

### Control

The Control of the simulation was done using the chimney described in Chap. 3 and is shown in Fig. 9.10, possessing identical dimensions and boundary conditions with the sole difference of the absence of wire mesh screen at the top exit. This configuration is also used to verify and validate the result of this simulation with Chu et al. (2016). The Control simulation's results are shown in Table 9.8 and Figs. 9.20 and 9.21.

**Table 9.8** Data for mode 0, control simulation

| Mass flow rate (kg/s) | Outlet velocity (m/s) | Temperature change (°C) | Pressure loss across wire mesh (Pa) | Heat gained by air (MW) |
|---|---|---|---|---|
| 55,964 | 6.52 | 16.98 | −7.8 | 0.958 |

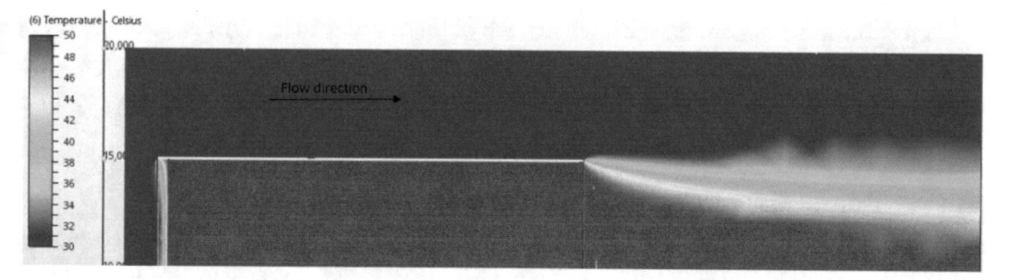


**Fig. 9.20** Temperature profile of mode 0, control (without wire mesh)

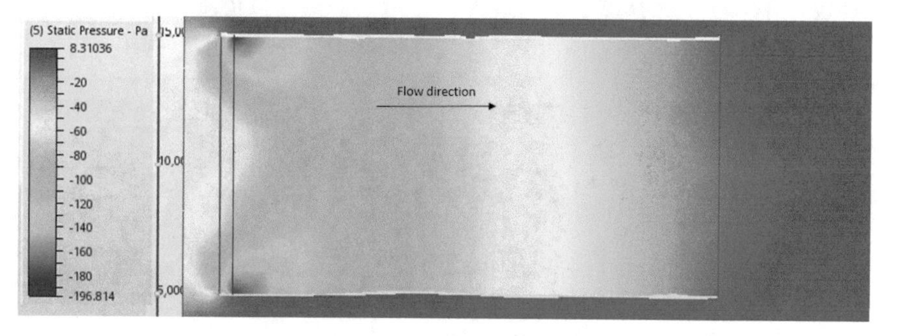


**Fig. 9.21** Pressure profile of mode 0, control (without wire mesh)

Although the chimney was not equipped with wire mesh screen, cold inflow is not significant in chimneys with inlet height of 14.68 m, staying consistent with Chu et al. (2016). The absence of wire mesh screen causes no change in pressure at the top exit of the chimney. Even though there was no wire mesh screen installed that would otherwise cause a pressure drop as the fluid flows through the wire mesh screen, it is suspected that the high velocity of the fluid flow was enough to prevent any significant cold inflow.

## Configuration 1

This configuration uses the same model and boundary conditions as other sets of simulation configurations. The wire mesh screen that this configuration uses was theoretically calculated to cause a decrease in pressure of the fluid flowing across by 14.55 Pa. The wire mesh screen responsible for the pressure drop has the following specifications:

i. Wire diameter, $d = 0.5$ mm
ii. Horizontal and vertical wire separation = 10 mm
iii. Number of stacks, $n = 5$
iv. Stack distance, $\sigma = 5$ mm.

Simulation results for Mode 0, Configuration 1 are shown in Table 9.9 and Figs. 9.22 and 9.23.

Configuration 1 equipped with wire mesh in theory would cause a pressure drop in the fluid flow across the chimney by 14.55 Pa. However, the simulation result shows that the pressure drop across the wire mesh screen was only 10.46 Pa. In theory, fluid

**Table 9.9** Data for Mode 0, Configuration 1 simulation

| Mass flow rate (kg/s) | Velocity before wire mesh (m/s) | Velocity after wire mesh (m/s) | Temperature change (°C) | Pressure loss across wire mesh (Pa) | Heat gained by air (MW) | Improvement in heat gain (%) |
|---|---|---|---|---|---|---|
| 54,819 | 6.43 | 6.39 | 19.9 | 10.46 | 1.0996 | 14.8 |

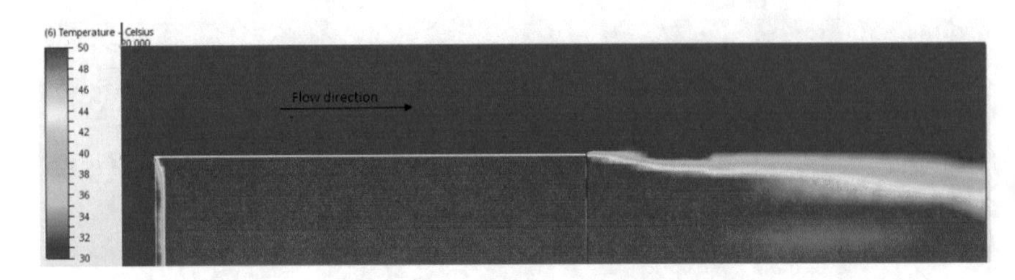


**Fig. 9.22** Temperature profile of mode 0, configuration 1

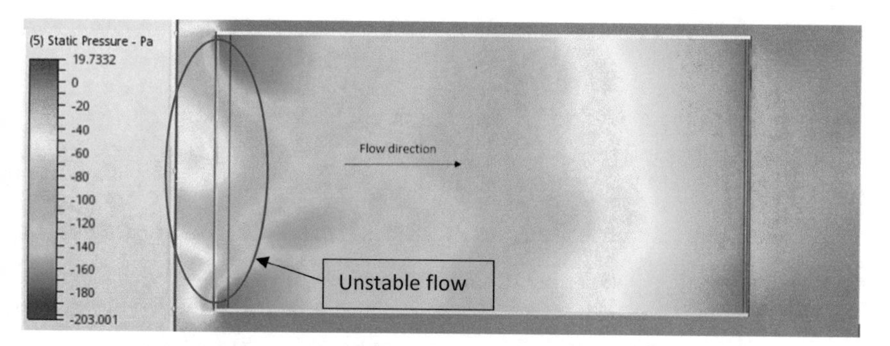


**Fig. 9.23** Pressure profile of mode 0, configuration 1

flow that experiences pressure drop should have an increase in velocity; in contrary, the result shows that the velocity decreases across the wire mesh screen.

These abnormal behaviours might be caused by the unstable flow at the bottom exit of the chimney. When the unstable flow reaches the wire mesh screen, the flow velocity is not uniform across the entire chimney area, causing some areas to have higher velocity than the other regions. This can further be proven by the revising Eq. (9.1), where the pressure drop caused by the wire mesh assumes that the velocity of the fluid flow is uniform across the whole area.

## Configuration 2

Configuration 2 uses the same chimney model and boundary conditions as other configurations. The difference between this configuration and the others was the wire mesh screen that was theoretically calculated to be able to cause a pressure drop within the fluid flowing across by 30.46 Pa. The wire mesh screen responsible for the pressure drop has the following specifications:

i. Wire diameter, $d = 0.5$ mm
ii. Horizontal and vertical wire separation = 5 mm
iii. Number of stacks, $n = 5$
iv. Stack distance, $\sigma = 5$ mm.

Simulation results for Mode 0, Configuration 2 are shown in Table 9.10 and Figs. 9.24 and 9.25.

**Table 9.10** Data for mode 0, configuration 2 simulation

| Mass flow rate (kg/s) | Velocity before wire mesh (m/s) | Velocity after wire mesh (m/s) | Temperature change (°C) | Pressure loss across wire mesh (Pa) | Heat gained by air (MW) | Improvement in heat gained (%) |
|---|---|---|---|---|---|---|
| 51,275 | 5.98 | 6.04 | 19.9 | 30.46 | 1.0285 | 7.38 |

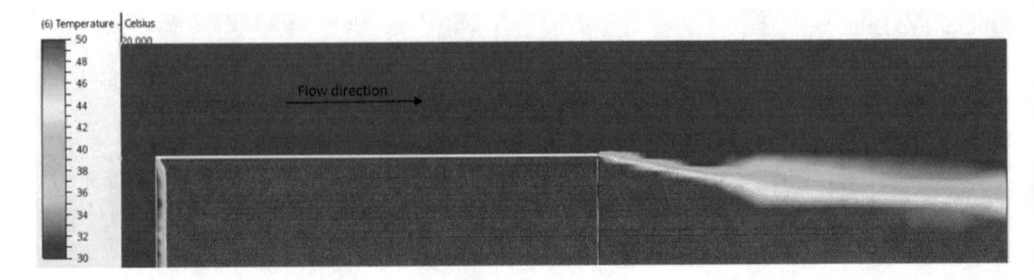


**Fig. 9.24** Temperature profile of mode 0, configuration 2

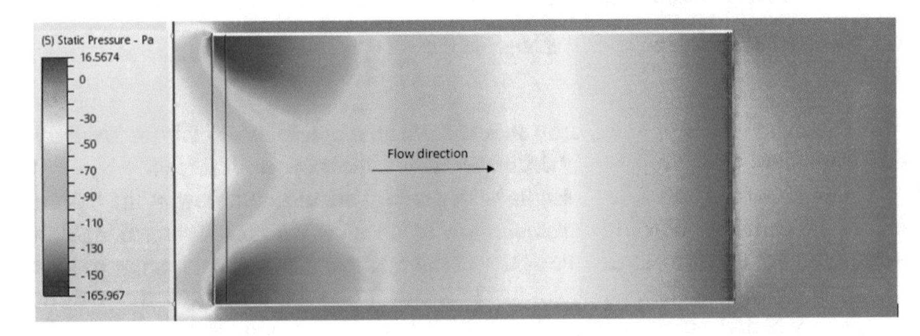


**Fig. 9.25** Pressure profile of configuration 2

Unlike Configuration 1, this configuration with wire mesh screen of supposed 30.46 Pa pressure drop was achieved; however, the velocity of the fluid across the wire mesh does increase, though by a very small amount. Moreover, when compared to the Control, the mass flow rate decreases in comparison.

## Configuration 3

Identical to other configurations, Configuration 3's model and boundary conditions are the same as the other configurations. The difference between this configuration and the others was the wire mesh screen. The wire mesh screen that this configuration uses was theoretically calculated to be able to cause a pressure drop within the fluid flowing across by 105.11 Pa. The wire mesh screen responsible for the pressure drop has the following specifications:

i. Wire diameter, $d = 5$ mm
ii. Horizontal and vertical wire separation = 10 mm
iii. Number of stacks, $n = 5$
iv. Stack distance, $\sigma = 5$ mm.

Simulation results for Mode 0, Configuration 3 are shown in Table 9.11 and Figs. 9.26, 9.27 and 9.28.

**Table 9.11** Data for mode 0, configuration 3 simulation

| Mass flow rate (kg/s) | Velocity before wire mesh (m/s) | Velocity after wire mesh (m/s) | Temperature change (°C) | Pressure loss across wire mesh (Pa) | Heat gained by air (MW) | Improvement in heat gained (%) |
|---|---|---|---|---|---|---|
| 52,169 | 5.79 | 6.08 | 19.9 | 99.2 | 1.0465 | 9.25 |

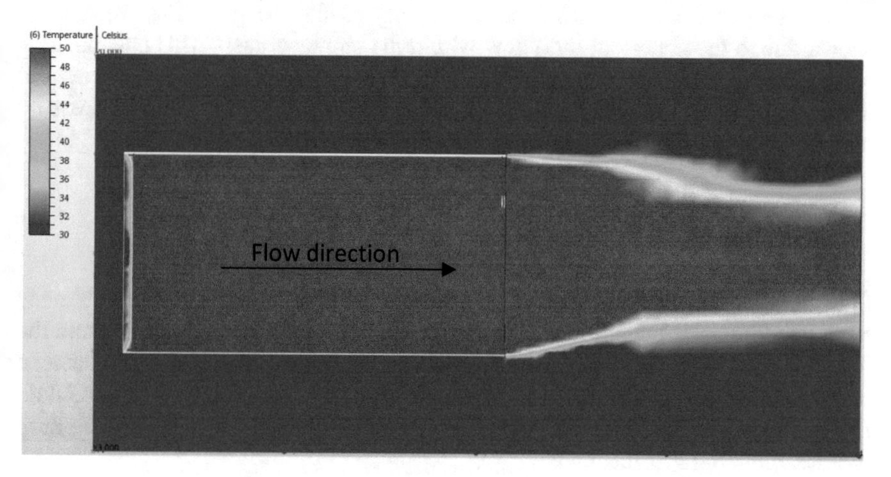


**Fig. 9.26** Temperature profile of mode 0, configuration 3

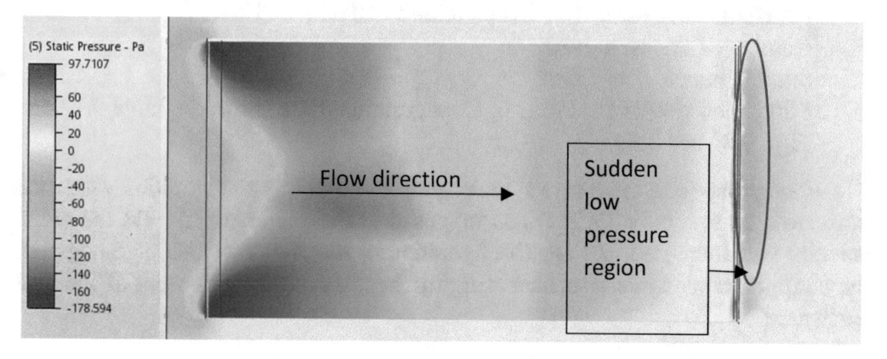


**Fig. 9.27** Pressure profile of mode 0, configuration 3

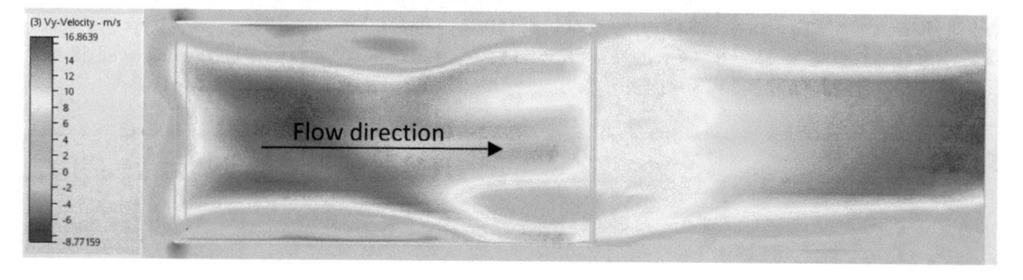


**Fig. 9.28** *Y*-axis velocity profile of mode 0, configuration 3

Configuration 3 is different than the other configurations as the air flow at the outlet is unstable, unlike other temperature profiles, Configuration 3's temperature profile may not be shown symmetrically along the middle of the chimney due to said reason. This might be caused by the sudden decrease in pressure across the wire mesh screen, where a decrease of 99.2 Pa is recorded evident by the green region directly at the outlet of the top exit in Fig. 9.27.

For this particular configuration, the velocity profile along the *Y*-axis has to be shown due to the abnormal fluid flow within the chimney itself. This phenomenon was suspected to be the result of the sudden and steep drop of pressure across a small distance at the top exit of the chimney, causing an instability in the flow, decreasing the mass flow rate and thus the heat gained by air.

### Configuration 4

Configuration 2 is virtually identical to the other configurations, in terms of chimney model and boundary conditions. The difference between this configuration and the others was the wire mesh screen, where this configuration uses a wire mesh screen that was theoretically calculated to be able to cause a pressure drop within the fluid flowing across by 58.20 Pa. The wire mesh screen responsible for the pressure drop has the following specifications:

i. Wire diameter, $d = 0.5$ mm
ii. Horizontal and vertical wire separation = 10 mm
iii. Number of stacks, $n = 20$
iv. Stack distance, $\sigma = 5$ mm
v. Simulation results for Mode 0, Configuration 4 are shown in Table 9.12 and Figs. 9.29 and 9.30.

Although the wire mesh screen under-performed in this configuration where the pressure drop was only 35.52 Pa, as supposed to the intended 58.20 Pa, the outlet velocity was increased. By far, Configuration 4 was the most stable amongst all the configurations while also increasing the heat gained by air output of the heat exchanger.

**Table 9.12** Data for mode 0, configuration 4 simulation

| Mass flow rate (kg/s) | Velocity before wire mesh (m/s) | Velocity after wire mesh (m/s) | Temperature change (°C) | Pressure loss across wire mesh (Pa) | Heat gained by air (MW) | Improvement in heat gained (%) |
|---|---|---|---|---|---|---|
| 56,732 | 6.32 | 6.53 | 19.9 | 35.52 | 1.1379 | 18.81 |

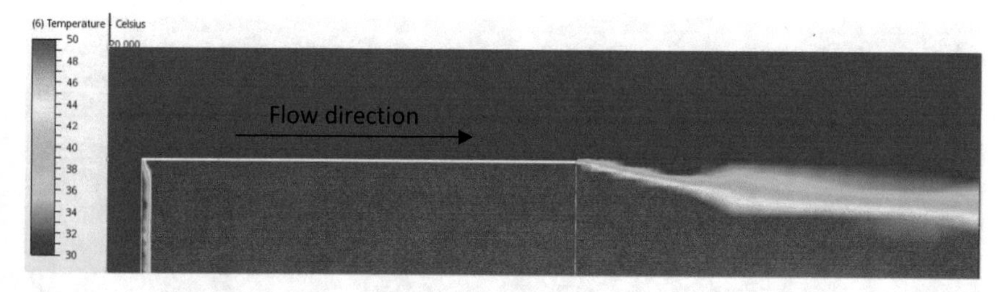


Fig. 9.29 Temperature profile of mode 0, configuration 4

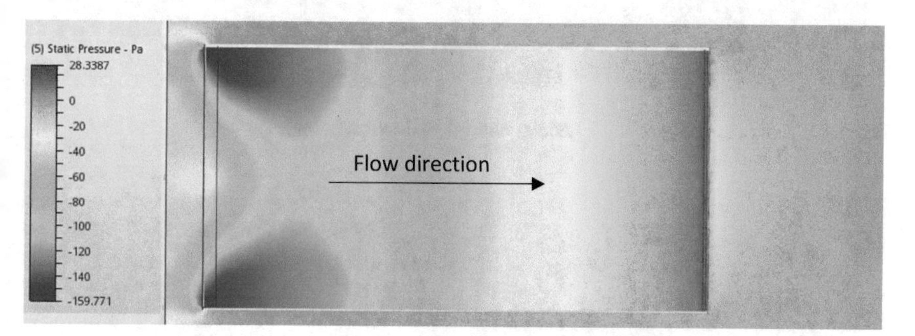


Fig. 9.30 Pressure profile of mode 0, configuration 4

## *Mode 1*

Mode 1 was a set of simulations with a Control and four configurations which were all simulated with the presence of a 10 m/s crosswind at the top of the solution domain in the negative *z*-direction, as illustrated in Fig. 9.11. Each configuration will have their heat gained by air compared with the Control of this mode.

### Control

The Control of Mode 1 was identical to the Control in Mode 0 in all aspects, including the model dimension and all but one boundary conditions which is the presence of a 10 m/s crosswind across the top of the chimney exit. The Control simulation's results are shown in Table 9.13 and Figs. 9.31 and 9.32.

**Table 9.13** Data for mode 1, control simulation

| Mass flow rate (kg/s) | Outlet velocity (m/s) | Temperature change (°C) | Pressure loss across wire mesh (Pa) | Heat gained by air (MW) |
|---|---|---|---|---|
| 43765 | 5.09 | 16.42 | 0 | 0.723 |

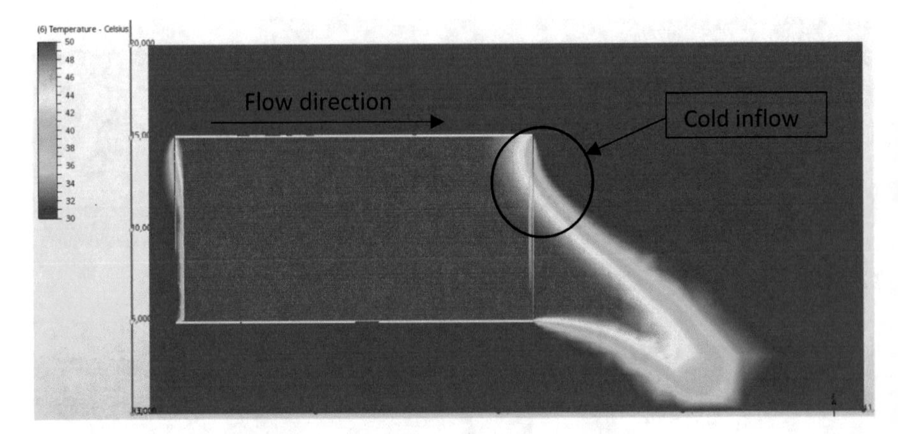


Fig. 9.31 Temperature profile of mode 1, control (without wire mesh)

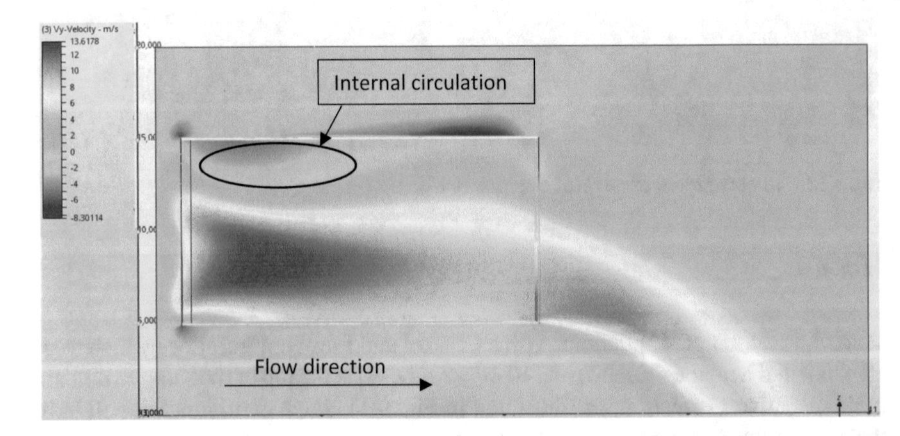


Fig. 9.32 *Y*-axis velocity profile of mode 1, control (without wire mesh)

Although the chimney, with inlet height of 14.68 m, was previously found to have no significant cold inflow in still air condition, the presence of a 10 m/s crosswind was able to cause a significant cold inflow at the top exit of the chimney, decreasing its mass flow rate, velocity, outlet temperature and heat gained by air when compared to Mode 0, Control. The absence of wire mesh screen causes no change in pressure at the top exit of the chimney and thus was unable to impede any flow reversal at the exit. If the ambient air flow has a higher velocity than the internal chimney air flow, cold inflow will occur.

Figure 9.32 clearly demonstrated the effect of cold inflow towards the internal air flow of the chimney, creating a circulating fluid flow near the inlet wall of the chimney, due the cold inflow at the top exit impeding the hot air to flow vertically outwards. This impedance of flow of hot air within the chimney greatly reduced the heat gained by air of the chimney when compared to the Control simulation from Mode 0.

## Configuration 1

With the exception of the presence of a 10 m/s crosswind, this configuration uses the same model and boundary conditions as Mode 0 and Configuration 1. The wire mesh screen that this configuration uses was theoretically calculated to cause a decrease in pressure of the fluid flowing across by 14.55 Pa.

Simulation results for Mode 1, Configuration 1 are shown in Table 9.14 and Fig. 9.33.

Configuration 1 was equipped with wire mesh which in theory would cause a pressure drop in the fluid flow across the chimney by 14.55 Pa. However, the simulation result showed that the pressure drop across the wire mesh screen was higher than the expected value, measuring at 19.06 Pa. In theory, fluid flow that experiences pressure drop should have an increase in velocity; however, the increase in this configuration was insignificant.

**Table 9.14** Data for mode 1, configuration 1 simulation

| Mass flow rate (kg/s) | Velocity before wire mesh (m/s) | Velocity after wire mesh (m/s) | Temperature change (°C) | Pressure loss across wire mesh (Pa) | Heat gained by air (MW) | Improvement in heat gained (%) |
|---|---|---|---|---|---|---|
| 47,201 | 5.43 | 5.49 | 17.12 | 19.06 | 0.813 | 12.45 |

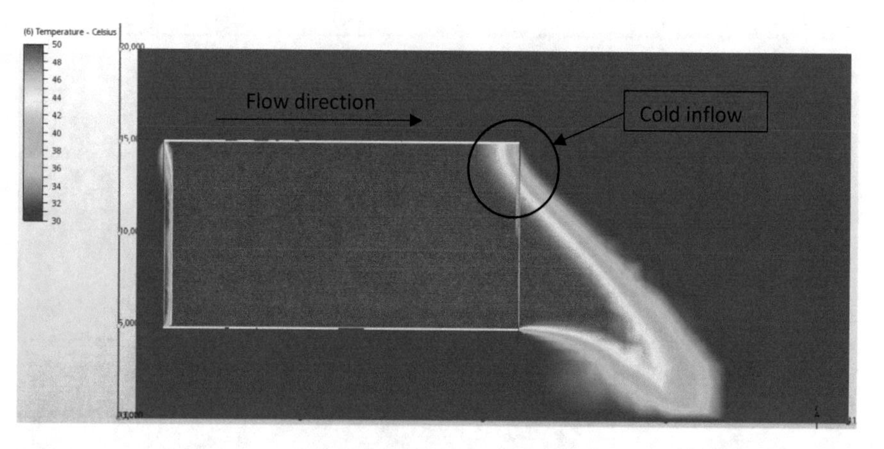


**Fig. 9.33** Temperature profile of mode 1, configuration 1

Cold inflow is still occurring in this configuration even with the presence of wire mesh screen. It is suspected that the mere 19.06 Pa decrease in pressure across the wire mesh screen was inadequate to impede the backflow of air at the chimney exit.

### Configuration 2

This configuration uses the same model and most boundary conditions as Mode 0, Configuration 2 with the sole difference being the addition of a 10 m/s crosswind. The wire mesh screen that this configuration uses was theoretically calculated to cause a decrease in pressure of the fluid flowing across by 30.46 Pa.

Simulation results for Mode 1, Configuration 2 are shown in Table 9.15 and Fig. 9.34.

The wire mesh screen installed in Configuration 2 can in theory cause a pressure drop in the fluid flow across the chimney by 30.46 Pa. The pressure drop measured across the wire mesh screen in this configuration was found to be 34.4, which is fairly consistent with the theoretical value. Although the air flow within the chimney experienced a decrease in pressure, the velocity increase was minimal if not insignificant. The negligible increase in flow velocity may be caused by the presence of crosswind in a direction perpendicular to the air flow coming out of the chimney, disrupting the intended flow direction and velocity.

**Table 9.15** Data for mode 1, configuration 2 simulation

| Mass flow rate (kg/s) | Velocity before wire mesh (m/s) | Velocity after wire mesh (m/s) | Temperature change (°C) | Pressure loss across wire mesh (Pa) | Heat gained by air (MW) | Improvement in heat gained (%) |
|---|---|---|---|---|---|---|
| 45,129 | 5.20 | 5.26 | 19.98 | 34.4 | 0.907 | 25.47 |

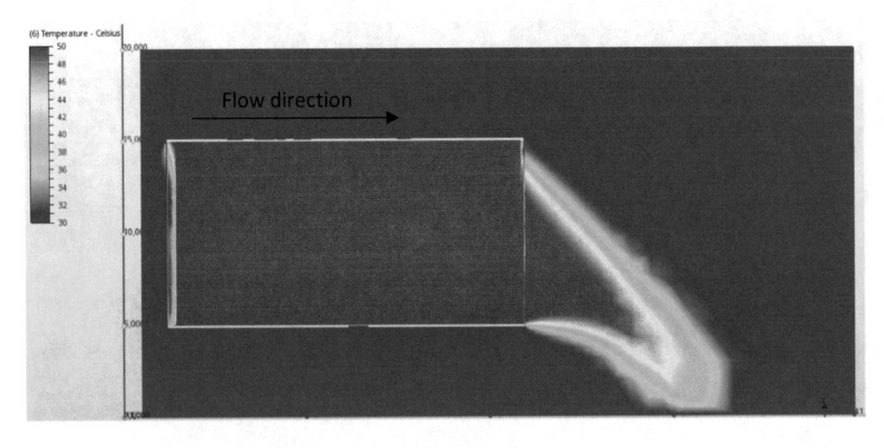


**Fig. 9.34** Temperature profile of mode 1, configuration 2

Although the crosswind remained an influence to the air flow above the chimney exit, cold inflow was seen to have completely been eliminated in this case, where it was highly suspected to be the pressure drop across the wire mesh screen that impedes the air flow reversal. This prevention of cold inflow had increased the amount of heat gained by air by 25.47% when compared with the Control simulation of Mode 1.

## Configuration 3

This configuration uses the same model and most boundary conditions as Mode 0, Configuration 3 except with the addition of a 10 m/s crosswind at the top portion of the solution domain. The wire mesh screen that this configuration uses was theoretically calculated to cause a decrease in pressure of the fluid flowing across by 105.11 Pa.

Simulation results for Mode 1, Configuration 2 are shown in Table 9.16 and Figs. 9.35 and 9.36.

Theoretical calculations done upon the wire mesh screen specifications installed in Configuration 3 suggested that a pressure drop of 105.11 was to be expected in the fluid flow across the wire mesh screen. However, the pressure drop measured across the wire mesh screen in this configuration was found to be only 74.3 Pa, which is very inaccurate. Even the temperature change in air saw a slight decrease when compared with the other configurations in Mode 1. It is suspected that there are correlations

**Table 9.16** Data for mode 1, configuration 3 simulation

| Mass flow rate (kg/s) | Velocity before wire mesh (m/s) | Velocity after wire mesh (m/s) | Temperature change (°C) | Pressure loss across wire mesh (Pa) | Heat gained by air (MW) | Improvement in heat gained (%) |
|---|---|---|---|---|---|---|
| 38,913 | 4.50 | 4.53 | 19.65 | 74.3 | 0.769 | 6.4 |

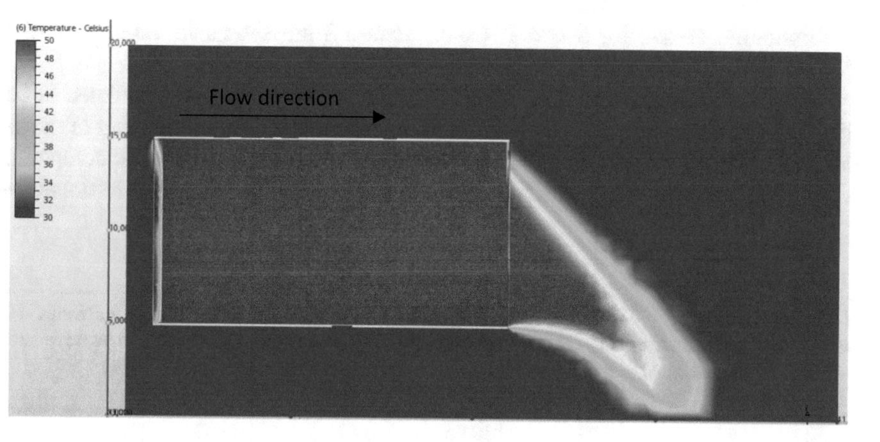


**Fig. 9.35** Temperature profile of mode 1, configuration 3

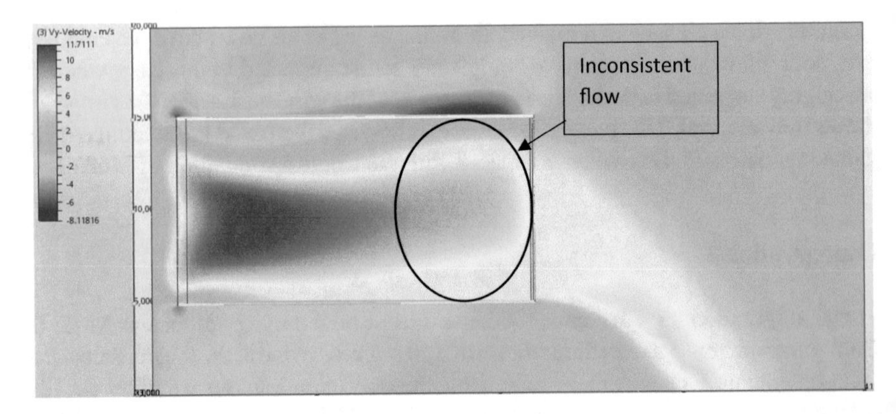


**Fig. 9.36** *Y*-axis velocity profile of mode 1, configuration 3

between the huge drop in pressure and the inconsistency of fluid flow as seen in Fig. 9.36. However, no prove was able to be produced with such little information on the configuration.

Although the wire mesh screen was able to impede the backflow of air at the top exit of the chimney, the improvement upon the heat gained by air was the least compared to the other configurations in Mode 1.

## Configuration 4

In this configuration, the same model and most boundary conditions as Mode 0, Configuration 4 were used with the only exception being the addition of a 10 m/s crosswind at the top portion of the solution domain. The wire mesh screen that this configuration uses was theoretically calculated to decrease in pressure of the fluid flowing across by 58.20 Pa.

Simulation results for Mode 1, Configuration 2 are shown in Table 9.17 and Fig. 9.37.

The theoretical pressure drop across the wire mesh screen was calculated to be 58.20 Pa for this configuration. This configuration produces no interesting result as the data obtained was consistent and showed not much difference when compared with the data from Configuration 2 of Mode 1. This configuration provides a quite

**Table 9.17** Data for mode 1, configuration 4 simulation

| Mass flow rate (kg/s) | Velocity before wire mesh (m/s) | Velocity after wire mesh (m/s) | Temperature change (°C) | Pressure loss across wire mesh (Pa) | Heat gained by air (MW) | Improvement in heat gained (%) |
|---|---|---|---|---|---|---|
| 41,911 | 4.85 | 4.88 | 19.96 | 47.32 | 0.842 | 16.41 |

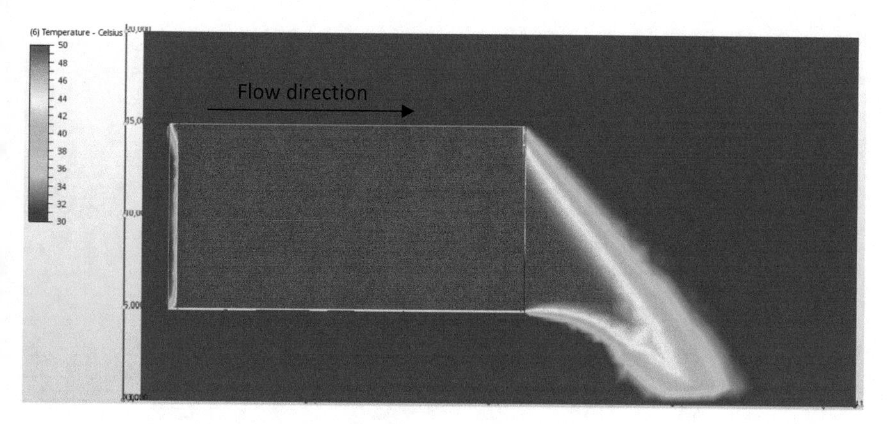


Fig. 9.37 Temperature profile of mode 1, configuration 4

significant increase in performance when compared with the Control but still felt a little short behind Configuration 2 of Mode 1, respectively.

## Discussion

The simulation data from Mode 0 agrees with the conclusion from Chu et al. (2016) where cold inflow is insignificant with natural draft chimney with inlet height of 14.68 m. However, Chu et al. (2016) demonstrated that the presence of wire mesh screen decreases the flow velocity while a research investigating the pressure drop across wire mesh screen done by Sun et al. (2015) showed that pressure drop due to wire mesh screen increases flow velocity. In this research, data that agrees with both contradicting researches may be concluded. However, the data collected for different wire mesh parameters in Mode 0 seems to show that the increase in pressure drop increases the velocity change, instead of directly affecting the velocity itself. Figure 9.38 is a plot illustrating said relation.

Figure 9.39 showed the increase in amount of temperature change as the pressure drop increases. Temperature increase is very significant when the pressure drop is within the range below 35 Pa. After 35 Pa of pressure decrease, the temperature change reaches it maximum possible amount and is stable afterwards. However, when the pressure drop is around 75 pa, the temperature increase experiences a slight decrease.

In both Mode 0 and 1, although there are no discernible trend in the percentage increase in heat gained by air as the pressure drop caused by the wire mesh screen increases, both Modes showed that the performance increase peaked when the pressure provided by the wire mesh screen is roughly at 35 Pa, as seen in Fig. 9.40.

Research is done by Seifi et al. (2017) with the goal of utilizing windbreakers to decrease the negative effects upon the performance of industrial-sized cooling

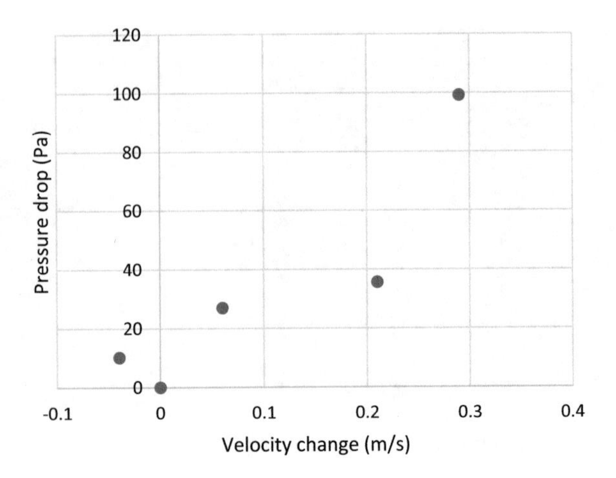


**Fig. 9.38** Pressure drop against the velocity change for different wire mesh size (mode 0)

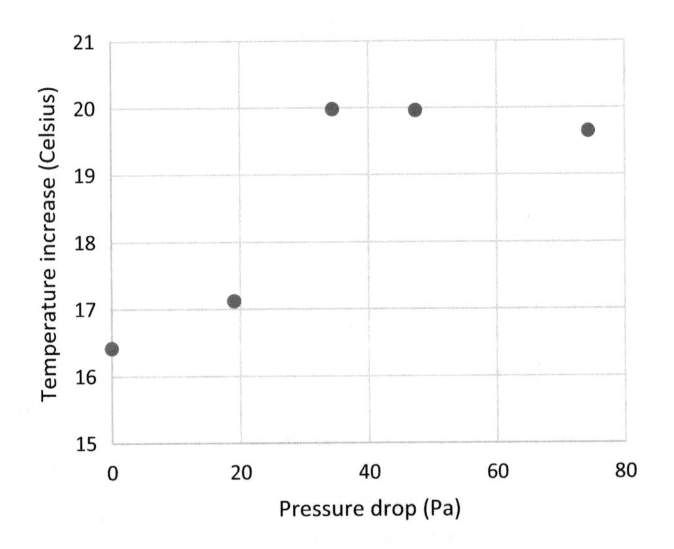


**Fig. 9.39** Temperature change against pressure drop plot (Mode 1)

tower due to crosswind. The maximum amount of increase in performance peaked at 10.5% in the heat gained by air when subjected to crosswind condition of 15 m/s with four windbreakers. Comparing the data to this research, the maximum increase in performance was recorded to be 25.47% when the model was subjected to a crosswind condition while using wire mesh screen that caused a pressure decrease 34.4 Pa. Besides having a higher increase in performance, the use of wire mesh screen is also more economically feasible when compared with the building windbreakers.

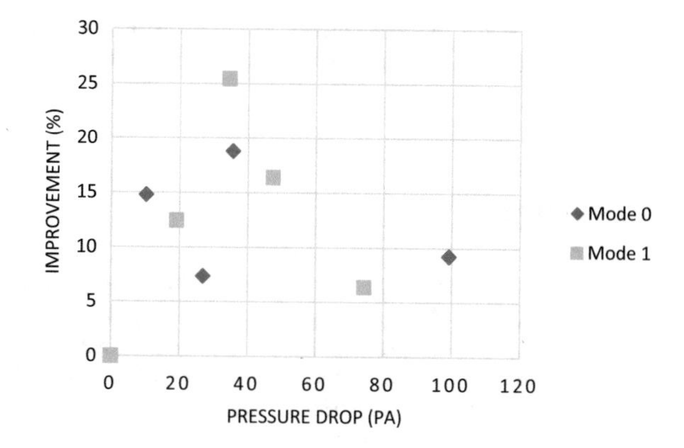


**Fig. 9.40** Percentage increase in heat gained by air against pressure drop

# Conclusion

Although many previous researchers (Andreozzi et al., 2010; Chu et al., 2012; Jörg & Scorer, 1967; Li et al., 2017) have stated that cold inflow could easily occur within the NDC, this research had failed to replicate the cold inflow in the NDC even with a straight chimney where Andreozzi et al. (2010) deemed vulnerable to cold inflow. The presence of a strong crosswind (10 m/s) was required to effectively induce a cold inflow effect in an NDC of this design.

For a NDC operating in still ambient air condition (Mode 0), the presence of wire mesh screen is able to increase the heat gained by air within the chimney, simply by increasing the temperature change of air. However, the increase in heat gain was not proportional to the pressure drop caused by the wire mesh screen, where after reaching a high pressure drop of 90 Pa, the fluid flow within the chimney becomes too unstable to provide an increase in heat gained by air. The wire mesh screen that was able to cause a pressure drop of 35 Pa was found to have the best result in this mode. However, the improvement towards the heat gained by air was only by 18%. Although the improvement is not entirely insignificant, in practice, the cost of installing wire mesh screen over a wide chimney exit area with diameter of 100 m might not outweigh the improvement it could potentially be brought.

However, for a NDC operating under crosswind condition, the introduction of wire mesh screen was found to be able to increase the overall temperature change of air within the chimney and thus the overall heat gained by air. Similar to Mode 0, this set-up found similar patterns where the increase in performance is not proportional to the pressure drop caused by a wire mesh screen. The increase in performance showed more promising results compared to Mode 0, which also happened at around 35 Pa pressure drop. Said configuration (Configuration 2) was able to increase the overall heat gain by air by 25.47% when compared to the Control from the same mode.

**Acknowledgements** The authors would like to express their gratitude to the Ministry of Education of Malaysia and Universiti Malaysia Sabah for their financial and facilities support through

Fundamental Research Grant (FRG0429-TK-1/2015). The authors are also thankful to insightful inputs from other researchers of the similar field as well.

## References

Al-Badri, A. R., & Al-Waaly, A. A. (2017). The influence of chilled water on the performance of direct evaporative cooling. *Energy and Buildings, 155,* 143–150.

Andreozzi, A., Buonomo, B., & Manca, O. (2009). Thermal management of a symmetrically heated channel–chimney system. *International Journal of Thermal Sciences, 48,* 475–487.

Andreozzi, A., Buonomo, B., & Manca, O. (2010). Thermal and fluid dynamic behaviors in symmetrical heated channel-chimney systems. *International Journal of Numerical Methods for Heat and Fluid Flow, 20,* 811–833.

Arie, M. A., Shooshtari, A. H., & Ohadi, M. M. (2017). Experimental characterization of an additively manufactured heat exchanger for dry cooling of power plants. *Applied Thermal Engineering, 129,* 187–198.

Auletta, A., & Manca, O. (2002). Heat and fluid flow resulting from the chimney effect in a symmetrically heated vertical channel with adiabatic extensions. *International Journal of Thermal Sciences, 41,* 1101–1111.

Boomsma, K., Poulikakos, D., & Zwick, F. (2003). Metal foams as compact high performance heat exchangers. *Mechanics of Materials, 35,* 1161–1176.

Bouchair, A. (1994). Solar chimney for promoting cooling ventilation in southern Algeria. *Building Service Engineering, Research and Technology, 15,* 81–93.

Busch, D., Harte, R., Krätzig, W. B., & Montag, U. (2002). New natural draft cooling tower of 200 m of height. *Engineering Structures, 24,* 1509–1521.

Çengel, Y. A., & Boles, A. M. (2015a). *Thermodynamics, an engineering approach* (8th ed.). McGraw-Hill Education.

Çengel, Y. A., & Ghajar, A. J. (2015b). *Heat and mass transfer: Fundamentals & applications* (Vol. 5). McGraw-Hill Education.

Chu, C.-M., Rahman, M. M., & Kumaresan, S. (2012). Effect of cold inflow on chimney height of natural draft cooling towers. *Nuclear Engineering and Design, 249,* 125–131.

Chu, C.-M., Rahman, M. M., & Kumaresan, S. (2015). Improved thermal energy discharge rate from a temperature-controlled heating source in a natural draft chimney. *Applied Thermal Engineering, 98,* 991–1002.

Chu, C.-M., Rahman, M. M., & Kumaresan, S. (2016). Improved thermal energy discharge rate from a temperature-controlled heating source in a natural draft chimney. *Applied Thermal Engineering, 2016.*

Ding, E. (1992). Air cooling techniques in power plants. *Water and Electric Power Press, 1.*

EPRI. (2013). *Informational webcast for EPRI/NSF joint solicitation on advancing power*. Electric Power Research Institute.

Fisher, T. S., Torrance, K. E., & Sikka, K. K. (1997). Analysis and optimization of a natural draft heat sink system. *IEEE Transactions on Components, Packaging, and Manufacturing Technology—Part A, 20,* 111–119.

Fu, S., & Zhai, Z. (2001). Numerical investigation of the adverse effect of wind on the heat transfer performance of two natural draft cooling towers in tandem arrangement. *Acta Mechanica Sinica, 17,* 24–34.

Fu, Y., Wen, J., & Zhang, C. (2017). An experimental investigation on heat transfer enhancement of sprayed. *International Journal of Heat and Mass Transfer, 112,* 699–708.

Hajidavalloo, E., Shakeri, R., & Mehrabian, M. A. (2010). Thermal performance of cross-flow cooling towers in variable wet bulb temperature. *Energy Conversion and Management, 51,* 1298–1303.

Hosseini, S. S., Ramiar, A., & Ranjbar, A. A. (2017). Numerical investigation of rectangular fin geometry effect on solar chimney. *Energy and Buildings, 155,* 269–307.

Jaber, H., & Webb, R. (1989). Design of cooling towers by the effectiveness-NTU method. *Journal of Heat Transfer, 111,* 837–843.

Jörg, O., & Scorer, R. S. (1967). An experimental study of cold inflow into chimneys. *Atmospheric Environment, 1,* 645–654.

Khanal, R., & Lei, C. (2012). Flow reversal effects on buoyancy induced air flow in a solar chimney. *Solar Energy, 86,* 2783–2794.

Kim, S., Paek, J., & Kang, B. (2000). Flow and heat transfer correlations for porous fin in a plate-fin heat exchanger. *Journal of Heat Transfer, 122,* 572–578.

Kong, Y., Wang, W., Huang, X., Yang, L., & Du, X. (2017). Annularly arranged air-cooled condenser to improve cooling efficiency of natural draft direct dry cooling system. *International Journal of Heat and Mass Transfer, 118,* 587–601.

Kurian, R., Balaji, C., & Venkateshan, S. P. (2016). Experimental investigation of near compact wire mesh heat exchangers. *Applied Thermal Engineering, 108,* 1158–1167.

Li, X., Gurgenci, H., Guan, Z., & Sun, Y. (2017). Experimental study of cold inflow effect on a small natural draft dry cooling tower. *Applied Thermal Engineering, 128,* 762–771.

Liu, N., Zhang, L., & Jia, X. (2017). The effect of the air water ratio on counter flow cooling tower. *Procedia Engineering, 205,* 3550–3556.

Ma, J., Lv, P., Luo, X., Liu, Y., Li, H., & Wen, J. (2016). Experimental investigation of flow and heat transfer characteristics in double-laminated sintered woven wire mesh. *Applied Thermal Engineering, 95,* 1–9.

Mansour, M. K., & Hassab, M. A. (2014). Innovative correlation for calculating thermal performance of counterflow wet-cooling tower. *Energy, 74,* 855–862.

Merkel, F. (1925). Evaporative cooling. *Z. Verein Deutscher Ingenieure (VDI), 70,* 123–128.

El-Wakil, M. M. (1985). *Power plant technology.* McGraw-Hill Book Company.

Naik, B. K., & Muthukumar, P. (2017). A novel approach for performance assessment of mechanical draft wet cooling towers. *Applied Thermal Engineering, 121,* 14–26.

Seifi, A. R., Akbari, O. A., Alrashed, A. A., Afshary, F., Shabani, G. A., Seifi, R., et al. (2017). Effects of external wind breakers of heller dry cooling system in power plants. *Applied Thermal Engineering, 129,* 1124–1134.

Shah, R., McDonald, C. F., & Howard, C. P. (1980). Compact heat exchangers—History. *Technological advancement and mechanical design problems.*

Smrekar, J., Oman, J., & Širok, B. (2006). Improving the efficiency of natural draft cooling towers. *Energy Conversion and Management, 47,* 1086–1100.

Sun, H., Bu, S., & Luan, Y. (2015). A high-precision method for calculating the pressure drop across wire mesh filters. *Chemical Engineering Science, 127,* 143–150.

Sun, Y., Guan, Z., Gurgenci, H., Hooman, K., & Li, X. (2017). Investigations on the influence of nozzle arrangement on the pre-cooling effect for the natural draft dry cooling tower. *Applied Thermal Engineering, 130,* 979–996.

Xu, P., Ma, X., Zhao, X., & Fancey, K. S. (2016). Experimental investigation on performance of fabrics for indirect evaporative cooling applications. *Building and Environment, 50,* 104–114.

Zhai, Z., & Fu, S. (2006). Improving cooling efficiency of dry-cooling towers under cross-wind conditions by using wind-break methods. *Applied Thermal Engineering, 26,* 1008–1017.

Zivi, S. M., & Brand, B. B. (1956). An analysis of the cross-flow cooling tower. *Refrigeration Engineering, 64,* 31–34.

Zou, Z., Guan, Z., & Gurgenci, H. (2013). Optimization design of solar enhanced natural draft dry cooling tower. *Energy Conversion and Management, 76,* 945–955.

# Chapter 10
# Economic Analysis of Solar Chimney: Literature Review

**Rezwan us Saleheen, Md. Mizanur Rahman, Mohammad Mashud, and Sajib Paul**

**Abstract** Power generation utilizing solar chimney (SC) appears to be a radical perspective for both residential and industrial applications. Such proposition includes safeguard of mother nature reducing the ecological combustion of fossil fuels. Solar chimney (SC) initiates an opportunity in the convenience of local and isolated communities for the augmentation of industrial efforts and hence expands a probability in economic growth sustaining ecological balance. Nowadays, large-scale solar chimneys can be manufactured without any technical problems and at defined costs. Moreover, the accommodation requirements of rapidly growing population lead to the vertical growth of the buildings. The significance of electrical utilities intensifying for the necessity of proper ventilation and daylighting in place of nourish natural air ventilation system. Solar chimney also devised a driveway for natural ventilation structure for high-rise buildings. This chapter imparts a brief outline of research and development of SC power technology in the past century along with an economic analysis and cost effectiveness of solar chimney power plant (SCPP).

**Keywords** Ventilation · Solar chimney power plant (SCPP) · Thermal comfort · Solar chimney · Levelized electricity cost (LEC) · Additional review

## Introduction

Globally, the demand of electricity is steadily growing up due to the economic development of the world as well as advanced in technologies. The increasing trend of electricity demands is not same in any one of the country in the world. Some of the developed countries are having very sophisticated technology that required continuous supply of electricity. Alternatively, in the developing countries, both domestic

R. Saleheen (✉) · Md. M. Rahman · S. Paul
Department of Mechatronics Engineering, Faculty of Engineering, World University of Bangladesh, Uttara, Dhaka, Bangladesh
e-mail: saleheen1@mte.wub.edu.bd

M. Mashud
Department of Mechanical Engineering, Khulna University of Engineering and Technology (KUET), Khulna, Bangladesh


Md. M. Rahman and C.-M. Chu (eds.), *Cold Inflow-Free Solar Chimney*,
https://doi.org/10.1007/978-981-33-6831-6_10

and industrial sectors are the main consumers of electricity. Therefore, the demand of electricity is increasing significantly in the developing countries like Bangladesh, India, Malaysia, etc. It is estimated that the demand of electricity will increase more than 1.4 folds in the year 2025 compared to the demand of 2012. The demand of electricity in 2012 was 19, 562 TWh and it is expected that the demand will reach 26,761 TWh by 2025. At the same time, to fulfil this demand of electricity, the capacity of power generation system needs to be increased from 6117 GW in 2014 to 8370 GW in 2025 which is also increased by 1.4 times between these years. It is expected that the capacity of power generation will increase more in the developing countries compared to developed country that is estimated about 70%. The storage capacity of the world fossil fuels is now under threat; therefore, the excess demands of power are need to be fulfilled by using efficient technologies and smart grid system or different types of renewable energy sources (Hong and Lee, 2018).

In countries like China and India, the growing demand of electricity is very high due to population growth as well as for the technologies development. The coal-fired power generation system plays an important role to meet the increasing capacity in the power sector. Recently the mandate of air pollution from coal-fired power plant make the situations more complicated than before. Therefore, the compound effect of environmental pollution due to conventional fuels and the variations on fossil fuel prices put together the escalating global energy requirements despaired. Developed countries' economies have the suspicion about energy supplies of fossil fuels, with consideration the fluctuation of oil, petrol prices, the global warming effects that represent a substantial threat of economic and environment as well (Papageorgiou, 2007). Consequently the search of convenient alternative energy sources has become an imperative concern. To reduce dependence on traditional sources that produce a big amount of gas emission, the most effective and realistic way to find out electricity to fulfil the demand is harvest energy from renewable energy resources. These endeavours may reduce the necessity of human's migration in search of healthy and virtuous environment (Zhou & Yang, 2012). Moreover, global safety and better life style for the habitats of this earth, a clear environment, renewable energy resources and highly efficient technology are needed to develop and to implement all over the wold.

Wind and solar energy are most favourable among the renewable energy resources (Mohammed et al., 2019). Renewable energy includes tidal, wave, biofuel and geothermal resources. The uses of these sources are rapidly increasing due to strong legislation in opposition of air pollutions. The applications of solar and wind resources are large and on- and offgrid-connected considering with other resources due to scarcity of efficient technology. At present, solar chimney plants may be a distinguished solution for all the technical and environmental requirements. Many factors encourage using solar energy to obtain clean energy, have a lower maintenance cost, robustness of system (Vlachos et al., 2002). Utilization of solar chimney system will bestow greatly to reduction of toxic materials, gas emission which has direct impact on the environmental, health and adoption as an alternative green energy source (Mohammad, 2011).

Recently many researchers and experts have introduced different models to take advantage of solar chimney power plant technology. A new approach, hybrid system for power generation from renewable energy sources, is showing higher reliability and lower per unit generation cost (Muselli et al., 1999; Billinton, 2005). According to the definition of hybrid system, it is a power generation system from two or more sources. The examples of hybrid systems are wind and solar, wind and diesel, pv and wind or it can be combination of any two resources. A hybrid system is not a new power generation system or a technology that can be used for power generation, it is considered as combination of existed systems used in this field more than two decades (Yang et al., 2009).

## Inception of Solar Chimney

The generation of electricity using solar chimney technology was persuaded at the beginning of twentieth century by diverse engineering explorers. In 1903s, 1931s, both of Cabanyes, Gunther initiate an investigation about study the concepts of solar chimney power plant and basic description as well (Cao et al., 2011a, b). The solar chimney power technology was initiated by Cabanyes (1903) for maintaining regulating of warm air inside a house. Implementation of a wind blade inside the house attained electricity generation along with warm air. During 1975, several engineering pioneers in Australia, Canada and the USA privilege several research outcomes carrying out the Cabanyes research. Meanwhile in 1926, Prof. Engineer Bernard Dubos demonstrated a solar aero-electric power plant on the slope of an adequate height. Prof. Bernard constructed the solar chimney for the French Academy of Sciences in North Africa (Ley, 1954).

The very first framework of solar chimney power plant (SCPP) was illustrated by Schlaich and his colleagues in between 1981 and 1982. That solar chimney prototype of Schlaich and his team was constructed at Manzanares, Spain. In the year 1982, Prof. Dr. Ing. Jorg Schlaigh, Bergerman and Partners, got together to oversee the demonstration of the operating model of a SAEPP located at Manzaranes (Spain). Later that paradigm, having a solar chimney made out of steel tubes and a power rating of 50KW, was discovered by the German Government. The design incorporated a solar chimney of 10 m diameter and 195 m of height. The chimney was in functioning order for approximately six years spreading a greenhouse surface of 46000m$^2$. During the operation of that chimney optimization, data was recorded by Schlaich (1995). Schlaigh (1996) narrates the massive expenditures for the construction of huge reinforced concrete solar chimneys having heights of 500–1000 m. Therefore, a comprehensive cost analysis (Kasaeian, 2017) comes to an end finding the similarities between the investment costs per KWh production on the concrete solar chimney technologies in the similar range. Although the CSP has nearly same generation cost for the comparative generated KWh, the CSP power plants can be upgraded by rupturing into smaller sections utilizing rational recourses. However, there are some significant outcomes of the proposed solar chimney technology corresponding to other considerable renewable technologies (Wind, SCP and PV). This technology is capable of furnishing its solar collectors with thermal storage facilities. Moreover, this upgradation can be done in almost negligible cost providing a successive generation capacity throughout the whole year. This plant illustrated a noble method of power generation for researchers.

Prolonging the previous illustration several researchers around the world proposed their solar chimney designs and buildings. Inspired by the research outcomes on solar chimney a 200 MW project was initiated by the Australian government in Mildura. The chimney requirements for the Australian government's project were 1000 m in height and 7000 m in the collector diameter. Mildura project was assumed to be sufficient for fulfilling the power requirements for 200,000 households (Nizetic et al., 2008; Enviromission Limited, 2006). Indian Government was supposed to construct a 100 MW SCPP in Rajasthan, India which was called off because of the political rivalry between India and Pakistan (Jiakuan et al., 2003). A 40 MW SCPP, having a chimney parameter of 750 m height and 3.5 km collectors, was presented in Ciudad Real, Spain. It was actually a building known as the Ciudad Real Torre Solar (Sagasta, 2003). During 2008, Namibian government initiated a 400 MW SCPP to act as a greenhouse for the implantation of agribusiness (Cloete, 2008). The building was titled as the 'Green Tower' having a solar chimney 1500 m high, 280 m chimney diameter and 37 km collector. In Shanghai, China a building designed for power generation along with tourist interest. The 1000 m high solar chimney in Shanghai was propounded and executed by the HUST team (Zhou et al., 2010). A solar power plant has capacity of supplying power for 200,000 households (Nizetic et al., 2008; Enviromission Limited, 2006). Zhou et al. (2010) demonstrated an intuition of that solar chimney of 1000 m in height and 7000 m in the collector diameter.

## Background

Zhou et al. (2010) provided a more empirical perception on the physical process of solar chimney illustrated up to 2010. After the initiation the researchers reviewed the experimental and theoretical studies on the physical process of solar chimney along with efficiency. Moreover, the reviews provide a comprehensive economic analysis of some different types of solar chimneys. The relative analysis involves predominantly on floating solar chimney, solar chimney with sloped collector and mountain-laid chimneys. Focusing the improvement techniques of SCPP a review of previous studies was illustrated by Chikere et al. (2011). The illustration put forward a profound alternative enhancement techniques utilizing waste thermal energy in the flue gas as the energy input in a solar chimney collector. The estimation of solar chimney in building was an innovative initiative by Zhai et al. in (2011). The appraisal of a combined structure of solar chimney condensed by its applications focused on roof and walls of buildings. The Zhai et al. (2011) estimation was prolonged by Dhahri and Omri (2013) in their review paper. The review paper evaluates building applications of solar chimney and characterized in details along with the exploration of the components of that structure. Dhahri and Omri (2013) categorized solar chimney analysis into fragments involving the solar chimney projects, the numerical studies and the unconventional solar chimneys until 2013. The future aspects of solar chimney and its limitations to the desert prone villages were estimated by Olusola Olorunfemi (2014) and Bamisile. Their estimation was a momentary of solar chimney with a focal point on the northern regions of Nigeria.

During the past several years, technologists carried out a variety of researches focusing on the aspects of solar chimney. Based on the outcomes of that work, distinct business plans were proposed and executed. Among the statements of the researches, Floating Solar Technology has appreciated and capitalized with several reviews and franchises (Papageorgiou, 2004a, b, 2009). Common structure of the floating solar chimneys is frame with free standing consisting of some lifting balloon tube rings filled with a gas. The best option as lifting gas for FSC structure is $NH_3$, which is insubstantial and available in a nominal price. Hence, we can consider FSCs structures consist of an affordable cost structures comparing with conventional solar chimneys. While the electricity generation using concrete solar chimneys is extortionate due to their concrete chimney constructions, the height of FSCs is analogous to the yearlong power production by the SAEPs. Moreover, FSCs solar collector surface area along with the annual horizontal irradiation at the plant site is also consistent to the power generation capacity. Again the gross construction cost per produced KWh of the plant can be reduced using a single solar chimney unit of specific structural parameters. While contingent of floating solar chimney SAEPs, a group of similar SAEPs having the same dimensions can be used for producing the same annual amount of electricity. Hence, the FSC farm has greater efficiency and generates more electricity than the concrete solar chimney SAEP for the same solar collector area. Moreover, the maintenance of FSC is insubstantial and effortless corresponding to equivalent air fabric structures when their diameters are smaller. The optimum dimensions and power ratings of floating solar chimney SAEPs are significantly trivial relative to concrete solar chimneys. During 2005, Schlaigh et al. estimated the construction costs of a 30 MW concrete solar chimney power plant. The solar chimneys were labelled as solar updrafts towers by them. These solar updrafts towers were assumed to have a manufacturing expenditure of 145 million EURO (2005 market price estimations), while it was rationally presumed that both the electricity generating power plants will have the generation capacity of 99 million KWh/year with these approximations. But the FSC farm are capable of generating 30% additional electricity with an approximated construction cost of about 6 million EURO (2010 market price estimations) for per capita floating solar chimney SAEP of the farm. Hence, the gross construction cost for the entire FSC farm will be 54 million EURO. Prolonging a comprehensive study between the working capital for both floating solar chimney farm and the concrete solar chimney solar updraft tower having same energy production ratings was demonstrated. Papageorgiou (2010) demonstration shows that the capital expenditures for FSCs are 3 to 5 times smaller than the concrete solar chimney.

## Economic Analysis of Solar Chimney

Economic analysis anticipates an opportunities to assess the competitiveness of different types of Solar Chimney Power Plants. Bernardes (2004) estimated the cost of a 100 MW plant with plastic collector roof, which was carried out with a comprehensive calculation by Fluri et al. (2008, 2009). Carbon credits, which is one of the great environmental criterions, was included as an additional revenue, for the first time by Fluri et al. (2009). Papageorgiou estimated the cost of a 100 MW SCPP through which ruled out the carbon credits. Carbon credits come up with ecological balance, that might acknowledged as an additional review. In Northwest China, the electricity cost of potential SCPPs was estimated by Zhou and Yang (2009), which includes the support of high mountains in the deserts. Zhou et al. illustrated a comprehensive economic analysis about the electricity production among several types of floating solar chimney power plant (FSCPP) throughout the year 2009. Researches evaluated the economic analysis among discrete types of SCPPs. They are:

- The Floating Solar Chimney Power Plants (FSCPP), Zhou et al. (2009a, b).
- The sloped solar chimney power plant (SSCPP), Bilgen and Rheault (2005), Cao et al. (2011b).
- The double cover solar chimney power plant (DCSCPP), (Pretorius, 2007).
- The SCPP using the mountain cave as the chimney, Zhou et al. (2009a, b).

The SCPP with the collector sited on the horizontal direction proposed by Schlaich is proclaimed as the conventional solar chimney power plant (CSCPP) (Haaf et al., 1983a, b, 1984). Schlaich's proposal was recommended by Zhou et al. (2009a, b) by conducting a relative study among different types of SCPPs. However, several Chinese researchers estimated the economic analysis for the CSCPPs, while the economic evaluation of the SSCPP is infrequently publicized. Furthermore, there are diverse review on comprehensive techno-economic analysis on 100 MW CSCPPs, (Schlaich et al., 1995, 2005; Fluri et al., 2009; Papageorgiou, 2004a, b) rather qualified researches with different power capacities having focal point on economics of CSCPPs are barely disclosed. Cao et al. (2013) executed such remunerative survey on generation capacity in between the SCPPs operating in individual generation capacities. The Fei investigation was performed among the power capacity of 5 MW, 30 MW and 100 MW, respectively. The additional revenue generated by carbon credits was included in the analysis. A comprehensive inspection on economics of the CSCPP and the SSCPP were executed in Lanzhou, China. The analysis, which was basically a case study, involves their cash flows during their service period.

Distinct investigations were carried out to evaluate the lucrative potentials along with the competency in generations of distinct SCPPs (Schlaich, 1996; Schlaich et al., 2004; Bernardes, 2004). There were several estimations regarding the relative installation costs for all plant constituents among various plant ratings. Schlaich et al. (2004) appraised the levelized electricity cost (LEC) and its sensitivity to the interest rates along with the unit costs among dissimilar plants. The competitive investigation was carried out for several plants under firm viable frameworks including the length of the depreciation period. Bernardes (2004), executed a similar exploration between the SCPPs under different sizes. His exploration involves the vulnerability interpretation of LEC to the moneymaking criterions along with the component costs and LEC. Moreover, Bernardes demonstrated a specified installation cost replica involving solar collector, solar chimney and Power Conditioning Unit. Fluri et al. (2009) cost model was the foremost details tariff dummy for the PCU. The model introduced carbon credits as an additional achievement along with LEC. The model also brings to light an exhaustive outcomes of the economic analysis of Schlaich and Bernardes. Accompanying with the inspiration of that comparison, further investigations were carried out considering two SCPPs as a footnote having same size and a generation capacity of 100 MW, respectively. However, the comprehensive price replica of the solar chimney involves the following specifications:

(1) Constituents price (2) Manufacturing price (3) Lifting price (4) Conveyance price.

And the solar collector installation price includes the followings:

(1) Constituents price (2) Manufacturing price (3) Conveyance price.

While the Power Conditioning Unit price incorporated the cost of the following entities:

(1) Stability of station (2) Generators (3) Turbines (4) Ducts (5) Power electronics (6) Main structures (7) Controls (8) Supports.

Relative costs of components of the two reference SCPPs were estimated using the model. These comprehensive investigations of the relative values stated in the review of Schlaich et al. (2004) and Bernardes et al. (2004) were demonstrated in Figs. 10.1, 10.2 and Table 10.1 (Zhou et al., 2010).

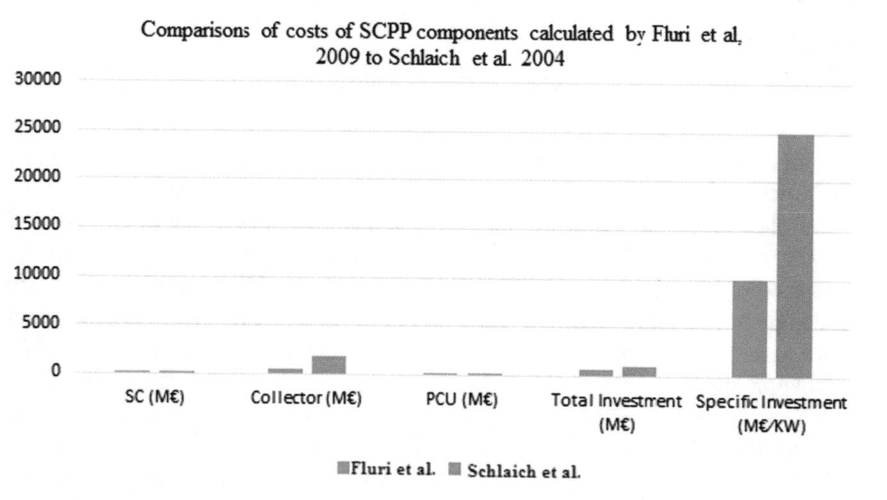


**Fig. 10.1** Relative costs of SCPP components reviewed by Fluri et al. (2009) to Schlaich et al. (2004) (Zhou et al. 2010)

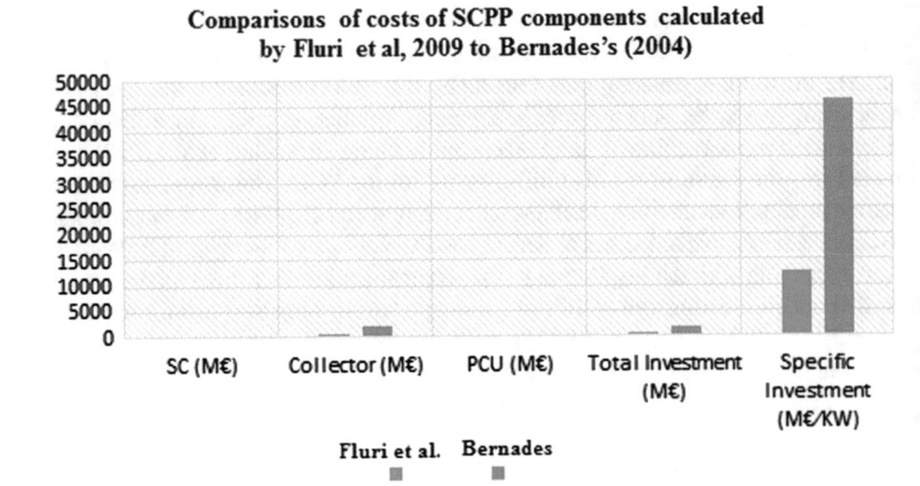


**Fig. 10.2** Relative costs of SCPP components calculated by Fluri et al. (2009) to Bernades (2004) (Zhou et al. 2010)

**Table 10.1** Relative LEC calculated by Fluri et al. (2009) to Schlaich et al. (2004) and Bernades's (2004) calculations (Zhou et al. 2010)

| | Fluri et al. | Schlaich et al. | Fluri et al. | Bernardes |
|---|---|---|---|---|
| Yearly power rating (GWH) | 190.4 | 320.0 | 181.3 | 281.0 |
| Reserve funds (M€) | 668.4 | 402.0 | 791.5 | 325.4 |
| Aggregate PV of OM (M€) | 38.9 | 34.9 | 923.2 | 379.5 |
| LEC (€ / KWh) | 0.270 | 0.1 | 15.59 | 15.59 |
| LEC involving carbon credits (€/KWh) | 0.232 | N/A | 0.430 | 0.125 |

Fluri et al.'s (2009) appraisal used Pretorius's thermodynamic model (Pretorius, 2007), which included the generation output of the plants considered for investigation. Simulation outcomes show a low-ranking generation output of 66 MW and 62 MW, which was cited from Schlaich and Bernardes's researches, respectively. Both the estimation exclude 100 MW reference plants. LEC was estimated with the same remuneration constants like profit rate = 6%, inflation rate = 3.5%, and active period = 30 years for both the reference plants having capacity of 100 MW. Estimation shows a better LEC outcome for both the reference plants that were considered for relative analysis shown in the above table. The LEC was re-evaluated using Bernardes's model with increased profit rate of 8% along with an inflation rate = 3.25% for the same working duration of 30 years. The manufacturing period estimated for that assumption was 2 years. Fluri et al. (2009) reviewed a reduced LEC, which was literally cited by Bernardes (2004) on account of a mistaken assumption. The SCPP excludes the distinctive $CO_2$ discharge of 0.95 kg per generation output from coal-fired power plant during the operation (US Energy Information Administration, 2006). These observations acquired enormous proportion of carbon credits for SCPP. Undisputedly the manufacturing of SCPP requires negligible fossil fuels consumption in comparison with conventional coal-fired power plants. Coal-fired power plants consume more fossil fuels because of long lasting service period of reinforced concrete solar chimney. The fossil fuel consumption of such plants is proportional to the aggregate of service life of two or more coal-fired power plants. Moreover, this model comprises the prospective consequence of carbon credits on LEC, which decrease the LEC. Generally, the lifespan of a reinforced concrete SC is about a centennial that inclines to additional moderation in SCPP LEC.

## Additional Revenues

Substantial use of lands is a significant agitate with solar technologies. Necessity of large coverage area appears because of less sensitivity on energy accumulation of sunlight. The installation of solar collector for SCPP covering a large area that intensifies the installation price very enormously was discussed in previous section. Utilization of large solar collector may be appreciated as an additional revenue sources. Ground covered by the solar collector acts as greenhouse that can be used for cultivating vegetables or fruits. This provides an opportunity to optimize the disadvantage of large installation cost to a source of possible additional revenues. But in order to cultivate foods using the ground covered by the solar collector, proper irrigation with fresh water is required. However, insufficiency of water source all over the place could be a complication for the cultivation of foods. Substantially the prospective sites of SCPPs are selected in deserts, where land is reasonable and sunlight is ample. Hence, for maintaining the consistency in agricultural activities along with the power generation, land selection becomes an optimum challenge. Collaboration between power productions and agricultural might appear to be a vital issue because of a damp cultivated ground is often darker than a dry flat. Moreover, evaporation may diminish the power output significantly on account of the conversion of segments of solar heat to latent heat. South African researchers demonstrated a combined project of a SCPP and a large profitable greenhouse. During 1998, the project titled as 'Greentowers' was initiated proposing a remedy for evaporation principle. Researchers implemented some black 'shadowing nets' involving multiple purposes (Bonnelle, 2008). During daylight solar radiation will be subsumed by those black shadowing nets which preserve the focal origin source of sensible heat to the moving air. Hence, no convection will occur and a hot environment than the agricultural greenhouse air will be maintained, whereas nightfall, the ground is warmer than the black shadowing nets. That emphasizes the air convection to prevail towards the black shadowing nets. The agricultural greenhouse gentle balmy air along with the cold solar collector develops a thermal contact point that will be performed by the black shadowing nets. Throughout the evaporation process, the continuous water consolidation coming from the ground with cooler bottom surface may introduce some dew on account of heat exchange which will retreat the humus. 'Shadowing nets' will avert the gateway for the escaping the solar chimney eliminating the tendency of wasting the fresh water and latent heat. The entire system will develop a very dark albedo by using the 'shadowing nets' (Zhou et al., 2010). Usually a slow photosynthesis phenomenon may occur due to shadowing a greenhouse. But, the light ambiences in bright sunlight regions can be accommodated using those black shadowing nets, where preservations of temperature and humidity are an additional advantage. Moreover, the superficial area of the solar collector can enhance the agricultural production significantly. Space covered by 2/3 solar collector can assemble a supplementary production about 170%. Meanwhile, these applications of heating collector operating air imply an effective cultivation including the subsidiary portion to the existing land. Hence, cultivation of foods in the solar collector also could enhance heat and generation capacity. (Greentowers, 2000).

## Conclusion

Flourishing energy demand has become distressing worrying widespread due to the fluctuating fossil fuel prices. Besides, deterioration of ecological balance is caused by the use of conventional fuels as the toxic emissions and global warming. Solar chimney power plants are a compulsive substitute to centralized electricity generation power plants. Solar chimney power plant is an innovative kind of solar thermal applications consisting of technological simplicity and simple operation of the installation. Developing countries having lack a sophisticated technical infrastructure can accommodate solar chimney power plants along with their conventional energy production systems. Large-scale solar chimneys can be built now without any technical problems and at defined costs. Installation costs are high but cost estimations made for chimneys of higher size show that energy cost decreases as size increases and the simultaneous use of greenhouses collector in drying or in agriculture.

Countries having high population growth emphasizes on high-rise building construction for providing adequate accommodation, thus involving interruption of natural air ventilation, which impels the people to use more energy consuming equipment's like AC, fan, etc. The solar chimney system is one of the distinguished techniques introduced in building ventilation for sustainable initiation. These provide an opportunity to reduce energy demand in domestic and commercial buildings. Inadequate authentic and rational outcomes originate an obstacle for hypothetical studies along with practical implementations. Moderate research and subsidy are required to cost optimization and improving the effectiveness and to make it in fascinating design. Such efforts emphasize the feasibility of providing a sensible technology, accessible to the technologically less developed countries based on environmentally sound production from renewable materials.

## References


Bernardes, M. A. D. S. (2004). Ph.D. dissertation, Universitat Stuttgart, Germany.

Billinton, R. (2005). Evaluation of different operating strategies in small stand-alone power Systems. *IEEE Transactions on Energy Conversion, 20*(3), 654–660.

Bonnelle, D. (2008). Private communication.

Bilgen, E., & Rheault, J. (2005). *Solar Energy, 79*(5), 449.

Cabanyes, I. (1903) Las chimeneassolares (Solar chimneys). La ernergiaeléctrica.

Cao, F., Zhao, L., & Guo, L. J. (2011a). *Energy Conversation Management, 52*(6), 2360.

Cao, F., Zhao, L., & Guo, L. (2011b). Simulation of a sloped solar chimney power plant in Lanzhou, China. *Energy Conversion and Management* 2360–2366.

Cao, F., Li, H., Zhao, L., & Guo, L. (2013). Economic analysis of solar chimney power plants in Northwest China. *Journal of Renewable and Sustainable Energy, 5*(2), 021406.

Chikere, A. O., Al-Kayiem, H. H., & Karim, Z. A. A. (2011). Review on the enhancement techniques and introduction of an alternate enhancement technique of solar chimney power plant. *Journal of Applied Sciences., 11,* 1877–1884.

Cloete, R. (2008). Solar tower sheds light on little-used technology. Engineering News Online. https://www.engineeringnewscoza/article.php.

Dhahri, A., & Omri, A. (2013). A review of solar chimney power generation technology. *International Journal of Engineering and Advanced Technology (IJEAT).*

Fluri, T., & Vonbackstrom, T. (2008). *Solar Energy, 82*(11), 999.

EnviroMission Limited-Technology (2006). Technology Overview. https://www.enviromission.com.au.

Fluri, T., Pretorius, J., Dyk, C., Backstrom, T., Kroger, D., & Zijl, G. (2009). *Solar Energy, 83*(3), 246.

Greentower. https://www.greentower.net.

Haaf, W., Friedrich, K., Mayr, G., & Schlaich, J. (1983a). Solar chimneys part I: Principle and construction of the pilot plant in Manzanares. *International Journal of Solar Energy., 2,* 3–20.

Haaf, W., Friedrich, K., Mayr, G., & Schlaich, J. (1983b). *International Journal of Solar Energy, 2,* 3.

Haaf, W. (1984). *International Journal of Solar Energy, 2,* 141.

Hong, C. S., & Lee, E. B. (2018). Power plant economic analysis: maximizing lifecycle profitability by simulating preliminary design solutions of steam-cycle conditions. *Energies, 11*(9), 2245.

Hamdan, M. O. (2011). Analysis of a solar chimney power plant in the Arabian Gulf region, United Arab Emirates. *Renewable Energy* 2593–2598.

Jiakuan, Y., Jin, L., & Po, X. (2003). A novel technology of solar chimney for power generation. *ACTA Energiae Solaris Sinica., 24,* 570–573.

Kasaeian, A. B., Molana, S., & Rahmani, K. et al. (2017). A review on solar chimney systems. *Renewable and Sustainable Energy Reviews, 67*, 954–987. ISSN 1364-0321.

Ley, W. (1954). *Engineer's dream.* Viking Press.

Mohammed, H., Alktranee, R., Yaseen, D. T. (2019). Evaluating the performance of solar chimney power plant. *International Journal of Contemporary Research and Review, 10*(6)

Muselli, M. N. G. L. A., Notton, G., & Louche, A. (1999). Design of hybrid-photovoltaic power generator, with optimization of energy management. *Solar Energy, 65*(3), 143–157.

Nizetic, S., Ninic, N., & Klarin, B. (2008). Analysis and feasibility of implementing solar chimney power plants in the Mediterranean region. *Energy, 33,* 1680–1690.

Olusola Olorunfemi, B. (2014). A review of solar chimney technology: Its' application to desert prone villages/regions in Northern Nigeria. *International Journal of Scientific & Engineering Research, 5,* 1210–1216.

Papageorgiou, C. D. (2004a). *Proceedings of ISES Asia-Pacific solar energy conference*, Gwangju, Korea, 17–20 October, pp. 763–772.

Papageorgiou, C. (2004b). Efficiency of solar air turbine power stations with floating solar chimneys. In *IASTED proceedings of power and energy systems conference Florida*, pp. 127–134.

Papageorgiou, C. D. (2007). Floating solar chimney versus concrete solar chimney power plants, Athens. *IEEE, 2007,* 761–765.
Papageorgiou, C. (2009). Floating solar chimney. E.U. Patent 1618302.
Papageorgiou, C. D. (2010). Floating Solar Chimney Technology.
Pretorius, J. P. (2007). Ph.D. dissertation, University of Stellenbosch, South Africa.
Sagasta, F.M. (2003). Torre solar de 750 metros de alturaen Ciudad Real (España). ed2003.
Schlaich, J. (1995). *The solar chimney: Electricity from the sun.* Stutgart: Axel Mengers Edition.
Schlaich, J. (1996). The solar chimney: electricity from the sun. Stuttgart, Germany: A. Menges (1995).
Schlaich, J., Bergermann, R., Schiel, W., & Weinrebe, G. (2004). Sustainable electricity generation with solar updraft towers. *Structural Engineering International.*
Schlaich J. et al. (2005). Design of commercial solar updraft tower systems-utilization of solar induced convective flows for power generation. *Journal of Solar Energy Engineering, 127*, 117–124R.
Solar Energy. ISBN 978-953-307-052-0
US Energy Information Administration (2006).
Vlachos, N.A., Karapantsios, T.D., Balouktsis, A.I., & Chassapis, D. (2002). Design and testing of a new solar tray dryer. *Drying Technology, 20*, 1243–71.
Yang, H., Wei, Z., & Chengzhi, L. (2009). Optimal design and techno-economic analysis of a hybrid solar–wind power generation system. *Applied Energy, 86*(2), 163–169.
Zhou, X. P., Yang, J. (2009). *Heat Transfer Engineering, 30*(5), 400.
Zhou, X. P., Yang, J. K., Wang, F., & Xiao, B. (2009a). *Renewable Sustainable Energy Reviews, 13*(4), 736.
Zhou, X. P., Yang, J. K., Wang, J. B., & Xiao, B. (2009b). *Energy Conversion and Management, 50*(3), 84.
Zhou, X., Wang, F., & Ochieng, R. M. (2010). A review of solar chimney power technology. *Renewable and Sustainable Energy Reviews, 14,* 2315–2338.
Zhou, X., & Yang, J. (2012). *A novel solar thermal power plant with floating chimney stiffened onto a mountainside and potential of the power generation in China's Deserts Wuhan*, China, pp. 400–407.
Zhai, X. Q., Song, Z. P., & Wang, R. Z. (2011). A review for the applications of solar chimneys in buildings. *Renewable and Sustainable Energy Reviews, 15*(8), 3757–3767.

# Chapter 11
# Technical Barriers Analysis for Solar Chimney Power Plant Exploration

**Md. Tarek Ur Rahman Erin, Mohammad Mashud, Fadzlita Mohd. Tamiri, and Md. Mizanur Rahman**

Over the past century, for electricity generation, the main energy sources used are fossil fuels, hydroelectricity and, since the 1950s, nuclear energy. Despite the strong growth of renewable energy sources over the last few decades, fossil-based fuels remain dominant worldwide. Over the time, new power-generating sources are added in power generation technology, from water and coal to oil and gas to the atom and, more recently, the wind and solar. Improvements in old technologies continued to be made, while new power sources were added, targeting cleaner-burning coal and performance enhancement. Until now, electricity generation using thermal power plants is about 40, and 25% of it is generated by combusting natural gas or oil petroleum. Electricity production by nuclear technology is around 13–14%, and hydropower is 16%. Other resources like wind, solar PV, solar thermal and geothermal are producing 4%, while biomass contributes by 2% for electric power generation. However, the natural gas power sector, which today dominates the power generation sector, was slower to replace the existing coal and oil fuel-based power generation. On the other hand, solar energy seems the main focus as a clean renewable energy source for the 2050 and above energy demand. Because of high energy demand globally, increased use of fossil-based fuels causes threat to our environment. On the contrary, pollution free energy source like solar chimney power plant

Md. T. U. R. Erin (✉) · Md. M. Rahman
Department of Mechatronics Engineering, Faculty of Engineering, World University of Bangladesh, Uttara, Dhaka, Bangladesh
e-mail: erin@mte.wub.edu.bd

M. Mashud
Department of Mechanical Engineering, Khulna University of Engineering and Technology (KUET), Khulna, Bangladesh

F. Mohd. Tamiri (✉)
Department of Mechanical Engineering, Faculty of Engineering, Universiti Malaysia Sabah, Kota Kinabalu, Sabah, Malaysia



Md. M. Rahman and C.-M. Chu (eds.), *Cold Inflow-Free Solar Chimney*,
https://doi.org/10.1007/978-981-33-6831-6_11

(SCPP) offers great opportunities for future power generation technology. To overcome the barriers and challenges in solar technologies, including PV, bio-solar and thermal conversions, the upcoming years are representing the research challenge. This paper presents some of the humble efforts paid to resolve some of the technical barriers in solar technologies. Many researchers around the world have introduced various projects of solar tower. Around 1500, a sketch of a solar tower called a smoke jack was made by Leonardo Da Vinci (Wengenmayr and Bührke, 2011). In 1902, Spanish engineer Isodoro Cabanyes first proposed the concept of employing the solar chimney technology in power generation. In 1931, German science writer Hans Gunther elaborated another solar chimney technology (Ngo and Natowitz, 2009). In 1903, Gunther proposed a design issue of "La Energia Eléctrica", entitled "Projecto de motor solar". In this design, a collector carries heated air to upwards toward a pentagonal fan. This arrangement covered by rectangular brick structure resembling a fireplace. Fan rotates as the heated air moves upward and generates electricity. After that, heated air escapes through a chimney of 63.87 m tall (Meyer & Mancha, 2008). In 1926, after studying several sand whirls in the southern Sahara, Prof Engineer Bernard Dubos proposed the construction of a Solar Aero-Electric Power Plant in North Africa (Hamilton, 2011).

The author claims that an ascending air speed of 50 m/s can be reached in the chimney, whose enormous amount of energy can be extracted by wind turbines (Ley, 1954). The academy suggested to implement the idea of Dubos in French North Africa where there were no fossil-based fuel sources. For that reason, Dubos developed his plans considering the North African Atlas Mountains in mind (Ley, 1954). In 1956, Dubos filed his first patent in Algeria. In this concept, a sort of round-shaped Laval nozzle generating ancestry atmospheric vortex artificially is mentioned. A French patent was received by Nazare for his invention in 1964 (Hamilton, 2011).

In 1975, the American Robert Lucier filed a patent request based upon a more complete design. This patent was granted in 1981 (Von Backström et al., 2008). In 1970s, Professor Schlaich brings forward this concept in some conferences and built a prototype of SCPP in Manzanares, 150 km south of Madrid, Spain. The height of SCPP chimney was 194.6 m, the diameter of the chimney was 10.8 m, and the collection radius was 122 m. For improving the design, researchers spent continuously 9 years. The power plant operated for approximately 7 years. In 1989, the tower was decommissioned due to corrosion, rust and storm winds (Von Backström et al., 2008). In 2002, this project was identified as one of the best inventions of the year by Time Magazine. The concept behind this revolutionary technology is that warm air moves upward (Ngo and Natowitz, 2009). After that, the research on SCPP has been attracted worldwide. The solar chimney power plant is a naturally driven power-generating system. It is a natural power generator that uses solar radiation to increase the internal energy of flowing air. In this technology, first solar energy is converted into thermal energy then into kinetic energy finally into electrical energy. It combines the concept of solar air collectors and a central updraft chimney to generate a solar induced convective flow which drives turbines to generate electricity. For generating electricity using solar radiation, the solar chimney power plant (SCPP) technology is a renewable source of energy. This technology is comprised of four sections: chimney, collector, energy storage layer and power conversion units (PCU).

The collector's main function is to heat up the air inside with the aid of solar radiation. As at the same height, the air density inside the system is less than that of the environment, buoyancy as the driving force comes into existence. This cumulative buoyancy results in a large pressure difference between the system and the environment. The chimney is erected in the middle of the collector; the heated air then moves upward to the chimney. If a turbine is placed where there is a huge pressure drop (at the bottom of the chimney or near the outlet of the collector), the potential and heat energy of the air can be converted into kinetic energy which then turns into electric energy.

Power generation based on renewable energy resources will definitely become a new trend of future energy utilization. Solar chimney power plant (SCPP) technology is one of renewable energy sources utilizing solar radiation. The solar chimney power plant (SCPP) technology that has a long life span is one of the most promising approaches for future large-scale solar energy applications. This paper presents a brief analysis of technical barriers of solar chimney power plant technology.

Al-Kayiem (2019), completed studies on the technical challenges and solutions for power generation using solar chimney power plant. In this article, they identified three major technical barriers in the solar-to-power energy conversion. First, the main technical setback is the low conversion efficiency of the solar-to-power energy conversion. Second is that large size land is required for harvesting solar energy to become feasible. And the third issue is the interruption during the night and cloudy days. In addition, there are some other setbacks like demand/production mismatch, and integration with energy storage arises in solar chimney power production technology (Al-Kayiem, 2019).

Ming et al. (2010), established a simple analysis on the performance of solar chimney power plant (SCPP). According to their study, it is found that a very wide space is required for producing a sufficient amount of energy due to low overall energy conversion efficiency of the solar chimney power plant (SCPP). On the other hand, the size of the plant (height of the tower and the diameter of the collector) becomes the major criteria for large capacity SCPP power generation technology, really considerable. Besides, the capital cost of the power plant cannot be underestimated. And it might be risky for investor to invest for this technology. The overall efficiency of the SCPP depends on the chimney's updraft efficiency, collector's greenhouse efficiency, efficiency of the turbines and generators. Chimney plays an important role for increasing the overall efficiency of the SCPP. As the chimney height increases, the overall efficiency also increases. Besides, the collector collects heat by absorbing sunlight. System output power and energy being stored increases with larger collector diameter. Considering the commercial use with an output power up to 100 MW of an SCPP, the collector diameter should be several kilometers, and the chimney should be about 1000 m. It will cause difficulty in cleaning the collector when collector size is large. At the same time, it will be difficult for constructing large chimney (Ming et al., 2010).

Studies conducted by Pasumarthi and Sherif (1997) that study showed that the collector performance can be modified by extending the collector base and by introducing an intermediate absorber. The modification helps to increase power output by increasing air temperature and mass flow rate inside the chimney (Pasumarthi & Sherif, 1997).

In the year 2010, Koonsrisuk et al. analyzed the effect of changing the tower area on both efficiency and the mass flow rate through the plant in a solar chimney power plant. Their study showed that, at the top of a convergent tower, velocity increases, but the mass flow rate remains same. At the same time, velocity increases near the base of the chimney (Koonsrisuk et al., 2010). Hamdan (2010) published results on a thermodynamic study of steady airflow inside a solar chimney. In order to predict the performance of the solar chimney power plant, a simplified Bernoulli's equation combined with fluid dynamics and ideal gas equation using EES solver is used. The study showed that both the chimney height and chimney diameter of the solar chimney are the most important design variables for solar chimney power plant. However, from the study, it is found that collector area has small effect on second-law efficiency but strong effect on harvested energy (Hamdan, 2010). Von Backström and Gannon (2004) presented an air standard cycle analysis of the solar chimney power plant to show the relationship between different variables to evaluate its performance. Their study includes chimney friction, system, turbine and exit kinetic energy losses in the analysis (Von Backström & Gannon, 2004).

In the year 2007, Pretorius has developed a numerical simulation model to examine the improvement and control of solar chimney power plant. The effects of ambient wind speed, temperature, lapse rates and night time energy storage system temperature variation on plant performance are observed. The work indicated that, the plant performance is highly affected by the collector reflectivity, emissivity, ground surface absorptivity (Pretorius, 2007). Backström et al. (2000), have been completed an analysis of the cycle of a chimney with the simple air standard cycle. Considering a simple air standard cycle as it determines the upper performance limits of the ideal solar chimney power plant. In the year 2002, Von Backström et al. developed a simple thermodynamic cycle analysis that includes system losses in the solar chimney power plant. Gannon and Von Backström (2000) analyzed and investigated that the chimney height and the solar collector temperature rise are the two major dominant parameters in terms of power output in the solar chimney power plant. Study also showed that the chimney height, collector height, collector diameter have a great impact on the performance of solar chimney power plant (SCPP). These technical parameters have effect on the power output and efficiency of the SCPP (Ahmed & Hussain, 2018). Ngala et al. (2013) have been traced out some technical obstacle constructing SCPP. Reinforced concrete is used for constructing conventional solar chimney. Although the reinforced concrete solar chimney has a long operating life and can be constructed as high as possible for improving the efficiency of SCPP, it has some demerits. High construction cost and limited height of the solar chimney due to the technological constraints are the major demerits. There is also risk of earthquakes, which can diminish solar chimney (Ngala et al., 2013).

In the year 2005, Schlaich et al. published an article and introduced collector roof and additional closed water-filled thermal storage system for monitoring and increasing power output from solar chimney power plant (SCPP). From the study, it is found that rather than single roof, intermediate secondary roof offers much more uniform output profile. At the same time, incorporating of additional closed water-filled system enhances the opportunities for monitoring and increasing power output from SCPP (Schlaich et al., 2005; Ling et al., 2017).

A study was done by Somsila et al., in the year 2010 and mentioned that in solar chimney technology energy is produced by using convection of hot air by passive solar energy. Air will flow through wind turbines by creating temperature differences between the collector and the ambient due to buoyancy force which would allow turbines to produce electricity. Solar chimney is defined as low temperature solar thermal power plants, which use the atmospheric air as a working fluid, where only one part of the thermodynamic cycle within the plant is utilized. From their work, it is found that temperature difference between the collector outlet and the ambient is a dominant parameter for the performance of SCPP. As SCPP needs high density solar radiation, at the same time, it makes decrease in temperature difference between the collector and the ambient, which results in less power output and less efficiency of SCPP. For this, under very high temperature, it may need to be employed with a cooling system in order to avoid loss in mass flow rate of air to the upright direction (Somsila et al., 2010; Rahman et al., 2018). Suhendri et al. (2018) discussed about the relation between average air velocity and chimney height. Higher air velocity goes into chimney outlet and air with lower velocity air stacked at the bottom because of stack effect, air movement that goes inside and outside the chimney because of buoyancy that occurs due to the different of air density inside and outside the chimney because of temperature and moisture differences (Suhendri et al., 2018). Bansod et al. (2014) published an article and suggested to use asphalt concrete as solar collector due to its high absorbing capacity of solar radiation. In this process, they used integrating pipes conducting liquid, through the structure of the asphalt concrete. In this process, one of the main identified problems is that if any of the pipe break down then liquid will spill out; eventually, it will cause damage to the asphalt concrete. From this article, it is also found that solar chimney size affects the output of plant. To generate 200–400 MW output, chimney tower should be 1000–1500 m tall, 160 m diameter (Bansod, et al., 2014). Agelaridou-Twohig and his team proposed to use reinforced concrete (RC) chimneys with fiberglass reinforced plastic (FRP) liners considering burning characteristics of the liner material. A parametric study is carried out to identify the chimney residual strength. Their study shows that structure failures would not occur after the chimney's post-fire structural capacity if any high scale lateral loads do not affect the chimney at the same time (Agelaridou-Twohig et al., 2014). A parametric study is carried out in order to find the relation between power generated per unit of land area and length scale of the power plant. After analyzing the geometry of a solar chimney for increasing the power generation capacity of solar chimney power plant, it is found that the pressure drop in the chimney depends on the friction loss in the collector, cold inflow in the chimney exit as well as losses at the entrance and exit of the plant (Koonsrisuk et al., 2010; Chu et al., 2012). Another researcher Ninic analyzed various collector types using dry and humid air and published the experimental results in 2006. The focused was on the effects of air flow into the air collector, air humidity and atmospheric pressure on various chimney heights. From this study, it is observed that under pressure the vortex motion moving downstream of the turbine can be controlled and acted like solid structure chimney (Ninic, 2006).

In the year 2008, Fluri printed an article and proposed a method to discover the layout and the number of turbines required for solar chimney assisted power plant for generation of electricity at low cost. In this work, the researcher conducted a numerical investigation on turbine design. Their objective is to identify a commercial application program as a tool in context with solar chimney turbines. A detailed cost model for the power conversion unit is also analyzed (Fluri, 2008). Bonnelle, in the year 2004, proposed some technical advancements of the solar tower to increase the overall efficiency. Their objective is to develop a larger solar collector with bigger towers having much technical advancement (Bonnelle, 2004). Atit (2009) has proposed theoretical, experimental and numerical approaches to increase the efficiency of a solar chimney. Determination of the scaling law for the flow in solar chimney systems using dimensional analysis was carried out and examined by using the computational fluid dynamics technique (CFD). Their work suggested that the flow area ratio can increase the plant performance (Atit, 2009).

Pretorius performed a sensitivity analysis presented in the year 2007 by using numerical simulation model on the most well-known operating and technical plant conditions. They analyzed the effects of ambient wind, temperature lapse rates and nocturnal temperature inversions on plant operational efficiency. For monitoring plant output according to specific demand patterns, new technologies are implemented. Their work showed that plant performance can be improved by the modification of the collector roof reflectance, collector roof emissivity, ground surface absorptivity or ground surface emissivity. Their work also suggested that as a base or peak load power station, solar chimney power plant can be utilized (Pretorius, 2007). Papageogiou et al. (2011) have been suggested to use low-cost greenhouse for applying in floating solar chimney (FSC) technology. The floating solar chimney can be raised anywhere and is lighter than air structure, and its cost is as low as 2% of the cost of the respective concrete chimney. In desert installations of the FSC technology, dust is a major problem. Their work also suggested that plastic covered low-cost greenhouse could solve this problem. A scale analysis of the FSC technology was also carried out according to the construction cost and the electricity generation capacity (Papageorgiou et al., 2011). Bernardes et al. (2003) have been developed a comprehensive analytical and numerical model to evaluate power output of solar chimney as well as to examine the effect of various ambient conditions and structural dimensions on the power output. Their work showed that important parameters for the design of solar chimneys are height of chimney, factor of pressure drop at the turbine, the diameter and the optical properties of the collector. Their study proved that by increasing the chimney height, collector area and the transmittance of the collector, the power output of SCPP can be improved (Bernardes et al., 2003). Von Backström and Fluri (2006) have been completed a study to evaluate the optimal ratio of turbine pressure drop to available pressure drop and thereby developed a method to maximize the fluid power in a solar chimney power plant. In their work, estimation of the performance of a solar chimney power plant was also carried out (Von Backström & Fluri, 2006).

Koonsrisuk and Chitsomboon (2013) predicted the performance characteristics of large-scale commercial solar chimney using mathematical model. Their work showed that important factors for performance enhancement of solar chimney power plant are the plant size, the factor of pressure drop at the turbine and solar heat flux. In their study, they also proposed a simple method using dimensional analysis to examine the turbine power output for solar chimney systems (Koonsrisuk and Chitsomboon, 2013). Harte et al. (2013) studied the working procedure of solar updraft power plants. Their work also analyzed the durability requirements of SCPP for operation in extreme desert climates (Harte et al., 2013).

In 2003, Chena et al. carried out experiments for different chimney gap-to-height ratio, variable heat flux and inclination angles of a model solar chimney. From their experiment, it is found that at an inclination angle around 450 for a 200 mm gap and 1:5 m high chimney, maximum air flow rate was attained. They also predicted the airflow rate for the vertical chimneys geometry using the prediction method available in the literature. In their work, they also evaluate the pressure losses at the chimney outlet by using loss coefficients obtained for normal forced flows (Chen et al., 2003). Maia et al. performed an analytical and numerical study of the unsteady airflow inside a solar chimney in 2009. The results indicate that rather than operational and geometric configurations found in the experimental prototype, they proposed a model to simulate airflow in solar chimneys. The research work showed that for solar chimney design, height and diameter of the tower are the most valuable technical factors (Maia et al., 2009). In 2007, Kasayapanand presented the results after analyzing the mechanism of natural convection inside the inclined solar chimneys using numerical simulations. The research work showed the effect of volumetric flow rate, pressure drop, heat transfer rate inside the solar chimney for optimization and economy design of solar chimney (Kasayapanand, 2008). Nizetic and Klarin (2010) approached a simplified analytical method to evaluate characteristic factor (pressure drop ratio in turbines) in SCPP. In their work, for the evaluation of the optimal pressure drop ratio, a simplified analytical approach is carried out. It is found that air flow rate and the velocity at the solar chimney inlet are the important factors that dominate the performance of solar chimney power plant (Nizetic & Klarin, 2010).

Tingzhen et al. (2008) evaluated the maximum power output of the solar chimney power plant prototype using numerical simulations. Under this research work, analyzed the chimney outlet parameters on the basis of turbine rotational speed. Numerical simulation result has been presented for a MW-graded solar chimney power plant system (Tingzhen et al., 2008).

Harris and Helwig (2007) investigated the performance of a solar chimney by using computational fluid dynamics (CFD) simulation software. It is found that slope of the chimney affects the performance of solar chimney. The focus of the study is also extended to improve performance as well as reduce the risk of overheating by varying the slope angle of solar chimney (Harris and Helwig, 2007). Another study is done by

Jing-yin Li et al., in the year 2012. The aim of the study is to analyze solar chimney performance at unloaded condition; the installation of the turbine in the solar chimney power plant (SCPP) has significantly lessened the power output of the SCPP. Their study proved that for generating maximum power output of SCPP, a certain amount of solar radiation is required. The critical value of the solar radiation depends on the mass flow rate and turbine generator, but in this study, the analysis was done for the solar radiation 500 w/m$^2$ to 1000 w/m$^2$. In addition, to achieve a longer uninterrupted power output, the left-hand-side of the maximum power line would be suggested to select the operating points of the turbine and the SCPP system. This research work is also identified that collector radius has some maximum limiting value 380 m beyond which power output of the SCPP increases slightly. On the other hand, in case of chimney height, no such findings are obtained (Li et al., 2012). Zandian and Ashjaee (2013) designed a model incorporating a thermal steam power plant dry cooling tower with a solar chimney to evaluate the effects of environmental temperatures and solar irradiations on the generated turbine power. The effects of chimney diameter on the hybrid system (HCTSC) power output and plant efficiency have been analyzed in their research work. The results indicated that for the same environmental conditions output power of the hybrid system is many times greater than the experimental results for the conventional solar chimney power plant prototype with similar geometrical dimensions in Manzanares (Zandian and Ashjaee, 2013). In the year 2014, a model is proposed by Guo et al., which is incorporating the radiation, solar load and turbines models. The effects of solar radiation, turbine pressure drop and ambient temperature on system performance are investigated in their work. From the simulation results, it is found that for preventing the overestimation of energy absorbed by the solar chimney power plant, the radiation model is essential. From their research, it is found that ambient temperature, radiation heat transfer, heat losses, the effects of solar radiation and turbine pressure drop affect the performance of SCPP considerably (Guo et al., 2014). Gholamalizadeh and Kim (2014) developed a three-dimensional unsteady CFD model simulating the greenhouse effect to evaluate the solar chimney power plant system. Their analysis predicted that characteristics of the flow and heat transfer in the system are influenced by the greenhouse effect through the collector. Finally, their results showed that for describing all the phenomena occurring in SCPP systems exactly, simulating the greenhouse effect is a significant parameter (Gholamalizadeh and Kim, 2014). Shen et al. (2014) proposed recommendation on the solar updraft power plant (SUPP) specifies that the effect of ambient crosswind (ACW) is very composite on SUPP. Demonstration from the Spanish prototype cites the numerical analysis on either chimney outlet and collector inlet. Apart from the impact of ACW, utilization through chimney outlet and collector inlet autonomously on the entire execution of SUPP is still ambiguous. Two geometric prototypes are considered for numerical simulation throughout the research. Those prototypes were on industrial-scale SUPPs in territory of 10 MW. The outcome of the explorations demonstrates that, along with cold ambient air into the collector emerging adjustment to fluid distribution and deterioration of buoyant driving force, implies the negative effect of ACW at the collector inlet. Whereas the positive effect that occurs at the chimney outlet causes entrainment of buoyant airflow with strong ACW passing through the chimney outlet. The overall power generation from SUPPs can be improved by avoiding deterioration with effective measures that can be apprehended to prevent ACW from entering the collector inlet. Hence, the beneficiary effects of high altitude strong ACW blowing across the chimney outlet will be persuaded (Shen et al., 2014; Chu et al., 2012; Rahman et al., 2014).

Andrezzi et al. (2009) completed and presented a parametric study on a channel-chimney system was conducted in order to evaluate some geometric optimal configurations in terms of significant dimensionless geometric and thermal parameters. The study illustrates that the channel walls are symmetrically heated at uniform heat flux and temperature wall profiles as a function of axial coordinate. Assunta also evaluated the thermal performances of the channel-chimney system in terms of maximum wall temperatures for different expansion ratios, which was determined as a function of the chimney aspect ratio. The difference between the highest and the lowest maximum wall temperature will enhance channel aspect ratio considering Rayleigh number values. The study also demonstrates the correlations between dimensionless mass flow rate, maximum wall temperature and average Nusselt numbers. Moreover, the correlation was expressed in terms of Rayleigh number and dimensionless geometrical parameters (Andreozzi, et al., 2009; Rahman et al., 2014).

In the year 2007, a renowned researcher Bacharoudis and the teams carried out a comprehensive study of the thermo-fluid phenomena occurring inside wall solar chimney that have been constructed and put at each wall and orientation of a small-scale test room. A qualitative analysis was performed between the buoyancy-driven flow field and heat transfer that take place inside the wall solar chimney. A control volume method was also implemented for solving the governing elliptic equations are solved in a two-dimensional domain. This method is appropriate for the simulation of solar chimneys of different aspects ratios and conditions. Several methods were used for a relative study between the numerical simulations of the turbulent flow inside the wall of solar chimney. Simulation results also illustrate the significance of the predictions achieved by the study for distinct environmental conditions, while they support the air mass flow rate that can be achieved through this system and the turbulence effects (Bacharoudis et al., 2007). Zhou et al. in 2009 demonstrated the maximum chimney height for convection avoiding negative buoyancy and the optimal chimney height for maximum power output using a theoretical model. The model was supported utilizing the measurements of the only one prototype in Manzanares that shows the standard lapse rate of atmospheric temperature. An optimal chimney height of 615 m is used for the maximum power output of 102.2 kW, which is lower than the maximum chimney height with a power output of 92.3 kW. The results obtained from the study illustrate that maximum height gradually increases with the increasing lapse rate and reach to infinity at a value of around 0.0098 $km^{-1}$. That the maximum value for the convection and the optimal heights for maximum power output increase with large collector radius. Furthermore, the power plant would obtain the maximum energy conversion efficiency if chimney height is equal to the optimal height (Zhou et al., 2009; Rahman, et al., 2014).

Zamora and Kaiser completed a numerical study in 2009 on the laminar and turbulent flow by natural convection in channels was initiated along with solar chimney configuration. The study involves a relative wall-to-wall spacing and different heating conditions for a wide range of Rayleigh numbers. Outcomes that obtained from the study also include numerical results for the average Nussle number and the non-dimensional induced mass flow rate for symmetrical isothermal heating. A relative analysis for the thermal optimum aspect ratio has been demonstrated for this heating condition. Zamora and Kaiser's study subleases to optimize the inter-plate spacing that maximizes the induced mass flow rate or the heat transfer within the chimney for a given condition (Zamora and Kaiser, 2009). Saifia et al. studied a detailed mathematical simulation in 2012 that was illustrated for the airflow in solar chimneys. Later, the observations from the simulations were experimentally carried out depending on the parameter of the design and the thermal performances for different geometrical configurations. The investigation revealed that the width of the channel and also the angle of inclination of the chimney have contributions on the field speeds in the chimney. Prolonging the research, the effect of inclination on the performance of solar chimney was evaluated in Ouargla Province, Algeria. The conservation equations of mass continuity and energy are solved by the finite volume method providing a good consistency between the experimental results and simulation. Experimental study was performed to determine the variation in temperature between the absorber and the pane varies according to indent solar flow under various chimney slopes. High air flow rate at chimney outlet can be obtained, which may be useful for exploitation in natural ventilation. For various chimney inclination, numerical simulation permits discovering temperature contours and velocity profile inside solar chimney with Rayleigh number Ra = 109. This study can be narrated as the variation of air blade thickness plays a very important effect to increase significantly air flow utilizing Boussinesq approximations. This work is also originating an inclination angle 450 for optimal thermal pulling (Saifi et al., 2012).

In the year 2014, Zhang and the team demonstrated an enhanced research work for improving the thermal environment in the data center utilizing a solar chimney integrated with under-floor air distribution (UFAD). The demonstration was validated using the system in the computational fluid dynamics (CFD) software Airpak. A relative study between the model calculation results shown in different solar chimneys was used for providing a better temperature and airflow distribution. The study illustrates that the temperature in upper zone of cold aisle can be decreased by 130 °C and the temperature field inside the rack can be developed significantly without any additional power. Solar chimney is a remarkable way to refine the thermal environment of the data center with UFAD system. Three distinct kinds of typical solar chimney were engaged utilizing the model, which furnished the power to exhaust air along with an effective distribution of temperature and airflow in both the room and racks. The relative study between the models reveals that the solar chimney introduced above the cold aisle is more effective to this system. Moreover, the solar chimney used in data center with (UFAD) system can achieve an improved cooling effect by the way of the distribution of temperature and airflow rather than increasing cooling load. Hence, the waste of energy and the burden of power system can be eliminated (Zhang et al., 2014).

In 2011, Yan et al. reviewed ventilation properties of solar chimney with vertical collector. Research reveals that the factors to affecting the solar chimney ventilation include heat collection height and weight, solar radiation intensity, inlet and outlet area ratio of chimney and air inlet velocity, etc. An optimal ratio between heat collector height and width was obtained for making the ventilation largest. The best air layer thickness is between 0.2 m and 0.4 m considering the urban architecture image. Furthermore, in certain solar radiation intensity, the airflow temperature in solar chimney is proportional to the chimney height (Yan et al., 2011; Rahman et al., 2014). Haghighi and Maerefat (2014) analyzed the capabilities of solar chimney were investigated to obtain the required thermal and ventilation in winter days. The study involves the heat transfer by natural convection along with a surface radiation in a 2D vented room. Room considered for numerical analysis was in contact with a cold external ambient. Depending on the performance evaluation for determining the appropriate operation conditions, regarding thermal comfort criteria, findings show that the system is capable of providing good indoor air condition at daytime in a room. The condition was applicable for the room even with poor solar intensity of 215 W/m$^2$ and low ambient temperature of 5 °C (Haghighi & Maerefat, 2014).

## Conclusions

Flourishing energy demand has become distressing worrying widespread due to the fluctuating fossil fuel prices. Besides deterioration of ecological balance caused by the use of conventional fuels as the toxic emissions and global warming, it also causes consequential degradation of the marine environment. Solar chimney power plants are a compulsive substitute to centralized electricity generation power plants. Solar chimney power plant is an innovative kind of solar thermal applications consisting of technological simplicity and simple operation of the installation. Developing countries having lack a sophisticated technical infrastructure can accommodate solar chimney power plants along with their conventional energy production systems. Large-scale solar chimneys can be built now without any technical problems and at defined costs. Installation costs are high, but cost estimations made for chimneys of higher size show that energy cost decreases as size increases and the simultaneous use of greenhouses collector in drying or in agriculture.

Countries having high population growth emphasize on high rise building construction for providing adequate accommodation. Thus, it involves interruption of natural air ventilation, which impels the people to use more energy consuming equipment's like AC, Fan, etc. The solar chimney system is one of the distinguished techniques introduced in building ventilation for sustainable initiation. These provide an opportunity to reduce energy demand in domestic and commercial buildings. Lacking of authentic experimental validation stands as an obstacle for both theoretical and modeling studies. Moderate research and subsidy are required to cost optimization and improving the effectiveness and to make it in fascinating design. Such efforts emphasize the feasibility of providing a sensible technology, accessible to the technologically less developed countries based on environmentally sound production from renewable materials.

## References


Agelaridou-Twohig, A., Tamanini, F., Ali, H., Adjari, A., & Vaziri, A. (2014). Thermal analysis of reinforced concrete chimneys with fiberglass plastic liners in uncontrolled fires. *Engineering Structures, 75,* 87–98.

Ahmed, O. K., & Hussein, A. S. (2018). New design of solar chimney (case study). *Case Studies in Thermal Engineering, 11,* 105–112.

Al-Kayiem, H. (2019). Solar thermal: Technical challenges and solutions for power generation'. *Journal of Mechanical Engineering Research and Developments (JMERD), 42*(4), 269–271.

Andreozzi, A., Buonomo, B., & Manca, O. (2009). Thermal management of a symmetrically heated channel–chimney system. *International Journal of Thermal Sciences, 48*(3), 475–487.

Atit, K. (2009). Analysis of flow in solar chimney for an optimal design purpose (Doctoral dissertation, School of Mechanical Engineering Institute of Engineering Suranaree University of Technology).

Bacharoudis, E., Vrachopoulos, M. G., Koukou, M. K., Margaris, D., Filios, A. E., & Mavrommatis, S. A. (2007). Study of the natural convection phenomena inside a wall solar chimney with one wall adiabatic and one wall under a heat flux. *Applied Thermal Engineering, 27*(13), 2266–2275.

Bansod, P. J., Thakre, S. B., & Wankhade, N. A. (2014). Solar chimney power plant-A review. *International Journal of Modern Engineering Research, 4*(11), 18–33.

Bernardes, M. D. S., Voß, A., & Weinrebe, G. (2003). Thermal and technical analyses of solar chimneys. *Solar Energy, 75*(6), 511–524.

Bonnelle, D. (2004). Solar chimney, water spraying energy tower, and linked renewable energy conversion devices: presentation, criticism and proposals. PhD, University Claude Bernard-Lyon.

Chen, Z. D., Bandopadhayay, P., Halldorsson, J., Byrjalsen, C., Heiselberg, P., & Li, Y. (2003). An experimental investigation of a solar chimney model with uniform wall heat flux. *Building and Environment, 38*(7), 893–906.

Chu, C. M., Rahman, M. M., & Kumaresan, S. (2012). Effect of cold inflow on chimney height of natural draft cooling towers. *Nuclear Engineering and Design, 249,* 125–131.

Fluri, T. P. (2008). Turbine layout for and optimization of solar chimney power conversion units (Doctoral dissertation, Stellenbosch: Stellenbosch University).

Gannon, A. J., & von Backström, T. W. (2000). Solar chimney cycle analysis with system loss and solar collector performance. *Journal of Solar Energy Engineering, 122*(3), 133–137.

Gholamalizadeh, E., & Kim, M. H. (2014). Three-dimensional CFD analysis for simulating the greenhouse effect in solar chimney power plants using a two-band radiation model. *Renewable Energy, 63,* 498–506.

Guo, P. H., Li, J. Y., & Wang, Y. (2014). Numerical simulations of solar chimney power plant with radiation model. *Renewable Energy, 62,* 24–30.

Haghighi, A. P., & Maerefat, M. (2014). Solar ventilation and heating of buildings in sunny winter days using solar chimney. *Sustainable Cities and Society, 10,* 72–79.

Hamdan, M. O. (2010) Analysis of a solar chimney power plant in the Arabian Gulf region. *Renewable Energy*, 1–6. https://doi.org/10.1016/j.renene.2010.05.002.

Hamilton, T. (2011). Mad like Tesla: Underdog inventors and their relentless pursuit of clean energy. ECW Press.

Harris, D. J., & Helwig, N. (2007). Solar chimney and building ventilation. *Applied Energy, 84*(2), 135–146.

Harte, R., Höffer, R., Krätzig, W. B., Mark, P., & Niemann, H. J. (2013). Solar updraft power plants: Engineering structures for sustainable energy generation. *Engineering Structures, 56,* 1698–1706.
Kasayapanand, N. (2008). Enhanced heat transfer in inclined solar chimneys by electrohydrodynamic technique. *Renewable Energy, 33*(3), 444–453.
Koonsrisuk, A., & Chitsomboon, T. (2013). Mathematical modeling of solar chimney power plants. *Energy, 51,* 314–322.
Koonsrisuk, A., Lorente, S., & Bejan, A. (2010). Constructal solar chimney configuration. *International Journal of Heat and Mass Transfer, 53*(1–3), 327–333.
Li, J. Y., Guo, P. H., & Wang, Y. (2012). Effects of collector radius and chimney height on power output of a solar chimney power plant with turbines. *Renewable Energy, 47,* 21–28.
Ling, L. S., Rahman, M. M., Chu, C. M., bin Misaran, M. S., & Tamiri, F. M. (2017, July). The effects of opening areas on solar chimney performance. *IOP Conference on Series: Materials Science and Engineering, 217,* 012001.
Maia, C. B., Ferreira, A. G., Valle, R. M., & Cortez, M. F. (2009). Theoretical evaluation of the influence of geometric parameters and materials on the behavior of the airflow in a solar chimney. *Computers & Fluids, 38*(3), 625–636.
Meyer, C. M., & Mancha, L. (2008). Towers of power: The solar updraft tower. Energize [online], pp. 51–54.
Ming, T. Z., Zheng, Y., Liu, C., Liu, W., & Pan, Y. (2010). Simple analysis on thermal performance of solar chimney power generation systems. *Journal of the Energy Institute, 83*(1), 6–11.
Ngala, G. M., Sulaiman, A. T., & Garba, I. (2013). Review of solar chimney power technology and its potentials in semi-arid region of Nigeria. *International Journal of Modern Engineering Research (IJMER), 3,* 1283–1289.
Ngo, C., & Natowitz, J. (2009). Our energy future resources, alternatives and the environment: Peat. Wieley Survival Guides in Engineering and Science. Section Peat, pp. 184–186.
Ninic, N. (2006). Available energy of the air in solar chimneys and the possibility of its ground-level concentration. *Solar Energy, 80*(7), 804–811.
Nizetic, S., & Klarin, B. (2010). A simplified analytical approach for evaluation of the optimal ratio of pressure drop across the turbine in solar chimney power plants. *Applied Energy, 87*(2), 587–591.
Papageorgiou, C. D., Psalidas, M., & Sotiriou, S. (2011). Floating solar chimney technology scale analysis. *Proceedings of IASTED International Conference on Power and Energy Systems Crete, Greece, 24,* 55–59.
Pasumarthi, N., & Sherif, S. A. (1997). *Performance of a demonstration solar chimney model for power generation.* California State University, Sacramento, CA, (USA), pp. 203–240.
Pretorius, J. P. (2007). Optimization and control of a large-scale solar chimney power plant (Doctoral dissertation, Stellenbosch: University of Stellenbosch).
Rahman, M. M., Chu, C. M., Kumaresen, S., Yan, F. Y., Kim, P. H., Mashud, M., & Rahman, M. S. (2014). Evaluation of the modified chimney performance to replace mechanical ventilation system for livestock housing. *Procedia Engineering, 90,* 245–248.
Rahman, M. M., Misaran, M. S. B., Jamanun, M. S. B., & Jawad, A. (2018). Estimate the ventilation effect from wire mesh screen assisted solar chimney. *Journal of Energy and Power Engineering, 12,* 127–131.
Saifi, N., Settou, N., Dokkar, B., Negrou, B., & Chennouf, N. (2012). Experimental study and simulation of airflow in solar chimneys. *Energy Procedia, 18,* 1289–1298.
Schlaich, J. R., Bergermann, R., Schiel, W., & Weinrebe, G. (2005). Design of commercial solar updraft tower systems—Utilization of solar induced convective flows for power generation. *Journal of Solar Energy Engineering, 127*(1), 117–124.
Shen, W., Ming, T., Ding, Y., & Wu, Y. (2014). Numerical analysis on an industrial-scaled solar updraft power plant system with ambient crosswind. *Renewable Energy, 68,* 662–676.
Somsila, P., Teeboonma, U., & Seehanam, W. (2010, June). Investigation of buoyancy air flow inside solar chimney using CFD technique. In *Proceedings of the international conference on energy and sustainable development: issues and strategies (ESD 2010)* (pp. 1–7). IEEE.

Suhendri, K. M. D., & Alprianti, R. R. (2018, July). Solar chimney as a natural ventilation strategy for elementary school in urban area, Vol. 1984, No. 1. In *AIP conference proceedings* (p. 030007). AIP Publishing LLC.

Tingzhen, M., Wei, L., Guoling, X., Yanbin, X., Xuhu, G., & Yuan, P. (2008). Numerical simulation of the solar chimney power plant systems coupled with turbine. *Renewable Energy, 33*(5), 897–905.

Von Backström, T. W., & Fluri, T. P. (2006). Maximum fluid power condition in solar chimney power plants–an analytical approach. *Solar Energy, 80*(11), 1417–1423.

Von Backström, T. W., & Gannon, A. J. (2004). Solar chimney turbine characteristics. *Solar Energy, 76*(1–3), 235–241.

Von Backström, T. W., Harte, R., Höffer, R., Krätzig, W. B., Kröger, D. G., Niemann, H. J., & Van Zijl, G. P. A. G. (2008). State and recent advances in research and design of solar chimney power plant technology. *VGB Powertech, 88*(7), 64–71.

W. Ley. (1954) Engineers' dreams. Viking Press.

Wengenmayr, R., & Bührke, T. (Eds.). (2011). *Renewable energy: sustainable energy concepts for the future*. John Wiley & Sons.

Yan, Z., Guang-e, J., Xiao-hui, L., & Qing-ling, L. (2011). Research for ventilation properties of solar chimney with vertical collector. *Procedia Environmental Sciences, 11*(Part C), 1072–1077.

Zamora, B., & Kaiser, A. S. (2009). Optimum wall-to-wall spacing in solar chimney shaped channels in natural convection by numerical investigation. *Applied Thermal Engineering, 29*(4), 762–769.

Zandian, A., & Ashjaee, M. (2013). The thermal efficiency improvement of a steam Rankine cycle by innovative design of a hybrid cooling tower and a solar chimney concept. *Renewable Energy, 51,* 465–473.

Zhang, K., Zhang, X., Li, S., & Wang, G. (2014). Numerical study on the thermal environment of UFAD system with solar chimney for the data center. *Energy Procedia, 48,* 1047–1054.

Zhou, X., Yang, J., Xiao, B., Hou, G., & Xing, F. (2009). Analysis of chimney height for solar chimney power plant. *Applied Thermal Engineering, 29*(1), 178–185.

# Summary

The novelty of this book is it is showcasing the benefits of adverse cold inflow prevention and the bonus stack effect of plume chimney above chimneys. The solar chimney applications are evaluated through investigations by theoretical, experimental and computational fluid dynamics (CFD), where it is demonstrated convincingly that cold inflow or flow reversal at the chimney exit is an undesirable phenomenon and its prevention brings back efficiency and stability to natural convection flows. While the theory behind solar chimney is relatively simple, the experiments can be challenging due to the transient nature of the experiments and the adverse effects brought by cold inflow. A study on the use of wire mesh blocker to block the cold inflow is reported. The measurement of the stack effect of plume chimney, the "invisible chimney", is very challenging due to instrumentation limits at low velocities. This was in a large part alleviated by CFD simulation which yielded insight into the stack effect. In the design of solar chimney, the basic is to obtain the simultaneous steady-state solution of hydraulic (pneumatic) balance and heat transfer from solar irradiation. The effect of geometry in two chapters and the hybrid turbo-solar chimney is given as a solution to weather and daytime–nighttime interruptions.

While the energy source of the solar chimney is free, it has very low efficiency (R0.1%) in converting it to electricity which renders it to be necessary to consider the economics of capital and operating costs, and an economic analysis is carried out on its viability. The commercialization of solar chimney as a power generator with the chimney height up to 1 km tall is still not realized to date. Various explanations are explored in the final chapter on the technical barriers that exist and what the way forward should be in harnessing a promising energy source for power generation.

Solar chimney as an air-mover, replacing fan and compressor, however, has much more potential for commercial applications, such as natural ventilation for thermal comfort, drying for processing farm and marine produce and small cooling towers for renewable energy. Unlike in power generation, the need for a tall chimney and turbine to convert the thermal energy into motion does not exist, and this has saved much of the capital and operating costs, in that a fan or a compressor does not need to be purchased or maintained, and the chimney at the same height of a building is normally sufficient to generate the required flow rate. Residences and shopping complexes are


Md. M. Rahman and C.-M. Chu (eds.), *Cold Inflow-Free Solar Chimney*,
https://doi.org/10.1007/978-981-33-6831-6

now known to have air wells and atriums that ventilate by solar chimney effect albeit, not very efficiently, and many journal papers have been written to report the analysis for deriving the optimum formula. This is where the prevention of cold inflow and the additional stack effect of plume chimney really come into their prominence by providing stable and efficient ventilation that to date has eluded many designers of natural draft chimney systems. Chimneys of large cross-sectional area in comparison with the slender and tall chimneys in the past can now be built for lower cost of construction and easier maintenance and operation.

Printed in the United States
by Baker & Taylor Publisher Services